Law for the Construction Industry

2nd edition

Stephanie Owen

LONGMAN

The CHARTERED
INSTITUTE OF
BUILDING

Addison Wesley Longman Limited
Edinburgh Gate
Harlow
Essex CM20 2JE
United Kingdom
and Associated Companies throughout the world

Co-published with The Chartered Institute of Building through
Englemere Limited
The White House
Englemere, Kings Ride, Ascot
Berkshire, SL5 8BJ, England

First published 1987
This edition 1997

ISBN 0 582 28708–1

British Library Cataloguing-in-Publication Data

A catalogue record for this book is available from the British Library

Set by 35 in 9/12pt Palatino
Produced through Longman Malaysia, LSP

For Barry, Edward and Chloë

Contents

Preface to the second edition

I was very pleased to be asked to update my original book, 'Law for the Builder', published 10 years ago. Now in its new guise, 'Law for the Construction Industry', I hope that a new generation of building students will find it useful, not only in their academic work but also on site, as I know it has been in the past. Extensive revisions have been made, as can be seen from the text. I have tried to steer a middle course, providing sufficient detail and depth, where I have thought it important or necessary. Where only an overview has been given, I hope that the book will provide a useful springboard and the appropriate legal advice is sought. The revision has been rather like painting the Forth Bridge and whilst every effort has been made to bring the law as up to date as possible, events often overtake one and a keen eye should be kept on the construction journals and newspapers. My greatest piece of advice to students based on my recent teaching experiences, is 'Buy the book!' or certainly, at least a text-book. Student finances are precarious, I know, but too many are trying to get through examinations, with only lecture notes and the occasional borrowed library book. My other piece of advice is that the law is not difficult – there is just so much of it. Steady work right from the beginning of the course will pay good dividends. If you don't understand something, leave it till the next day when you are fresh and better able to absorb the information, and if you still don't understand, make sure you ask your lecturer for help.

I wish all readers every success in their studies.

Stephanie Owen, Ashtead 1997

Preface to the first edition

This book is not intended for lawyers. It has been written for the many thousands of builders who are faced with legal problems either in their daily work or when studying for their building examinations. After lecturing to building students for a number of years it became apparent that the average builder takes to the law like a duck to water, providing the information is shown to be relevant to the building industry and wherever possible plenty of diagrams are given. This book has been designed to convey the essential elements of English law in a relatively simple way, providing as much information and practical suggestions as possible and there are over 50 diagrams to assist. The index is detailed and should be used in conjunction with the text as a glossary.

Preferably the book should be read from the beginning as many of the mysteries of English law can be simply explained by reference to legal history.

In a book of this nature which covers so many areas of the law, I had to finally fix the law as best I could as at October 1985. I appreciate that there will be some areas in which there will be changes and I urge the reader to open a legal file into which he can start putting relevant cuttings, brief notes on radio programmes, articles from professional journals etc. so that he builds up a living picture of the English legal system as it affects him. I hope also that by the time he throws away this book (and buys a new edition) that it will be dog-eared with use, pencilled and underlined, which is how a law book should be especially for those studying for examinations. There is undoubtedly a correlation, in my experience, between high grades in examinations and 'dog-earedness' of books.

In a book of this nature, it would be extremely difficult and lengthy if I were to use both masculine and feminine examples at all times. For this reason I have adopted the usual law textbook approach i.e. by Section 6 of the Interpretation Act 1978, where words referring to the masculine gender are used, the female is included unless the contrary is shown.

Finally I hope that my book will encourage all builders to share my enthusiasm for the law. No other discipline is so relevant to everyday life and a well informed person gets much more satisfaction from their work.

Stephanie Owen
Ashtead, April 1986

Acknowledgements

We are grateful to the following for permission to reproduce copyright material;

The CIOB for figure 6.2 from their archives, the London Illustrated News Picture Library for figure 3.4 and The Solicitors' Law Stationery Society Limited for figures 1.3 and 4.9.

Author's acknowledgements

The miracles of the word processor have meant that I can now type my own manuscript, however, I am still indebted to Breda Dallimore, Margaret Gierlinska, Eileen Powell-Davies, Charlotte Johnson and my husband Barry for their help in proof-reading. I would also like to thank the library staff at NESCOT and the Law Society for their assistance with my research, and Valerie Leach and Ann Davidson for helping to keep my family ticking over. My one regret is that my great friend Leslie Bell, who proof-read virtually the whole of the first edition, is no longer alive. He is sadly missed in this community.

Whilst every effort has been made to ensure accuracy, any errors remaining are my own. In practical legal difficulty, I cannot recommend too highly that professional advice is sought as soon as possible and before the situation gets worse.

Cases

Statutes

Statutory instruments

1 The nature of law and its sources

'To begin at the beginning' DYLAN THOMAS/UNDER MILK WOOD

What is law?

The law

Laymen often have a jaundiced view of the law, and no wonder. To them the 'law' means big bills, policemen, red tape, divorce, prison and American movies. What they forget is that the more harrowing aspects should only play a small part in the life of the average man. In other respects, however, the law can creep into every nook and cranny of his life. Nearly every important occasion is affected by legal considerations, whether it is registering a birth, celebrating a coming of age, marrying, buying a house, making a will or dying. Even the builder lives his life in this manner!

The law and the builder

At work, the builder is even more beset with the need to comply with legal requirements. Before he can build, he may have to buy land by the process known as conveyancing. He may enter into a contract to build on someone else's land and will have to enter into hundreds of contracts to buy all the necessary materials. He must make sure that he has planning permission and must deposit plans which satisfy the Building Regulations. He will have to enter into contracts of employment with his workers and may negotiate with trade unions. He may operate his business as a sole trader or in partnership or may form a limited company. He must build carefully to comply with the contract specifications, the Construction Regulations, the Defective Premises Act 1974, the Health and Safety at Work etc. Act 1974 and to avoid being sued for breach of contract or negligence. He must not cause nuisances to his neighbours whilst building. He should make satisfactory insurance arrangements to cover his liability for many occurrences. He must not trespass on the neighbour's land, build beyond building lines, cause obstructions of highways nor dismiss his workers without good reason. The list is endless!

Why do we need laws?

Even in the most primitive of societies, people live their lives by rules which they create. Some things they make compulsory, others they prohibit. Such rules are imposed to make people behave in a similar way, in order to provide a harmonious way of life. For example, it is undesirable to have uncontrolled violence or theft, and ancient British tribes would have had customary rules dealing with such matters, punishing those who transgressed.

Without such rules, each person would have to make individual choices concerning his behaviour, which, whilst being perfectly logical and moral for himself, would not benefit society as a whole. For example, it is logical and much more usual to drive on the right-hand side of the road, yet it is the law of England and Wales that we drive on the left. Anyone failing to do so would cause chaos. And so it is for the bulk of our law – it is there to prevent chaos, to create an ordered way of life.

1

What sort of laws do we need?

Obviously, laws are needed which support popularly held ideals. Thus, in the early days of English law, people would have been most concerned with protecting themselves, their families, land and property, and the laws that developed reflected this.

As society became more complex, and people stopped manufacturing everything on their own, they began to buy and sell things, and to get others, such as builders, to work for them rather than do the work themselves. As a result of these business dealings, a large body of mercantile law developed.

Following the Industrial Revolution, the state increasingly began to take a protectionist attitude towards the people, far different from the old *laissez-faire*, caveat emptor approach of nineteenth-century law. Much social and consumer legislation has been passed in the twentieth century, and today there is a high level of legal control in all walks of life.

Thus, the law has to a certain extent mirrored the changes in society, getting more complex and increasing in volume as society has grown more sophisticated. Similarly, the methods of law creation have grown more complicated. What originally would have been a mere customary rule must now be created either by Parliament or by judges making decisions on the finer points of law in court cases.

Who benefits from the law?

As stated before, the law is mainly needed to benefit society as a whole, but this should not be at the cost of the individual. A fine balance must be maintained. This is so in nearly every country. However, sometimes the balance is tipped in favour of the state, as in communist countries. Even in this country, many people feel that the law is unfair to them individually.

Substantive and adjective law

The laws necessary to achieve a harmonious way of life are substantive laws, i.e. the very substance of the law itself such as a law against killing people. Such law on its own, however, would fail through lack of procedure to follow the law through to prosecution and, if necessary, punishment. Thus, adjective law, which is concerned with procedural law, court practice and rules of evidence, is needed to make substantive law actually work.

Good law and bad law

What is good law and bad law is entirely subjective. What may be good for society may be bad for the individual. What is good and what is bad? Once again, the law can only reflect society's wishes, and if society does not meet with our personal conception of right or goodness, then we dub it bad law. Nazi Germany or Idi Amin's Uganda had laws which were perfectly valid during those regimes. To change such law one needs to change the society that created it. This can be done peacefully by evolution, as in most Western states today, or forcefully, by revolution, for example, as in Iran, where a modern legal system has been replaced by traditional Islamic law.

Even with an evolutionary process such as we now have in England and Wales, we have good and bad law, depending on our particular point of view. The way we think depends on our upbringing, religion, education, politics and many other external influences. Such factors also affect one's private morals, and for that reason morality is generally not the concern of the law.

Because we are not all alike, the law will not suit everybody. It is merely the lowest common denominator which will satisfy most people's requirements of a legal system, and achieve maximum stability in the state.

The 'law' compared with other types of law

- The long arm of the *Law*
- The *Law* of gravity
- Criminal *Law*
- French *Law*

The word 'law' of course, is used in other contexts, and thus it is important to define one's terms. Scientific laws, for example, are totally different from legal ones, for they describe things which must necessarily happen, such as the law of gravity. Legal laws can only lay down rules for people to follow. Whether they do or not is up to them. To encourage adherence to the rules, most have some element of coercion behind them, whether it is a threat of loss of liberty or some other right, or a financial penalty.

Every country in the world has law of some sort, and within each legal system there are many different areas covering all aspects of life. Fortunately, for the purposes of this book, we are 'only' going to examine English law, i.e. the law applying to England and Wales.

The composition of a legal rule

Every area of law is made up of many individual little laws. A complex example of this can be found in the law of contract, which has hundreds of thousands of legal rules.

Each legal rule consists of two parts: a **right** and a **duty**. Thus, there is a duty imposed on all people in this country – not to kill others. Conversely, everyone has a right to life, or a right not to be killed. If someone breaks this rule, or as we say **breaches** his duty, then the state, on behalf of the people for whom the rule was designed to protect, will punish the wrongdoer. Thus, all legal rules are rather like see-saws (see Figure 1.1).

This book is therefore concerned with many rights and duties under the law with which the builder must concern himself, and it is hoped that this introduction to the concept of law will help the reader with his studies.

English law

Legal systems in the British Isles

English law is the law of England and Wales – not of the United Kingdom, for in the United Kingdom of Great Britain and Northern Ireland there are three legal systems, Scotland and Northern Ireland having their own. The Channel Islands and the Isle of Man are merely direct dependencies of

the Crown and have their own legal systems. The reason for there being no uniformity is purely historical. The Scottish legal system, for example, developed in an entirely different way with much European influence, as it was a totally separate country until the Act of Union 1707. It would be foolish to imagine that two countries would ever develop in the same way through a period which knew no other forms of communication than that provided by horse or foot. Wales, on the other hand, was conquered by the English in the thirteenth century, at a time when English law was only in its infancy. The new overlords imposed the new developing law on the Welsh, and thus the laws of Hywel Dda are now lost in the mists of antiquity.

Common law

English law is a common-law system which is also found in many other countries in the world, mostly former British colonies. The words 'common law' have a number of different meanings. The first is of historical significance, and is the key to understanding what our law is all about. It is probably far easier to understand modern English law, with all its peculiarities, if one first looks at its historical development.

The development of English common law

Before the Norman Conquest
Before 1066, England was not as we know it today. It was merely a collection of many different kingdoms, e.g. Mercia, Wessex and Kent. These kingdoms were in perpetual disruption, because of fighting between their respective leaders and invasions from Europe of many warring tribes such as the Saxons.

Highly significant in the north-east was the arrival of the Vikings, who established their own kingdom called the Danelaw. All these tribes settled and interbred, so that old and new customs were intermingled. Some of their customs would obviously have concerned 'legal' ideas, i.e. not killing people or stealing their sheep. As few people could read or write, and the dialects

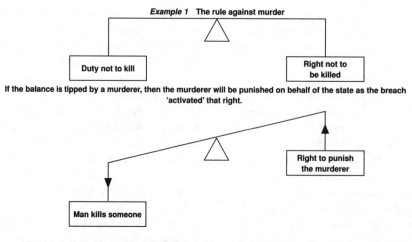

Example 1   The rule against murder

If the balance is tipped by a murderer, then the murderer will be punished on behalf of the state as the breach 'activated' that right.

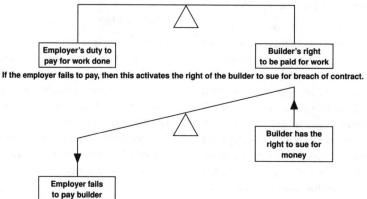

Example 2   In a building contract, the builder and the employer enjoy many different rights and are under many duties. At its simplest, the main object of the contract is that the employer wants building work done and the builder wishes to be paid for that work.

If the employer fails to pay, then this activates the right of the builder to sue for breach of contract.

A similar diagram relating to the builder's duty to carry out the work and the corresponding right of the employer to have the work done could be drawn.

Figure 1.1   Diagram to illustrate the concepts of rights and duties

were different, most laws tended to be unwritten and customary.

In 924, Athelstan was crowned on the Kings Stone at Kingston-upon-Thames, having invaded Scotland, defeated many tribes in the North of England and been paid homage by Welsh and Scottish kings. England was now united, but its laws remained much the same, and without an efficient system of communication it would have continued but for the Norman Conquest.

For governmental purposes, it was divided into shires, each with its own courts. Shires were sub-divided into hundreds which also had their own courts. The king's representative in the shire, the shire reeve (or sheriff) would sit as a judge in the shire courts with the gentry and clergy, for there were no lawyers as such.

The landowners also operated a sort of private enterprise court system, known as the manorial courts, which dealt with disputes between people living and working on their land. In all the courts, local customary law was used, supplemented by the occasional written law known as a Doom issued by the King and his advisors, the Witan.

This, then, was the legal picture at the time Harold lost the Battle of Hastings.

After 1066

William of Normandy had been promised the throne of England by Edward the Confessor, who had been brought up partly at the Court of the Duchy of Normandy. Unfortunately, Edward also named Harold II as his heir. To right this betrayal, William invaded England, and fought and won the Battle of Hastings.

Realising that to retain the throne of England would be a difficult task, he set out to overcome the Anglo-Saxon hatred of their Norman conquerors, and to establish himself as a strong ruler. In this seeking after complete power, he sowed the seeds of what was to become England's first unified legal system, or law – common to all men. However, he did not abolish the existing legal system based on custom, knowing that this would be too unpopular.

The development of modern English law occurred as a result of a number of innovations introduced by William and subsequent Norman kings.

The feudal system – a form of land holding

First, the legal structure of society was completely changed by the introduction of the Norman *feudal system*. This was a system of land tenure or holding, whereby the king confiscated all the land, declaring himself to be the only true **owner** of England. He then made grants of the land to those he wished to reward for past favours. These people merely held the land as tenants of the king. They in turn granted estates or interests in the land, and so on. All the tenants and sub-tenants owed their loyalty to their immediate landlord, and had to pay feudal dues to keep their land. In addition, all swore to be faithful to the king. This process of granting sub-tenancies was called *sub-infeudation* (see Figure 1.2). By this means, William repaid past loyalty, and ensured continued obedience, as land could be forfeited to the landlord in certain situations.

The feudal system also produced a form of communication from the king right down to the lowest of the low, the serfs. They were rather similar to slaves, in that they could be sold with the land, they could not hold an interest in land itself, and could not marry without their lord's permission.

In a time of poor communication, this system provided an adequate framework on which society in England could be unified. It also provided, by means of the feudal dues, money and armies to strengthen the monarchy, which were needed to impose the embryonic king's law on the antiquated, inefficient and regional legal system already in existence.

The Curia Regis – a forerunner of government departments, House of Lords and modern courts

Second, the Norman kings reinforced their rule with the introduction of the Curia Regis or King's Council, which was an advisory body, consisting of the tenants-in-chief, or barons. Initially they met only on formal occasions, but as life became more complex, many of the advisors took on special responsibilities for various aspects of government, such as taxation. Later, they gathered around them departments of suitably educated staff to help with administration, and from these beginnings, Parliament and the courts as we know them began to develop.

Circuit judges – unifying England by common legal rules

Third, the kings started to send Royal judges around the counties, giving them specified tasks known as commissions, e.g. the Commission of the Assize. They went on circuit, and took with them 'central' policy from the King's Council. At first they had to use the old customary laws at the trials, as there was no other law. These, they applied, but on their return to the king, those customs of which they disapproved would be rejected, and those they liked would be incorporated into a new set of rules, to be used all over England, the common law.

Thus, the English people now had two court systems, the old pre-1066 courts, and those which were modified or introduced by the Norman kings. The Church continued to deal with matters relating to marriage and the family. Gradually, the

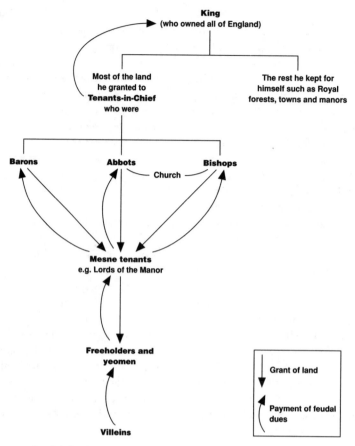

Figure 1.2 The process of sub-infeudation

people began to see the advantages of using the circuit courts of the king, for there was uniformity in treatment, little corruption and virtually no bias. The nobles who ran the manorial courts, and who derived a great income from them, were, of course, less than happy with Royal Law. Whenever a king was weak, the nobles would force him to do something which restricted his power in some way. This is well illustrated when the development of the writ system was hindered by the enactment of the Provisions of Oxford in 1258 (see p. 7).

At first, the new royal justice was concerned with law and order, so criminal law was important. Property, of course, was of extreme significance and under the sole jurisdiction of the Crown, and the Royal judges would be asked to deal with disputes concerning property rights. In order to seek help from the new system, a plaintiff would have to obtain a document from one of the new government departments. This document was aptly called a writ. It was because of the writ administration system that the common law developed in a strange way. Eventually, because of the shortcomings of the common law, a parallel set of rules was developed known as equity (see p. 7). In addition to the common law created by the circuit judges, the king would also enact statutes on very important matters such as the Provisions of Oxford.

The writ system

Even today, if you wish to sue someone in the High Court, you need to issue a writ, a written document containing particulars of the claim you are making against the defendant (see Figure 1.3). The modern writ is a printed form, which the plaintiff completes, to cover the circumstances involved. The old writs, on the other hand, were specially written by the clerks in Chancery which was one of the new government departments spawned by the Curia Regis. The head of the Chancery was the Lord Chancellor, who was one of the most important members of the Curia Regis. In the early days of common law, he was always a priest, and also the King's Confessor. This department was mostly made up of educated clergymen, as they were some of the few people who could read and write. The clerks drew up the writs, using a mixture of old English and Norman French. Each writ was unique, in that it only applied to the circumstances pleaded by the plaintiff. One can understand the delight of those people who had travelled sometimes hundreds of miles to obtain one of the new writs, for it meant that whatever their grievance, they would be able to take a private action in the modern king's courts.

This legal 'Shangri-La' was short-lived, for a side effect of allowing the issue of writs for every private legal grievance was that the people flocked to use the king's courts. As a result, the manorial courts, run by the feudal lords, began to lose revenue to the Crown. Even today one must pay court fees before a civil action can be started.

In 1258, the Crown was comparatively weak. The lords pressurised King Henry III to enact a statute, called the Provisions of Oxford. This dealt with many matters, but one matter of great significance was that the statute forbade the making of any **new writs**. This had a profound effect on the development of the common law, for unless you had a grievance already covered by the wording of an existing writ (which could be repeated, incorporating your name and other particulars) you had **no** remedy in the Royal courts. The result of this was that for many people, the only remedy open to them was to be found in the manorial courts – a retrograde step.

All was not lost, however. In 1285, King Edward I, a successful and strong king, enacted the Statute of Westminster II. Whilst he could not reverse the Provisions of Oxford, he could introduce a modification concerning the writ system. He decreed that, whilst no new writs could be issued, the wording of the old pre-1258 writs could be used to cover analogous situations. As a result, the civil side of common law continued to develop in an artificial fashion.

Equity

Development of equity

Thus, one of the main problems of the common law was the lack of freedom to start a private action, unless there was a suitable writ, or one that could be adjusted to fit your grievance. No writ – no right! There were other problems which stemmed from this restriction. The wording of the writ became sacrosanct. Indeed it was called the form of action, and had to be applied exactly in each subsequent case. One word wrong could lose you your case, an early example of 'sticking to the letter of the law'.

So, for people who wanted to go to law but who were unable to use the common law, because there was no writ available, there were two alternatives. Either they went back to the manorial courts, or they applied directly to the king, who was the 'fountain of all justice'. The second method was excellent, as the king could grant whatever remedy he wished. But it could involve following the king around, sometimes even behind battle lines, until one got an audience. Even so, applying directly to the king for justice became so popular that eventually the hearing of applications or petitions was taken on by the Lord Chancellor. As we saw before, although he was in charge of the Chancery, originally, his main job was to be the king's spiritual advisor. As a priest, it is understandable that he viewed the problems presented to him in an entirely different light from that of the king and the judges. His was a moral and religious viewpoint, and he made his decisions based on his own personal ideas of what was fair and just. He was not restricted by the writ system as, to start with, each appeal was made in person by

COURT FEES ONLY

Writ indorsed
with Statement
of Claim
[Unliquidated
Demand]
(O.6, r. 1)

IN THE HIGH COURT OF JUSTICE
Queen's Bench Division

[~~District Registry~~]

1997.— F .—No. 123

Between

BIGGLES BUILDING SUPPLIERS   Plaintiff
LIMITED

AND

BORIS MOLE   Defendant

(1) Insert name.   **To the Defendant** (¹)   Boris Mole

(2) Insert
address.   of (²)   16, Chemical Road, London. N. W. I

This Writ of Summons has been issued against you by the above-named Plaintiff in respect of the claim set out overleaf.

Within 14 days after the service of this Writ on you, counting the day of service, you must either satisfy the claim or return to the Court Office mentioned below the accompanying **Acknowledgment of Service** stating therein whether you intend to contest these proceedings.

If you fail to satisfy the claim or to return the Acknowledgment within the time stated, or if you return the Acknowledgment without stating therein an intention to contest the proceedings, the Plaintiff may proceed with the action and judgment may be entered against you forthwith without further notice.

(3) Complete
and delete as
necessary.   Issued   from   the   (³)   [Central   Office]   [~~Admiralty and Commercial Registry~~]
[~~District Registry~~] of the High Court
this   1st   day of   August   1997.

NOTE: —This Writ may not be served later than 4 calendar months *(or, if leave is required to effect service out of the jurisdiction, 6 months)* beginning with that date unless renewed by order of the Court.

IMPORTANT

Directions for Acknowledgment of Service are given with the accompanying form.

Figure 1.3   Writ endorsed with statement of claim for liquidated sum. High Court form A3B reproduced by kind permission of The Solicitors' Law Stationery Society Limited.

Statement of Claim

The Plaintiff claim is for £28,966.75 being the price of goods sold and delivered to the Defendant by the Plaintiff Company and for interest thereon pursuant to Section 35A Supreme Court Act 1981

Particulars

1995

April 22nd
to
December 27th

To goods comprising timber and bricks sold & delivered to the Defendant full particulars of which have been given to the Defendant.

£28,966.75

Figure 1.3 cont'd

(Signed) *Timothy Twitch for Fitch, Twitch & Smith.*

(1) If this Writ was issued out of a District Registry, this indorsement as to place where the cause of action arose should be completed.

(2) Delete as necessary.

(3) Insert name of place.

(4) For phraseology of this indorsement where the plaintiff sues in person, see *Supreme Court Practice*, Vol 2, para 1.

(¹) [(²) [~The cause One of the causes of action in respect of which the Plaintiff claim relief in this action arose wholly or in part at (³) in the district of the District Registry named overleaf]

(⁴) **This Writ** was issued by Messrs Fitch, Twitch & Smith of 213, High Street, London N. W. I.

[~Agent for~

~of~]

Solicitor for the said Plaintiff whose ~address (³) is are~] registered office is situated at Biggles House, The Street, Uptown, Surrey.

Figure 1.3 cont'd

1997.— F .—No.123

IN THE HIGH COURT OF JUSTICE
Queen's Bench Division

[~~District Registry~~]

Biggles Building SUPPLIERS LIMITED

V

BORIS MOLE

Writ of Summons

[~~Un~~liquidated Demand]

Date issued 1st August   1997

*Acknowledgment of Service
lodged*   19

Fitch, Twitch & Smith
                    Plaintiff's Solicitor

Solicitor's Reference HF/BBS **Tel. No.** 0181-111-2222

OYEZ The Solicitors' Law Stationery Society Ltd, 12.94 F28485
       Oyez House, 7 Spa Road, London SE16 3QQ 5044051
                                          ★ ★

High Court A3B

Figure 1.3 cont'd

the plaintiff. His remedies were based on concepts of conscience and fair play, and he would often order the defendant to perform his legal duty, which was fairer than requiring him to pay damages, which is all he would have got at common law.

Eventually, the Chancellor was given his own court, the Court of Chancery. Other judges were appointed, and a body of rules called **equity** developed, offering *not* an alternative system, but a supplement to 'stop the gaps' in the common law relating to private legal matters.

The nature of equity

Remedies

Equity thus developed its own remedies, such as **injunctions**. These may order people to stop doing something they should not be doing in the first place. Ignoring the order may result in imprisonment. Also, the decree of **specific performance** is used to order a person to perform his part of a contract, such as for the sale of a house.

Unlike legal remedies, i.e. the common-law remedies, which are yours **as of right**, once you have proved your case, equitable remedies were, and still are, at the discretion of the judge.

Maxims (guiding principles)

At first, there was no ordered way of coming to a decision in equity. Indeed, it was said 'Equity varied with the length of the Chancellor's foot'. Equity did not use precedents (see p. 21) but it did use **maxims**, which, even today, are at the root of equitable decisions. An example of one is, 'He who comes to Equity, must come with clean hands'. This means that a plaintiff asking for an equitable remedy, or for relief in equity in some way, must have behaved fairly throughout the affair and not merely complied with his legal duties.

Other equitable maxims include 'Delay defeats Equity', meaning that you cannot expect the court to help you in equity if you delayed seeking that help, as it may have adversely affected the opposition's position.

Jurisdiction

Because of the many deficiencies in the common law, equity developed its own jurisdiction in certain areas, totally separate from that of the common law. Thus, in some things, there were only rules in equity, and none at common law. One such area was that of **trusts**. The concept of a trust was totally beyond the scope of common law. How could anyone give property to someone else if that was not the end of the matter? Equity, however, acknowledged that it was possible to give property to others for the benefit of other people, even the donor or his family.

Equity also acknowledged that, as far as the law was concerned, the donees were the owners – in common law, but *not* in equity (see Figure 1.4).

It must be stressed, therefore, that in some areas of English law, equity never developed. Its main influence was and is in *private* or *civil* law. It gave *individual* rights and remedies in such areas as contract, tort, mortgages and trusts. Nevertheless, equitable ideas and concepts have been taken into other areas of the law by analogy, in court decisions and statute.

The problems caused by common law and equity

By the nineteenth century, the English law system was in chaos. There were still many outdated anachronistic courts dating back to pre-Conquest times, using customary law, which had virtually fallen into disuse. Within the main court system there were criminal courts using mainly common law and statute, such as the criminal assizes

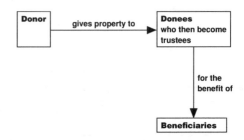

Figure 1.4 Diagram to illustrate a trust

(finally abolished in 1971); civil courts using common law and statute; and Chancery which used equity and statute.

Equity, of course, plugged the gaps in the common law, and was designed to achieve fairness within the total legal system. Unfortunately, one of the main problems with the system was that the rules were applied totally separately in the different courts. As each type of law had its own remedies a litigant, anxious to make sure of some remedy, would have to start his action in a common-law court and in Chancery – a ludicrous state of affairs.

Reform

As a result of centuries of dispute, compromise and court decisions, the matter was finally resolved by Parliament in the Judicature Acts 1873–75. These acts were of inestimable importance, for they did many things:

1. They unified the court system, so that there were no longer common-law courts and equity courts.
2. The rules of equity and common law were applied in all courts, so that only one action was necessary.
3. If there was a conflict between a rule of common law and equity, then the equitable rule prevailed. As equity had developed to remedy the injustices in the common law, this was only right.
4. They reorganised the structure of the courts, by creating a hierarchical ladder system. This made it easier to determine to which courts appeals should be made.

Is should be noted that the Acts did *not* combine the two sets of rules. Equity and common law remain like oil and water – they are present together in the same system, but they do not mix.

The court system has been altered greatly since 1875, but we still retain the courts in descending order of importance, as first introduced by the Judicature Acts, from the House of Lords right down to the magistrates' courts.

In conclusion if we examine how English law has developed **historically** we find that we have a legal system which consists of law derived from custom, the common law, equity and statute.

(Note: if one is not talking in historical terms, lawyers tend to refer to common law as non-statute law (see p. 21).)

Divisions of English law

If we examine the subject-matter of English law, we find that it falls into two broad categories – public and private law (see Figure 1.5).

Public law

This deals with matters which are of importance to the state, are for the well-being of society, and which give public rights and duties to the individual.

Criminal law
This encompasses not only the crimes with which one is familiar, such as murder, rape, bigamy and theft, but also matters which have been regarded by Parliament to be so detrimental to the state as to necessitate treatment by the criminal courts, e.g. breaches of the Construction Regulations, the Health and Safety at Work etc. Act or road traffic offences.

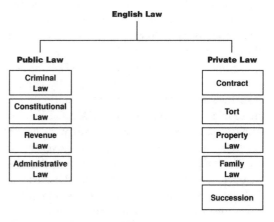

Figure 1.5 Divisions of English law

A person who fails to abide by a criminal rule will be prosecuted by the state or its agents, and will be punished accordingly, e.g. by imprisonment or fine. The most serious offences are prosecuted in the name of the Crown, and the law report will be headed *R* v. *Bloggs* – *R* standing for Rex (King) or Regina (Queen).

Constitutional law

This deals with the relationships between the individual as a citizen and the state itself. It covers such matters as the roles of the Crown, Parliament, the government, the judiciary (judges) and the citizens. It gives the citizen certain public rights, e.g. the right to vote at elections, and imposes certain public duties on the elector, e.g. the duty to sit on a jury if so required. (Note: you must perform duties, but you need *not* exercise a right.)

Failure to perform one's duties may result in prosecution and punishment.

Revenue law

The state cannot be run without financial resources and so taxes are levied on people and organisations. Failure to pay may result in prosecution. There are many different taxes, e.g. income tax, inheritance tax, stamp duty (paid on transfer of land), value added tax and corporation tax (company income tax).

Administrative law

As society has become more sophisticated and the state has played more of an interventionist role, a large body of law has had to grow to deal with the way in which the government interacts with government departments, such as the Department of Social Security, and how they actually administer their statutory functions. Thus, not only are there many Acts and pieces of delegated legislation (see p. 21) relating to social security, the health services, land compensation, local government etc., but a large body of case law has also developed. It also deals with nationalised industries, judicial control of public authorities and interpretation of statutes, tribunals and enquiries, and remedies in the courts.

Private law (civil law)

This deals with law relating to relationships between citizens. Whether a citizen has rights or duties depends on which branch of private law he is concerned with. For example, only parties to a contract will normally have rights under that contract. Should there be a breach of duty, then the party suffering will be able to sue in the **civil** courts, i.e. in the non-criminal courts.

The following are some branches of English private law.

Contract

This deals with legally binding agreements between people, e.g. a contract for the sale of land, a contract of employment, a contract to buy a toothbrush, or a building contract.

Each party can only expect from the other what has been agreed between the parties, or what is implied into the agreement by Act of Parliament or the common law. By S.14 of the Sale of Goods Act 1979, when buying materials from a builders' merchant, one expects, without having to express the wish, that the goods will be of satisfactory quality. If they are not, then the party in breach could be sued for damages, or the goods could be returned on delivery.

Tort

A tort is a civil and not a criminal wrong. It is not punishable by the state, unless the behaviour amounts to a crime as well. Each tort has grown up in a separate way, and has its own special requirements and remedies. Examples of torts include negligence, trespass, private nuisance and defamation (libel and slander) (see Ch. 7).

Law of property

This deals with the ownership or title to land and personal property, and the transfer of such property (see Chs 4 and 5).

Family law

This deals with all matters relating to the family, such as adoption, legitimacy, divorce, marriage and nullity. Such law is derived, almost entirely, from statute law.

Succession

This concerns the transfer of property on death, whether the deceased died testate, having left a will containing directions, or intestate, having left no will, when a statutory order of transferring the property to the next of kin is applied.

Other types of law affecting us

One or two other types of law should also be noted, as we must remember that English law is merely one example of national or municipal law.

International law

Public international law

This deals with the interrelationships between countries, on a political level, and is not concerned with the internal or municipal legal systems. Thus, instead of talking about for example the law of England and Wales or Nevada, we are now concerned with the general effect of international law on the United Kingdom and the United States of America.

Public international law is not all-embracing, but covers legal problems of international significance, such as the law of the sea. This deals with international fishing rights, territorial waters, oilfields and piracy. The law of the air is concerned with, among other things, violations of air space and hijacking.

Such law is created by political leaders signing treaties or conventions on behalf of their respective governments. Only the parties to the treaties, or subsequent signatories, become subject to the agreement thus created. Thereby hangs the problem. A country not signatory to a particular treaty or convention cannot be politically punished by the other signatories, because they were never party to it in the first place. Such punishment as can be meted out depends on the international standing of the signatory countries. If the recalcitrant country is a member of the United Nations, a complaint can be made there, or offensive action can be taken, as in the Falkland crisis in 1982 or the Gulf War in 1991, or economic sanctions can be taken. The content of a treaty or convention may be incorporated into a state's internal law using that state's legislative system, e.g. the European Communities (Amendment) Act 1993 ratified the Treaty on European Union (the Maastricht Treaty).

Private international law

This deals with the legal problems when two or more municipal (internal) legal systems are involved in one situation. Such a conflict between law occurs frequently in the building industry, when an English or Scottish company contracts to build in another country such as Iran. If there is a legal dispute, then where is the action heard? It depends on many factors. Different countries have different rules. Sometimes an English court applying English law will determine that it may hear a case or, if the case has been heard abroad, will recognise a foreign judgement or not. Since 1991 in England and Wales the question as to which country's law governs a contract will be determined by rules set out in the Rome Convention signed by the United Kingdom in 1981 and subsequent conventions referred to and incorporated into English law by the Contracts (Applicable Law) Act 1990 as amended. Often, where there is an American state involved, the party suing will try to have the case tried in that state as, on the whole, damages awarded in the United States are much higher than those awarded here.

On a more personal level, individuals often get involved in private international law, in divorce cases or on death, e.g. an Englishman marries a French girl and goes to live in Utah. She leaves him, and wants to get a divorce in California. Once again, the law is quite complex, and may involve (in England) the principle of an individual's **domicile**. This is not his residence, but rather relates to the legal system to which the individual owes his allegiance. So, although the Englishman is British by nationality, as far as the divorce is concerned he may have changed his domicile to that of Utah, if he no longer felt bound to the English legal system. If he was living there temporarily, then this is unlikely.

Note: when a person goes to a foreign state, he must while there abide by its criminal laws, or face the consequences. There is no conflict of law in such a situation, and no choice in the matter. Furthermore, in most countries, ignorance of the law is no defence.

European Union law

This is a relatively new source of English law; see p. 18.

Origins

The original European Economic Community (EEC) was set up in 1957 by the first Treaty of Rome, which was signed by the original six member states. Two other communities were also set up, the European Coal and Steel Community (ECSC) by the Treaty of Paris 1951, and the European Atomic Energy Community (Euratom) by a second Treaty of Rome 1957. The administrative aspects of the three communities were merged in 1967, but they remain separate legal entities. The first Treaty of Rome was amended by the Single European Act 1986. As a result the EEC was from then on referred to as the European Community (EC) and the first Treaty of Rome as the EC Treaty. The Treaty on European Union (the Maastricht Treaty) 1991 was finally adopted by all the member states in 1993. The Maastricht Treaty amended the EC Treaty and now the collective communities are referred to as the **European Union** (EU).

Objects

The objects of the EU are to create a huge common market for all member states, in which workers, capital, transport and goods can move about without restraint, thereby encouraging prosperity in Western Europe. In particular, as stated in Article 2 of the first Treaty of Rome (the Treaty) as amended, the main object of the Community is to 'promote throughout the Community, a harmonious development of economic activities, sustainable and non-inflationary growth respecting the environment, a high degree of convergence of economic performance, a high level of employment and of social protection, the raising of the standard of living and quality of life, and of economic and social cohesion and solidarity among Member States'.

In 1985 a further step was taken by all member states signing the Single European Act 1986 in order to create the Single European Market on 1 January 1993. With this move trade barriers were abolished, border controls for Union nationals were eliminated and employees were permitted to move freely between the states.

The Maastricht Treaty was also concerned with, among other things, a common foreign and security policy, the setting up of a European Central Bank, the proposition of a single currency by the end of the millennium and the right of all EU citizens to vote or even to stand as candidates in local and European Parliamentary elections.

Accession by the United Kingdom

In 1971, the United Kingdom, Ireland and Denmark signed the Brussels Treaty of Accession, by which they agreed to join the three communities. Like all treaties, this would have remained a mere political agreement had the United Kingdom's Parliament not passed the European Communities Act 1972. This came into effect on 1 January 1973. The Act ratified (confirmed) the treaty and incorporated the three treaties as part of the law of the **United Kingdom**, as well as all the case law from the European Court of Justice. At one fell swoop there was a new source of English, Scottish and Northern Irish law. The Single European Act 1986 and the Maastricht Treaty were ratified in the United Kingdom by the European Communities (Amendment) Act 1986 and the European Communities (Amendment) Act 1993 respectively. Thus, the English courts must apply community law, if it is applicable. If there is conflict, EU law prevails.

EU institutions

The EU is run by four main institutions, the Council of Ministers of the European Union, the European Parliament, the European Commission of the European Union and the European Court of Justice (ECJ). Each member state has voluntarily given up to these bodies some of its own rights to make law and decisions (sovereignty). This is because the communities are **supra-national**, having authority to make decisions for all the member states on certain community matters, e.g. the Common Agricultural Policy (CAP).

The **Council of Ministers** of the European Union, which sits at Brussels, is the principal executive body of the EU. The Council of Ministers contains one appropriate representative from the government of each member state. Thus, if an environmental matter is to be discussed then an environment representative is sent. The Council of Ministers has the final decision on proposals put forward by the Commission and legislation may be passed with a qualified majority (achieved by weighting the votes among the states), but complete unanimity is required in foreign and security policy, justice and home affairs, after consultation with the European Parliament. The Council also comprises a number of working groups of officials from the member states.

The European **Commission** is the executive of the EU. It sits in Brussels and consists of 20 independent members appointed by the member states, two from Germany, France, Italy, Spain and the United Kingdom and one from each of the other states. The president is appointed by common accord between governments after consulting the European Parliament, but there is a power of veto which applies to commissioners as well as to the President. Commissioners are nominated by governments in consultation with the new President and the resultant list is then sent for approval to the European Parliament.

The European Commission implements EU policy, and initiates and draws up proposals for legislation for the Council to approve. It also polices and enforces EU law and has extensive investigative powers. It can start proceedings before the European Court of Justice. Decisions are decided on by majority. Member states must inform the Commission of proposed domestic law. It is divided into a number of directorates. For example, Directorate General XI deals with the environment, consumer protection and nuclear safety.

The **European Parliament** sits at Strasbourg. It has evolved from a mere advisory assembly into a directly elected representative Parliament having equal rights to those of the European Council on budgetary matters. Its powers *vis-à-vis* the introduction of new legislation have been extended by the Single European Act and the Maastricht Treaty so that there is now a conciliation committee designed, where possible, to produce a joint statement from the Commission and Parliament. It can set up inquiry committees to investigate maladministration and any citizen of the EU can directly petition Parliament on community matters that affect him. He can do this whether he acts in a private capacity or as a member of an organisation, including companies. It has been suggested that this right is not used enough.

The member states

Since 1 January 1995 there are 15 member states: Spain, Portugal, United Kingdom, Denmark, Belgium, Italy, France, Finland, Ireland, the Netherlands, Germany, Luxembourg, Austria, Greece and Sweden (mnemonic SPUD BIFFING LAGS).

EU law

The three community treaties are obviously the main source of community law which the Single European Act 1986 and the Maastricht Treaty supplement, but the Commission and Council are allowed to make three types of law, called regulations, directives and decisions.

1. Regulations. These are rules which, once made, are immediately binding on all the member states, without reference to their legal systems.

   Examples are EC Council Regulation 1210/90 establishing the European Environmental Agency and EC Council Regulation 2062/94 which set up the European Agency for Safety and Health at Work.

2. Directives. These are orders or requirements, directed to all or some of the member states, the results of which are binding, but the means to achieve them are left to each member state. This is useful because of the difficulties in translating legal ideas between different legal systems.

An example of this is Directive 86/188/EEC. This was the Council Directive 'The protection of workers from noise' which in the United Kingdom was implemented under the Health and Safety at Work etc. Act 1974 in the Noise at Work Regulations 1989.

Proposals for directives may come from the Council of Ministers or the Commission. Preparation of drafts of directives is undertaken by the Commission who may set up technical working groups where necessary. Directives must be approved by the Council of Ministers. In some cases Ministers may have the power of veto but on others only a majority agreement is required. Implementation of a directive is normally within a specified time, usually two years. Once the domestic law has changed there has been formal compliance.

These two types of law and the treaties are binding on **all** the member states and their citizens – in effect it becomes new English law.

3. Decisions. These are decisions on some aspect of community life, which are binding on the person or body to whom they are addressed.

An example of these is EC Council Decision 91/690/EEC which approved the amendment of the Montreal Protocol on substances that deplete the ozone layer.

European Court of Justice

This court, situated at Luxemburg, is a court specially set up to deal with problems of community law. Each state sends at least one judge. The court rules on the meaning of the treaties and, under the Treaty of Rome, a national case must be sent to the court for clarification of the community law involved if there is any doubt as to the treaties' meaning (Art. 177 EEC Treaty).

The court also has jurisdiction over disputes involving aspects of community law between states, corporations and even individuals.

The court therefore provides another source of English judicial precedent binding national courts by its decisions. The ECJ itself is not bound by its own previous decisions (see p. 21).

An example of a case brought before the ECJ is Case C-415/93 *Union Royale Belge des Sociétés de Football Association ASBL* v. *Bosman; Royal Club Liégois SA* v. *Bosman; Union des Associations Européennes de Football* v. *Bosman* 1996. The ECJ was concerned with the fairness of the rules of the Belgian Football Association and the European governing body, payment of transfer fees, fielding of foreign players and restriction of employment as contrary to EC Treaty Art. 48.

In addition a Court of First Instance was established by the Single European Act 1986 to which cases can be brought directly to the ECJ without the necessity for a referral from a domestic court hearing.

Sources of English law – creating the rules

We have already seen how English law developed historically. Also, we looked at the different categories of English law – but how is English law created today?

There are two main methods (see Figure 1.6): legislation and judicial precedent.

(Note: EU law is an external source of English law but of fundamental importance.)

Legislation

In England and Wales, legislation is law made by the Houses of Parliament (the Commons *and* the Lords). As Parliament fulfils two roles in the United Kingdom, governmental and legislative, much of the legislation covers all of the country and not just England and Wales.

Legislation can *create* new law completely from scratch. It may also *repeal* old legislation when it is no longer relevant or useful, *amend* existing legislation to cover new situations or plug loopholes and *consolidate* piecemeal legislation into one comprehensive Act. Occasionally, an Act may be used to *codify* law which has been created by case law or case law and legislation (mnemonic; CRACC).

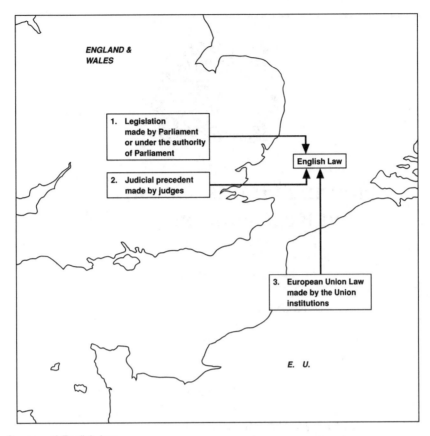

ENGLAND & WALES

1. Legislation made by Parliament or under the authority of Parliament

2. Judicial precedent made by judges

English Law

3. European Union Law made by the Union institutions

E. U.

Figure 1.6 Sources of English law

Development

There has been legislation since Anglo-Saxon times, when the king, having sovereign power, issued laws called Dooms on something of importance. The Domesday Book of 1086 was the result of such a doom. Other types of legislation were also made, e.g. statutes, provisions and charters (Magna Carta). Today we still have charters (see Ch. 6) and the word statute is an alternative word for Act of Parliament.

Legislation, thus, is a most sophisticated form of law-making, relying on the wishes of Parliament, hopefully reflecting the needs of the people. Once a piece of legislation has been made, it must be applied in all the courts and by judges and magistrates of whatever political persuasion. However, if a Bill has a political bias, then whether it goes through both Houses successfully will

depend on the strength of the party in government. A governing party is the one with the majority of seats in the House of Commons.

Direct and indirect legislation

Legislation nowadays takes two basic forms – direct and indirect. But all legislation stems from an Act of Parliament in some way.

If we look at Figure 1.7 we see the first page of a typical Act, the Housing Grants, Construction and Regeneration Act 1996. After the Royal Coat of Arms comes its year date and Chapter No. 53, i.e. the 53rd Act of the year. Following this is the preamble, which sets out briefly what the intentions of the Act are. The reference to the Queen indicates that the Royal Assent has been given, and that the House of Lords *and* the Commons have passed the bill. At the beginning of the Act,

Housing Grants, Construction and Regeneration Act 1996

CHAPTER 53

ARRANGEMENT OF SECTIONS

Part I

Grants, &c. for renewal of private sector housing

Chapter I

The main grants

Introductory

Section
1. Grants for improvements and repairs, &c.
2. Applications for grants.

Preliminary conditions

3. Ineligible applicants.
4. The age of the property.
5. Excluded descriptions of works.
6. Defective dwellings.

Renovation grants

7. Renovation grants: owner's applications and tenant's applications.
8. Renovation grants: certificates required in case of owner's application.
9. Renovation grants: certificates required in case of tenant's application.
10. Renovation grants: prior qualifying period.
11. Prior qualifying period: the ownership or tenancy condition.
12. Renovation grants: purposes for which grant may be given.
13. Renovation grants: approval of application.

Common parts grants

14. Common parts grants: occupation of flats by occupying tenants.
15. Common parts grants: landlord's and tenants' applications.

Figure 1.7 First page of Housing Grants, Construction and Regeneration Act

you will find the arrangement of the sections of the Act. Towards the end of each Part are miscellaneous and supplementary provisions, including definitions. For example, the expression 'construction operation' is defined **for the purposes of this Act**. Such a definition may then be adopted for other later Acts.

1. Direct legislation

Here, the law is actually contained in the new Act itself. This means that Parliament debated all the law found in the Act, and took it through the lengthy legislative process from Bill to final Royal Assent. Obviously, as Parliament has a limited amount of time to deal with all matters, pieces of direct legislation are usually going to be of vital importance to the country, or are considered to be so by Parliament, e.g. Finance Acts.

So, in the case of direct legislation, reference must be made to the Act to find out what Parliament enacted, and subsequent precedents, to see how the Act was applied in practice.

2. Indirect or delegated or subordinate legislation

Sometimes Parliament sensibly acknowledges that it neither has the time, expertise, or even inclination, to deal with certain problems requiring legislation. It therefore solves the problem by passing what is known as an **enabling Act**, setting out the aims of the Act and giving to other people or bodies the power to make the necessary legislation. Legislation made in this way is called **indirect** or **delegated** or **subordinate** legislation.

Despite these descriptions, however, such legislation has **all** the force of Parliament, because it could only be made following the transference of power from Parliament to the designated persons or bodies.

A classic example of an enabling Act is the Health and Safety at Work etc. Act 1974. The Act itself is relatively short because it transfers virtually all of the technical, policy-making and enforcement powers to others, mainly the Health and Safety Commission and the Health and Safety Executive.

Legislation made by such people as ministers of state or bodies such as the Health and Safety Commission are usually called **rules** or **regulations** and **orders**. In the case of local authorities who are empowered to make rules for their areas by virtue of the Local Government Act 1972, these rules are called **byelaws**. Most forms of delegated legislation are published in a document called a **statutory instrument** (instrument meaning a document). Thus, the Building Regulations are published in Statutory Instrument reference SI 1991 No. 2768 and the relevant enabling Act is the Building Act 1984.

Judicial precedent (case law or judge-made law)

If someone buys a round of drinks for his office colleagues because it is his birthday and no-one has done this before, then we might say that he has created a precedent. The others would then be expected to follow suit. It is this concept that is at the root of the legal doctrine of precedents. But, of course, it is not quite as simple as that.

The elements of a court case

When a civil or criminal case goes before a court, there are two elements involved in coming to a decision, the relevant *facts* and the relevant *law*.

The facts are relatively simple to determine on the evidence presented. (See the rules relating to evidence on p. 38.) But the law may be more difficult to apply and will even cover the acceptability of the evidence. This is due to the human element altering each factual situation, so that the law must be applied differently in each set of circumstances. This is so even if the facts of the case appear to be identical to those in another case.

Thus, in every court hearing, the judge or the magistrates must apply the appropriate legal rules which may have been created by statute or by judges in previous cases.

Precedents

A precedent (from the verb to precede) is a decision made by a judge on a point of law. There are four kinds of precedent: (1) original; (2)

declaratory, both of which may be (3) binding; and (4) persuasive, which cannot be binding.

Original precedents

If there is no relevant law applicable to a case being decided, because a particular aspect has not been considered judicially before or there is no relevant piece of legislation, then the judge must make a decision on what he thinks the law *should* be. He will come to his decision after listening to the arguments of the lawyers on both sides. Their arguments will be based on precedents gleaned from previous cases. For this reason, the judge cannot and should not come up with a startlingly unusual decision. He will take the other precedents and try to mould them to fit the case in hand. By doing this he is said to 'uncover' what the law is. His decision is called a precedent and if his decision appears to break new ground, even in the smallest way, then it is called an **original** precedent.

But precedents cannot be plucked out of thin air. There must already be some law on the subject-matter in existence. Significant and material changes in the law in a new direction can only be achieved by statute law.

Original precedents are of vital importance in the English legal system, as they create new laws to solve new problems posed in cases brought before the courts. This is an advantage over statute law as precedents can only be made to fit the facts of a case which have actually happened, however weird or fantastic. Statute on the other hand is made in a sort of vacuum and law may be created for situations which may never occur or which may never happen again.

Binding precedents

Precedents are even more important, however, if they are made by judges in one of the superior courts, e.g. the House of Lords, Court of Appeal or even the High Court, for their decisions **bind** the courts below. The lower courts, in cases dealing with the same legal situation at a later date, must apply the precedent, even if the judge privately disagrees with it. As the courts are operated as a hierarchy, it is relatively simple to determine which courts bind lower courts by their decisions (see Figure 1.8). Neither the magistrates' nor the county courts make binding precedents, nor does the Crown Court, unless it is sitting as an appeal

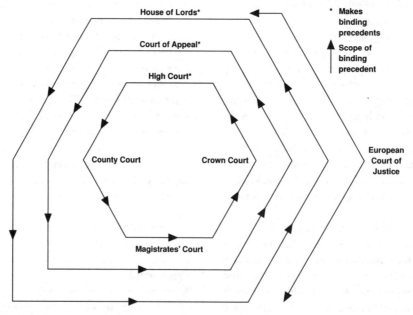

Figure 1.8 The English court system with special reference to those courts which bind others by their decisions

court. As a rule of thumb, only courts sitting as appeal courts make binding precedents, except for the High Court which can do so even in cases heard **at first instance** (for the first time).

The House of Lords does not bind itself by its own decisions, although it will endeavour to follow them in most cases.

Generally, other courts are bound by the decisions of courts higher up the ladder. The Court of Appeal (Criminal Division) is not necessarily bound by its own decisions, unlike the Court of Appeal (Civil Division) which binds itself.

(Note: the European Court of Justice can overrule the House of Lords in matters relating to the European Union. Also, legislation may override decisions of the Lords.)

Persuasive precedents
Most cases, however, are not taken on appeal to other courts. Thus, if the judge, during the case, makes an original precedent, it may not be binding, but it will be a **persuasive** precedent. This means that another judge, faced with the same legal problem, would undoubtedly be influenced by that decision. Unfortunately, not all such cases are reported in the law reports, and therefore finding details of relevant decisions is difficult. This is why one must keep abreast of the legal pages in professional journals, in which contributors give first-hand reports of cases to readers.

Persuasive precedents are also made by courts in other countries which have common-law systems, e.g. New Zealand and most states in Canada and Australia, and by the Judicial Committee of the Privy Council (JCPC). The JCPC hears appeals from certain Commonwealth countries.

Declaratory precedents
Where a judge is using a precedent that is already made, then he merely applies the precedent in a declaratory way, i.e. he declares how the law already stands. For example, in every criminal trial, the judge will tell the jury that they must only find the defendant guilty if they believe the evidence proved the case beyond all reasonable doubt.

Ratio decidendi and *obiter dicta*
In coming to a decision on a point of law, the judge must give his reasons for doing so. This is called the **ratio decidendi**. This is the only part of the decision which may be binding.

Occasionally, to illustrate his argument, he uses hypothetical examples to further explain his decision. This is called **obiter dicta** or words by the way. As the obiter dicta is not strictly based on the case in hand, it can never be binding – but it may be a **persuasive** precedent in future cases.

Judge v. jury
You will have noticed that no reference has been made to a jury in the above. This is because, first, juries are usually found only in serious criminal cases in the Crown Court. Generally, few building problems are dealt with in that court. Second, a jury's duty is to give a decision on the **facts** of the case. Thus, it is not concerned with legal arguments, and indeed, wherever a point of law is argued, the jury is asked to leave the courtroom.

Later cases on similar facts
When a judge is hearing a new case based on similar facts to those on which the binding precedent has been made, and that precedent is referred to him by one of the lawyers involved, he may do two things. He may agree with the precedent and **follow** it, or he may say that the facts of the current case are different and that he will not follow the precedent, and will therefore **distinguish** it.

Appeals
If a case goes on appeal, and the appeal court agrees with the court below, it is said to **uphold** or **affirm** the decision.

If it wishes to go against the lower court's decision, then it **reverses** it.

A decision may, or may not, involve a precedent. If the appeal court is considering a precedent set by the lower court, it may **approve** or **disapprove** it and **overrule** it. If the precedent referred to was set by a higher court than the one being appealed to in the current case, then it cannot **overrule** the precedent.

The law reports

The proceedings of courts of record, i.e. those courts which can make binding precedents, are reported in special journals called law reports. There are many different types. The law reports are published by the Incorporated Council of Law Reporting, representing the legal profession. These are either Appeal Cases (AC) or divisions of the Queen's Bench, Chancery or Family Divisions of the High Court (QB, Ch, Fam).

The Council also publishes the *Weekly Law Reports* (WLR), *All England Law Reports* (All ER), *Times Law Reports* (TLR) and *Building Law Reports* (Build LR).

The reports are prepared by qualified lawyers, usually barristers, who will have been present at the court hearing, and who will have access to the transcript of the case.

Reference to law reports is always made as in Figure 1.9.

Without the law reports, finding out about relevant precedents would be extremely difficult. Sometimes, one sees advertisements in legal journals for information on unusual cases. However, the courts are reluctant to take notice of cases

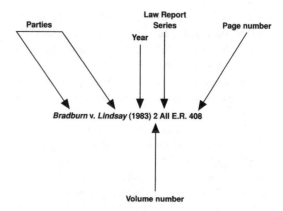

Figure 1.9 How cases are referred to in law books

which have not appeared in the major law reports mentioned above.

Nearly all the cases mentioned in this book contain at least one important precedent. Some are known as *leading* cases as they contain original and binding precedents of such significance that the law has taken a completely new direction (see Murphy v. Brentwood District Council 1990 (p. 170)).

2 The administration of English law

The legal profession

Barristers-at-law and solicitors of the Supreme Court

In England and Wales, the legal profession is divided into two completely separate branches, those of **barristers** and **solicitors**, which have developed in entirely different ways. Collectively, they and legal academics are known as lawyers.

Generally

The man in the street seeking legal advice would firstly need to see a solicitor, whose name could be supplied from the yellow pages of the telephone directory, or the lists kept at Citizens Advice Bureaux.

A solicitor could be described as the general practitioner of the legal profession. He will give general legal advice, and will undertake to do divorces, adoptions, conveyancing (buying and selling land), drawing up of wills, obtaining probate on the death of someone, and debt collection through the courts. Usually, he works at his office, but, if his work involves court appearances, then a solicitor is entitled to present the case (act as an **advocate**) in the magistrates' court, the county court, and in certain situations the Crown Court. If, however, the work involves a trial at the Crown Court, or litigation in the High Court, or there is an appeal to the Court of Appeal or House of Lords, then he must normally obtain the services of a barrister.

Under the Courts and Legal Services Act 1990, solicitors who are good at advocacy have the right to apply to extend their rights of audience to the higher courts. Such solicitors must undergo a further training course and examination in advocacy and, in addition, must obtain a certificate of eligibility based on having taken part in 50 contested hearings over two years. As a result, in 1996, only 420 solicitors out of a total of 70,000 had become solicitor advocates, and of those only 71 qualified via the course. The rest had been barristers originally.

Thus, the barrister's main function is to act as an advocate, and he has a right of audience in all English courts. Except in the magistrates' court, where robing is unnecessary, he will be dressed in the old-fashioned wig with pigtail, winged collar, tabs and a gown. Solicitor advocates are not permitted to wear wigs. Etiquette demands that his **instructing solicitor** be present at the same time in court (and at any time when the client is present), as he will be more familiar with the client as a result of frequent pre-trial meetings.

In addition to his function as an advocate, the barrister, who is always referred to in the profession as **counsel**, will often give specialist advice to solicitors, called **counsel's opinion**. Suppose your company wished to sue another organisation for something which turned on an undecided legal point; your company's solicitor would hesitate before advising the board of directors to commit vast sums of money on a court case, the outcome of which was very much in doubt. He would therefore prepare a **brief** containing all the relevant details of the dispute. In particular he would specifically ask a number of questions, e.g. 'Would counsel give his opinion on the merits of the case, and in particular, estimate the possible **quantum** (amount) of damages that may be awarded'. The brief, tied with pink tape, would then be sent to a barrister having known specialist knowledge in that particular field. If counsel's opinion was favourable, the solicitor would consult with his client and ask him, in view of what counsel said, what he wished to do.

This is where the analogy of general practitioner breaks down, for there are many barristers who could not be said to specialise in such work, and whose main tasks are to do the everyday criminal court work. 'Rumpole of the Bailey', John Mortimer's brilliant creation, comes to mind here. Other barristers, however, specialise in certain areas, such as taxation, matrimonial law, building contracts, and you would be unlikely to find them doing any criminal work. Similarly, some solicitors, especially in London, tend to specialise in certain areas, e.g. shipping, theatrical contracts, and would not wish to handle your divorce.

Solicitors in more detail

Solicitors are governed by the Law Society, which regulates the entrance examinations, articles (apprenticeships), and admits them, on qualifying, onto the **rolls of solicitors**. The Law Society also issues **practising certificates**, without which no solicitor is allowed to practise. Also, compulsory insurance must be obtained through the Law Society to cover against claims for professional negligence.

A client instructing a solicitor will have a contract with him, and therefore can sue him for breach of contract, and can himself be sued for unpaid fees. Indeed, a solicitor who has not been paid by a client has a **lien** over his clients' papers, i.e. he will not release possession of them until the bill has been settled.

Solicitors are not allowed to tout for business, nor could they at one time advertise, except to announce that they have opened a new office, and to give a list of the sort of work they will undertake. Now they are permitted to state their expertise in certain areas and name clients in their advertising provided they have the client's written consent. Also, they are allowed to put a similar list on the outside of their premises. This is because it was discovered that people did not understand the variety of work a solicitor is able to do. Once again, American films, with their divorce and real estate lawyers, where the legal

work is split up between types of law and not between professions as in this country, are to blame for such ignorance.

Solicitors have strict accounting rules, and must have two sets of bank accounts, the **office** account and the **client** account. If money from the client account is put into the office account without authority, whether deliberately or not, serious penalties can occur – the ultimate being **struck off** the rolls of solicitors.

As solicitors are court officers (unlike barristers), they are entitled to have oaths sworn before them. In the past, solicitors had to apply to become commissioners for oaths, but nowadays any solicitor with a current practising certificate is entitled to act as a commissioner. Most people are aware of the oaths sworn in court, e.g. 'I swear to tell the truth, the whole truth, and nothing but the truth' (not 'So help me God' which is American). Swearing an oath before a solicitor, on whatever religious book you require, has the same effect. So, you could be prosecuted for perjury if what you had sworn was knowingly false. If you are sent a form called an **affidavit** (a sworn statement), with a request to get it sworn, you will need to go to any solicitor in practice (but not the one who prepared the affidavit). Holding the Bible, or whatever religious book you adhere to, in your right hand, you say, 'I swear that this is my name and handwriting, and the contents of this, my affidavit, are true'. If you have no religion, you may **affirm**, i.e. promise, that it is true. The solicitor signs his signature, and dates the document.

Barristers

Barristers are governed by the Senate of the Inns of Court and Bar, which has under its control a number of committees dealing with discipline and professional etiquette. The Council for Legal Education deals with the examinations and educational requirements of prospective barristers.

All barristers must belong to one of the four Inns of Court, Gray's Inn, Inner Temple, Middle Temple or Lincoln's Inn. They are all situated close to the Royal Courts of Justice in the Strand in London. However, there are groups of barristers working all round the country, based not in offices, but in **chambers**, in which they 'rent' a room, and usually share both the services of a **clerk** who runs the chambers and the expenses of maintaining the establishment. Since the Courts and Legal Services Act 1990, there is no longer a legal requirement to have a clerk. For convenience, all sets of chambers tend to be very close to the town's courts.

After passing the required examinations, and eating a certain number of meals at his Inn, the prospective barrister is called to the **Bar**. If he intends to practise as a barrister, he must then do a year's **pupillage** (apprenticeship) with an experienced barrister.

There are a few aspects of the barrister's work which must be particularly noted.

First, there is no contract between the client and the barrister, for it is the solicitor who instructs him on behalf of the client. Strictly speaking, there is no contract either between the lawyers. A barrister's fees are called an **honorarium**, i.e. a voluntary payment for professional services. Therefore, a barrister cannot sue for his fees. In practice, this does not matter, for it is the solicitor who collects the fees with his bill, and he can sue the client for breach of contract. Also, if the solicitor does not pass on the fee, the barrister would not undertake any further work for that solicitor, and would undoubtedly tell his colleagues to do the same. Furthermore, the solicitor may be reported to the Law Society for professional misconduct. If the solicitor remains unpaid, he is professionally obliged to settle counsel's fee, however.

Second, it used to be a hard and fast rule that no one could sue a barrister for professional negligence. This was slightly modified by the case of *Saif Ali* v. *Sydney Mitchell (a firm) and others* 1978, in which it was stated that, as a point of law only, it should be possible to sue a barrister for pre-trial work, but not in relation to his work at the trial itself. There has yet to be a case tried on such facts, however. This professional immunity has been extended to solicitor advocates by S.62 Courts and Legal Services Act 1990.

If you ever visit London, a worthwhile trip can be made to visit the Law Courts in the Strand (see Figure 2.1). Around the corner, in Chancery Lane, is the Law Society and the Public Record Office, and close by are the Inns of Court. Up Ludgate Hill, on the left, is the Old Bailey, London's Central Criminal Court which is a Class 1 Crown Court. Details of conducted walks around this area can be obtained from tourist offices.

Queen's Counsel

Barristers who have been practising at the Bar for at least ten years may apply to the Lord Chancellor for a recommendation to be appointed a Queen's Counsel (QC). If his application is successful, he is said to **take silk**, because he exchanges his barrister's gown made of 'stuff' for a silk one. From then on, he takes fewer cases, and in many trials may not appear without another barrister, called a **junior**. Most QCs specialise in some particular area of the law, and most have chambers in London.

The courts

Introduction

In England and Wales, the court system, unified since the Judicature Acts 1873–75, is best studied in two parts, the criminal courts and the civil courts.

The criminal courts deal with pure crimes and behaviour treated by statute as if it were criminal (see Figure 2.3, p. 35).

The civil courts deal with disputes or matters between private citizens, such as breaches of contract, tort, divorce, adoption, trusts, bankruptcy, winding up of companies, administration of estates on death, mortgages, land actions and admiralty claims (see Figure 2.2).

The civil courts

The magistrates' courts

Jurisdiction (authority, and extent of that authority to administer law)
More properly designated a criminal court, the magistrates' courts have a certain amount of civil jurisdiction.

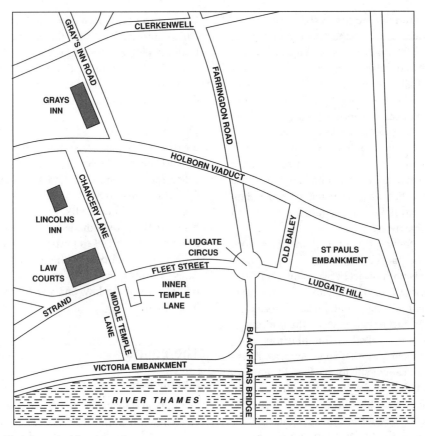

Figure 2.1 Sketch map of legal London

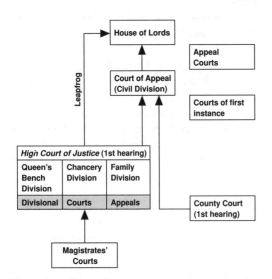

Figure 2.2 The civil court system in England and Wales

1. Domestic and family proceedings. They can deal with domestic disputes between husband and wife, the custody of children, commit children into the care of a local authority, consent to the marriage of minors over 16, decide on the paternity of a child of an applicant and adoptions. When sitting as such, it is called a Family Proceedings Court and will be scheduled so that only family business is dealt with and magistrates drawn from a special family panel are chosen.

2. Licensing sessions. Licensed premises must have current licences issued by the magistrates at special sessions of the court.

3. Civil debts. The magistrates can deal with the enforcement of tax arrears, social security contributions, council tax and gas and electricity supplied. All of these are called civil debts, as the debt is owed to an agency of the state, e.g.

the Inland Revenue, or a utility company, or local government. Because of the nature of the debt, the proceedings are to some extent criminal, and defaulters could be committed to prison.

Venue

Because of their great criminal jurisdiction, there are magistrates' courts in most towns in England and Wales. Whether a particular court has jurisdiction to hear a case will depend on the wording of the Act of Parliament governing the circumstances.

'Judges' (magistrates; addressed as 'Your Worship')

The 'judges' are not proper judges at all, and are called magistrates. They may either be **Justices of the Peace** (who are part-time) or **stipendiary magistrates**.

Justices of the Peace (JPs) are laymen who have been chosen, usually, because of their services to the community. They are appointed for a particular area, known as a **Commission**, by the Crown on the advice of the Lord Chancellor. He chooses them from lists drawn up by countrywide advisory committees. Suggestions for recommendations may be made by anyone to the local Clerk to the Justices. The JPs are not paid, but receive expenses. Because they are laymen, and have only received minimal training, they are advised in court by a **Clerk to the Justices** who sits in front of the **bench** (the collective term for JPs). The clerk is either a barrister or a solicitor who has been qualified for at least five years. When sitting in court, the usual number of justices is three, although it is quite proper for any number between two and seven to hear a case.

Stipendiary magistrates are professionally qualified magistrates, being either barristers or solicitors of at least seven years' standing. They are appointed in cities where the amount of work warrants the need for a full-time magistrate. Only one is required to hear a case.

Procedure

The procedure is less formal than in the county court or High Court.

Civil proceedings are started by a **complaint** to a single magistrate. This is usually done by the complainant visiting the magistrate court office. The complaint may be written or oral. A summons is then issued.

Practice

Details of jurisdiction, procedure and practice may be found in *Stone's Justices' Manual* (the Blue Books) which is published annually. This is a vital book for those who have to appear regularly in magistrates' courts.

County courts

Jurisdiction

These are governed by the County Courts Act 1984 as amended by the Courts and Legal Services Act 1990 and the High Court and County Court Jurisdictions Order 1991 (the 1991 Order). There are over 350 of these courts all over the country and they are of great value to the builder.

Rather like a junior version of the High Court, the county court has jurisdiction over many types of proceedings. All divorces have to be started there, and matters concerning adoptions and guardianship are also dealt with.

Before 1991 cases were brought before the county court or the High Court depending on the sums involved. However, since the 1991 Order made under the Courts and Legal Services Act 1990 and the County Courts Act 1984, the county courts now have **concurrent** (at the same time) jurisdiction with the High Court *whatever* amount is involved in cases of contract and tort, excluding libel and slander. An outline of the methods used in the county court to collect debts is given on p. 138. Other areas of concurrent jurisdiction of interest to the builder include actions to recover land, applications made by tenants to defend claims by their landlords to forfeit their leases for non-payment of rent, applications to renew business tenancies and applications under the Access to Neighbouring Land Act 1992 by someone requiring access to his neighbour's land in order to carry out necessary works on his own land (see p. 207).

Additionally, the county courts have **exclusive** jurisdiction in matters relating to claims for unlawful sexual or racial discrimination and mortgage possession proceedings.

County courts have **limited** financial jurisdiction in certain equity proceedings, e.g. where someone applies to foreclose or redeem a mortgage or enforce a charge or lien, provided the sum outstanding is not more than £30,000. (See p. 100 re mortgages and p. 26 re liens.) Other equitable matters include the dissolution or winding up of a partnership where the assets of the partnership are no more than £30,000 and actions for specific performance (see p. 79) or rectification of contracts where the sums involved are less than £30,000 (see p. 65).

Some county courts are designated as having jurisdiction in bankruptcy in which case they will also have jurisdiction over companies.

County court or High Court?

Commencement The practice involved is found in the 1991 Order. Obviously if the county court has exclusive jurisdiction you must start proceedings in the county court. If it only has concurrent jurisdiction then proceedings may be started in either court. But, if one is suing for damages for **personal injuries** the action must commence in the county court unless the claim is for more than £50,000 (at present), in which case the action may be started in the High Court and the writ must be endorsed with a special statement to that effect. Certain factors are taken into account when a decision is taken as to which court to use. One example is whether the county court has jurisdiction to grant the relief required.

Trial The rationale behind the 1991 Order was to recognise that some large claim actions nevertheless involved simple principles of law and could be easily tried in the county court, a cheaper and simpler procedure. Conversely, some relatively low claims involved difficult points of law and needed to be tried in the High Court. Thus, by Art. 7(3) actions for less than **£25,000** should be tried in the county court **unless** the county court decides to transfer the action to the High Court and the High Court agrees or the case is started in the High Court at the outset and the High Court agrees to try the case. The corollary is that actions for more than £50,000 should be started in the High Court unless it was started in the county court and that court sees no reason to transfer to the High Court or the High Court considers that it ought to transfer the case to the county court for trial (Art. 7(4)).

By Art. 7(5), when the courts are making the decision whether to try or transfer, they must take into account **four** criteria:

1. the *f*inancial substance of the action including the value of the counterclaim (F);
2. whether the action is important or raises matters of *p*ublic *i*nterest (PI);
3. the *co*mplexity of the facts, legal issues, remedies or procedures involved (CO);
4. whether transfer is likely to result in a more *s*peedy *t*rial (ST).

(mnemonic PIFCOST)

Merely trying to achieve a speedier trial is not a good ground for transfer.

Venue By Order 4 of the County Court Rules an action can be started in the local county court for the district where the defendant lives or carries on his business; or in the court in whose district the cause of the action wholly or partly arose, e.g. the car crash happened there; or, in the case of a default action, in any court. If the sum claimed is a liquidated amount (i.e. for a fixed amount) then the action will be automatically transferred to the defendant's home court on his filing a defence. This is to reduce costs and travelling times to a court should the case actually come to trial.

Judges (usually circuit judges; addressed as 'Your Honour')

Since the Courts Act 1971, England and Wales has been divided into six regions or Circuits. Within each region there are a number of High Court judges, circuit judges, recorders and assistant recorders, all of whom, together with Court of Appeal judges, may sit as judges in the county courts.

District judges (addressed as 'Sir' or 'Madam') appointed under the 1984 Act have limited rights in the county courts. They fulfil many functions in addition to their old function as Registrar. They deal with applications and preliminary matters in chambers prior to the court hearing before the judge. In certain cases where the claim has been admitted or the defendant fails to appear, the district judge can determine the action. He can grant injunctions and can fine or commit people to prison for contempt of court. Of most interest to the builder is that he can hear cases that have been referred to arbitration (Small Claims procedure) where the sum claimed is £3,000 or less, taking no account of the amount of counterclaim. In such cases he is sitting as an **arbitrator** and not as a judge and thus has fewer powers. Circuit judges can sit both on less serious Crown Court criminal cases as well as in the county courts. To be appointed, they must have been practising barristers for 10 years, or have been recorders for five years. (Solicitors may be appointed as recorders, so may ultimately become circuit judges.)

Procedure
Actions in the county court are started by **plaint**. Once the plaint has been filed at the court a **summons** is issued and served on the defendant. For an account of a default action, see p. 138. The parties are referred to as plaintiff and defendant.

Other types of proceedings called **matters**, e.g. adoptions, or windings up of companies or bankruptcies, are started either by **originating applications** or **petitions**, depending on the regulating Act and county court rules. In such cases, the parties are called applicant and respondent, petitioner and respondent or appellant and respondent respectively.

Practice
The 'bible' of the county court is the *County Court Practice* (the Green Book) published in three volumes. The rules are called the County Court Rules (CCR).

Other personnel
The administrative side is dealt with by a court staff headed by the district judge. Deputy district judges can also be appointed and the chief clerk may in some cases be able to carry out certain duties delegated to him by the judges. Additionally, the court now employs full-time bailiffs to serve summonses and to execute judgements.

(Note: both the magistrates' court and the county court are courts which have **original** jurisdiction only, i.e. they have no rights to hear appeals.)

The High Court
The High Court is split into three divisions – Queen's Bench, Family, Chancery – each having original and **appellate** jurisdiction, i.e. they can be sitting for the first time hearing a case, or may be hearing appeals from the magistrates' court.

When sitting as an appeal court, the court is called a **divisional** court (see p. 33).

Jurisdiction
See under each individual division.

Venue
Unlike the county and magistrates' court, the High Court has jurisdiction over the whole of England and Wales. The 'headquarters', however, are found at the Royal Courts of Justice in the Strand, London (see Figure 2.1) where the administrative headquarters called the **Central Office** of the Supreme Court may be found. In the county, High Court cases will usually be heard in the Crown Court building, but, of course, the court will be convened as a High Court. The provincial administrative offices are called district registries.

Judges (High Court judges; addressed as 'Your Lordship' – knighted on appointment)
They must either have been circuit judges or have been practising at the Bar for at least 10 years. They are called **puisne** (pronounced puny) judges, in that they are of a lesser standing than the senior judicial officers, such as the Lord Chief Justice.

Procedure
See under each individual division.

Practice
High Court practice and procedure is set out in the *Supreme Court Practice* (the White Books). The rules are called the Rules of the Supreme Court (RSC).

Other personnel
At the Central Office, administration is dealt with by masters and registrars. In the district registries, this is undertaken by district judges.

The Queen's Bench Division (Head: Lord Chief Justice)

Jurisdiction
There are two specialist sub-courts of the Queen's Bench Division: the Commercial Court and the Admiralty Court.

The Commercial Court has judges with a commercial background hearing the case and using a simplified procedure. It deals with such matters as commercial contracts, including building contracts.

Commercial Court judges, because of their expertise, may also act as arbitrators or umpires in commercial arbitrations (e.g. under a building contract) (see p. 40).

The Admiralty Court deals with many other matters, salvage claims and prize claims (property captured from an enemy at sea, as for example in the Falkland campaign). The judges are appropriately qualified.

Procedure
Actions are started by **writ**. For an account of an action begun by writ, see p. 138.

Other types of proceedings called matters may be started by originating summons.

Family Division (Head: President)
This is the most modern of the divisions, created in 1970 by the Administration of Justice Act 1970. Once again, it has similar jurisdiction to the matrimonial and family cases of the county court.

Jurisdiction
This deals with defended divorces and nullity actions, judicial separations, recognition of foreign decrees of divorce, ancillary financial matters relating to proceedings, custody and protection of children, protection from molestation and violence, wardship proceedings etc.

Procedure
Divorces and certain other proceedings are commenced by **petition**; other proceedings may be started by originating application or originating summons.

A subdivision of the Family Division deals with the granting of probate, or letters of administration on death. These are obtained, either from the **Principal Probate Registry** in London's Somerset House, or at one of the district probate registries. Applications for a grant are done by filing an executor's or administrator's oath.

Chancery Division (Head: Lord Chancellor – in practice the Vice Chancellor)

Jurisdiction
Obviously, this deals with predominantly equity matters, such as trusts, mortgages, rectification of deeds, specific performance of contracts, interpretation of documents such as contracts and wills.

In addition, it deals with revenue cases, company law in the Companies Court, as well as contentious probate (i.e. where there is a dispute) and patents in the Patents Court.

Venue
Until recently, nearly all chancery cases were heard in London, at the Royal Courts of Justice, which, whilst being convenient for the judges and the chancery barristers, was less convenient for the parties and their witnesses. Now, chancery cases can also be heard at certain designated regional centres (sitting in the local Crown Court building).

Judges
There are judges specially assigned to the Chancery Division created from the chancery Bar, which consists of barristers specialising in chancery matters.

Procedure
Originally, the Chancery Court used no writs. However, it is possible to use writs in certain

circumstances in the Chancery Division nowadays. Other modes of starting cases are by such means as originating summonses and **motions**.

The Divisional Courts

1. *Queen's Bench Divisional Court.* This hears criminal appeals on points of law from the magistrates' court or the Crown Court **by way of case stated** (see p. 36).

    It also has jurisdiction of a supervisory nature over lower courts and tribunals if they have failed to exercise their own jurisdiction properly. The court exercises this supervisory jurisdiction by the use of **prerogative orders** of **mandamus**, **prohibition** and **certiorari**.

    Mandamus means 'we command'. It is used to order the performance of a public duty, e.g. where a local authority has failed to do something required of it by law.

    The order of prohibition is used to prevent a lower court from exceeding its jurisdiction, acting against the rules of natural justice, and to prevent a government minister or public corporation abusing their quasi-judicial functions, e.g. where a government minister refuses to hold a planning enquiry.

    The order of certiorari is used to command a lower court or tribunal to certify, as to some aspect of a judicial matter, that it should be investigated by the High Court.

    Most appeals are heard by two judges.
2. *Chancery Divisional Court.* This deals with appeals from the county courts on questions of bankruptcy and land registration. Also, it hears appeals from the Commissioner of Inland Revenue.
3. *Family Divisional Court.* This hears appeals on family matters from the magistrates' court, county court and Crown Court.

Court of Appeal – Civil Division (Head: Master of the Rolls)

Jurisdiction

As its name implies, this is a court with pure appellate jurisdiction. It can hear appeals from the High Court, the Queen's Bench Divisional Court,

county courts, the Admiralty Court, the Commercial Court and the Employment Appeals Tribunal, the Land Tribunal, the Patent Appeal Tribunal, the Restrictive Practices Court, the Social Security Commission Tribunal, and from arbitration appeals which have gone to the divisional court, or from commercial judges sitting as arbitrators.

Judges

These are called Lords Justices of Appeal, and are appointed from the ranks of High Court judges.

They remain knights and do not become lords in the normal sense. But they are also made Privy Councillors on appointment.

The head of the civil division of the Court of Appeal is the Master of the Rolls.

Venue

Normally, the appeal is heard by two judges in the Royal Courts of Justice in London.

Procedure

Notices of appeal must be served on the other party within four weeks of the judgement in the lower court. The notice must contain sufficient details of the grounds for appeal.

The appeal is 'by way of rehearing' similar to that in the Crown Court (see p. 36), but the witnesses are not called, and transcripts of their evidence are read, and counsel for all parties put forward their arguments.

House of Lords

Jurisdiction

This is the highest court of appeal for all British courts in civil matters. (Note: its criminal jurisdiction is limited to England, Wales and Northern Ireland.)

It is part of the House of Lords, but when sitting in its judicial capacity, its lay peers, i.e. the hereditary life peers and bishops, are no longer allowed to sit as judges.

Generally, only cases which raise a point of law which is of public importance may be taken to the House of Lords. A statement to this effect may be made, either in the Court of Appeal or the High Court. In the latter case, it is possible, in cases of great public importance, to by-pass or **leap frog** the Court of Appeal, and take the appeal straight to the House of Lords. In such cases, however, all the parties to the action must agree, as must the House of Lords Appeal Committee. Furthermore, in this situation, only certain aspects of the law may be the subject of an appeal to the House of Lords, e.g. concerning a binding precedent (see p. 22) of the House of Lords itself or the Court of Appeal.

Judges
In the House of Lords, the judges are called Lords of Appeal in Ordinary (to distinguish them from the Lords Spiritual, who are the bishops, and the Lords Temporal who are the lay peers).

A law lord must have been at the Bar for at least 13 years, or have been a senior judge for at least two years.

Venue
The appeal is heard in a committee room at the House of Lords at Westminster.

Procedure
At least three lords called an Appellate Committee listen to counsel arguing their case. Afterwards, they vote whether to allow or dismiss the appeal. Their judgments take the form of written speeches, which often take some time to prepare.

Criminal cases

Introduction

All criminal cases begin in the magistrates' court. Whether the case is actually tried there depends on whether the offence is an **indictable** offence or a **summary** one. At the most basic, indictable offences are very serious crimes, summary being less so. But, unfortunately, there are some offences triable either way, which can be tried summarily or on indictment depending on certain circumstances. Indictable offences were originally created at common law, although nearly all of them are now statutory offences, and thus have the benefit of trial by jury. Summary offences and

offences triable either way, on the other hand, are purely the creation of Parliament and listed in Schedule 1 of the Magistrates' Court Act 1980, and this has increased the jurisdiction of the magistrates, especially in the twentieth century. Whether a trial is to be by magistrates or by jury in the Crown Court will depend on many factors set out in the Magistrates' Court Act 1980 as amended (see Ss. 17–27).

Both the prosecutor and the accused are asked which appears to be more suitable, summary trial or trial by jury on indictment. The court then decides which method is most appropriate. If the magistrates decide that summary trial is to be preferred, they tell the accused who is then given the opportunity to opt for jury trial. If the magistrates opt for jury trial then he is merely given that information. The magistrates are helped by the National Mode of Trial Guidelines 1995 issued by the Lord Chief Justice which suggest but do not lay down the mode of trial in appropriate cases, e.g. handling stolen goods should be tried summarily unless the offence has professional hallmarks or the property is of high value.

Summary offences

Examples of summary offences
Being drunk and disorderly; failing to pay social security contributions where liable to do so; obstructing the highway under the Highways Act 1980; using a television without a licence; leaving litter contrary to the Environment Protection Act 1990.

Magistrates' court
Procedure
Criminal proceedings are started either by arrest, charge and production to court of the defendant or by an **information** and not by complaint as in civil matters. In the case of an information this is **laid** before the magistrates in court by the informant, whether it is the police, or a member of the public, or a private individual. If a warrant needs to be issued for the arrest of the defendant, then the information must be in writing on oath. Where

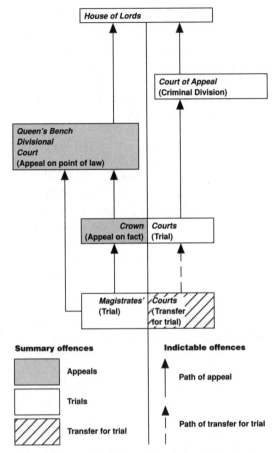

Summary offences

Appeals

Trials

Transfer for trial

Indictable offences

Path of appeal

Path of transfer for trial

Figure 2.3 The criminal court system in England and Wales

there is no need for a warrant for arrest a summons is then issued to summon the defendant to court to answer the charge (unless the charge is one where the defendant can plead guilty by post, e.g. for speeding). The information must contain sufficient particulars to deal with the case.

At the hearing, the Clerk to the Justices reads the charge and asks for the accused's plea – guilty or not guilty.

1. *Plea of guilty.* If he pleads guilty, then there is, of course, no trial, and the bench proceeds to sentencing immediately. The accused person's record is read and any relevant reports, such as social workers' and probation officers', are studied. If the convicted person is legally represented, then a plea in mitigation may be made. This is a speech to the bench in which the defendant's legal representative will try to explain, but not justify, the reasons for committing the offence, and hopefully secure a lesser punishment. Where the magistrates do not feel they have adequate sentencing powers (which have been restricted by statute) they can commit the convicted person to the local Crown Court, where the judges' powers are less restricted.

2. *Plea of not guilty.* If the accused pleads not guilty, however, then the case must proceed. The prosecution will outline the facts of the case briefly, and call the witnesses for the prosecution, i.e. all those witnesses whose testimony will support the *prosecution's* case. The Crown Prosecutor or policeman asks questions which will elucidate the supportive answers. This is called the **examination-in-chief**. At the end of this, the witness may then be cross-examined by the defendant's legal representative or, if unrepresented, by the accused himself. Should greater clarification be needed by the prosecution (to try and mend the holes picked in the case by the defendant), then he may request a re-examination of the witness. The defence then calls his witnesses. After hearing all the evidence, the Crown Prosecutor makes his final speech followed by the defence lawyers. The bench then retires to think about their decision. Should they require help on legal aspects, then the Clerk to the Justices must assist. If they decide the defendant is not guilty, then he is acquitted without a stain on his character. If, however, he is found guilty, then the magistrates proceed to sentencing, as described in the previous section.

Appeals – from the magistrates' court

To the Crown Court

An appeal on facts relating to the conviction, or against the sentence, may be made to the Crown Court. Generally, only the defendant may make such an appeal.

The case will be reheard by a judge and two or more justices (not the ones who heard the case in the magistrates' court). Of course, there is no jury. All the evidence is presented again.

To the Divisional Court of Queen's Bench

An appeal on a point of law relating to the case may be made to the Divisional Court of the Queen's Bench (see p. 33).

The appeal takes the form of being by way of case stated. This means that there is some legal aspect of the original trial which needs to be clarified. Any party to the proceedings may appeal, unlike appeals on fact or sentence in the Crown Court.

The party must apply to the magistrates' clerk in writing stating the question of law at issue. He then drafts the case and sends copies to both parties and magistrates. Following redrafting, if necessary, the case is then sent to the High Court.

There is no rehearing. Counsel for each side argue the points of law.

Appeal to the House of Lords

An appeal from the Divisional Court of Queen's Bench may be made to the House of Lords. But the court must first **certify** that a point of law of general public importance is involved. Either the Divisional Court or the House of Lords itself must give **leave** to appeal. Thus, the House of Lords only deals with the most important and significant cases.

Indictable offences

Transfer for trial procedure

By S.4 Magistrates' Court Act 1980 as substituted by the Criminal Justice and Public Order Act 1994 Schedule 4, where someone is charged before a magistrates' court with an offence which is triable on indictment only or one triable either way and the court has decided the trial is more suitable for trial by jury or the accused has not consented to being tried summarily, then the court and the prosecutor proceed to transfer the proceedings to the Crown Court. The prosecutor then serves a notice of his case on the magistrates' court specifying

the charges to be transferred for trial. The accused is also served with a copy. He may then make an application for the charge to be dismissed. The prosecutor can oppose the application. The court considers the written evidence and any oral representations and may dismiss a charge if it appears that there is not sufficient evidence to put him on trial. Dismissal of the charge or any of the charges against the accused has the effect of barring any further proceedings on that charge except in certain special circumstances.

Where no application for dismissal has been made during the time allowed or the application itself has been dismissed, then the court will transfer the proceedings to the appropriate Crown Court, taking into account the convenience of the defence, the prosecution and the witnesses and the expediting of the trial.

Until the time of the transfer for trial, the accused may be held in custody in jail. Each week he will have to be brought before the magistrates for a further period of being **remanded in custody**. In newsworthy criminal cases, such as the 'Yorkshire Ripper' case, television reporters may record such appearances. Reporting restrictions may be made.

In less serious cases, the accused will be allowed to remain at liberty, being **remanded on bail**. This means that if he fails to turn up at court for any appearance, be it for remand, committal or trial, he will suffer serious penalties. If others had to stand as **sureties** for the accused, they will lose the money they promised to pay if he lets the court down. If he stood as his own surety, then he loses his own money.

The trial at the Crown Court

Jurisdiction

The Crown Court has jurisdiction to hear cases in any part of England and Wales. There are Crown Court regional centres in most big towns. The Crown Court is divided into three tiers. The first has both High Court and circuit judges, and acts as provincial court rooms for High Court civil actions as well as for criminal cases. The second tier has High Court and circuit judges, but deals only with criminal cases. The third tier only has circuit judges.

Offences are similarly divided into four groups, Class I being the most serious, which can only be tried by a High Court judge, down to Class IV, which can be tried by a circuit judge or recorder.

Judges

1. High Court judges, see p. 31.
2. Circuit judges, see p. 30.
3. Recorders – addressed as '**sir**'. These are part-time judges in the Crown Court. Since the Courts Act 1971 they must be barristers or solicitors of 10 years' standing.

Procedure

Indictable offences are brought before the court by laying an information before the magistrates (see p. 34), or by arresting the accused and then bringing the accused before magistrates for re-mand within 24 hours of arrest.

Practice

The rules of the Supreme Court apply to the Crown Court (see p. 32), as the Crown Court, the High Court and the Court of Appeal together make up the Supreme Court.

The trial at the Crown Court

The accused will be taken to court and put in the cells if he has been remanded in custody. If he is on bail, he must surrender to his bail and wait in the cells to be called for trial; if he does not, a warrant for his arrest may be made.

The arraignment

The accused is asked to answer the indictment, i.e. plead guilty or not guilty to the charges made.

If he pleads guilty, there will be no trial, and the judge will proceed to sentencing in the same way as in the magistrates' court (see p. 35).

If he pleads not guilty, then the jurors will be summoned and sworn in. Under the Juries Act 1974, anyone on the electoral roll over 18, subject to certain exceptions, may be called to sit on a jury.

The course of the trial is much as described in the magistrates' court, except that the prosecution will be by Crown Prosecutors and the defence must be undertaken by barristers with instructing solicitors. Furthermore, it is, of course, the jury's job to judge whether the accused is guilty 'beyond all reasonable doubt' which is the standard of proof required in criminal cases. Any submission concerning the law, however, is heard by the judge in the absence of the jury.

The verdict

After the judge's summing up, addressed to the jury, the jury is sent out to the juryroom to consider its verdict. It is possible, where the jury has been out for a long period, that the judge will allow a majority verdict of either 11–1 or 10–2. But they must have been out for at least two hours.

The sentence

If the accused is found guilty, then the judge will then sentence him.

Types of punishment

The following are the types of punishment that may be meted out in the magistrates' and Crown Court.

- Imprisonment *The maximum sentence of life imprisonment is compulsory for murder and the maximum for manslaughter. For many offences, minimum and maximum sentences are laid down by statute*
- Fines *There may be statutory upper and lower limits*
- Detention and attendance centres *These are appropriate where young persons are involved*
- Probation
- Absolute and conditional discharges
- Binding over to keep the peace
- Community service orders
- Compensation orders
- Restitution orders (to return stolen property)
- Order of criminal bankruptcy (see p. 142)
- Deportation orders
- Care, supervision and guardianship orders of minors

- Deprivation of property
- Disqualification from holding licences
- Compulsory admission to hospital

Appeals

Appeals from the Crown Court go either to the Queen's Bench Divisional Court on a point of law or to the Court of Appeal, Criminal Division, and from there to the House of Lords. But an appeal to the House of Lords is only permitted if the Court of Appeal has given its certificate, stating that the case involves a point of law of great public importance, and either the Court of Appeal or the House of Lords itself grants leave to appeal.

Essentials of the law of evidence

Need for evidence

No action can be satisfactorily presented in England or Wales, whether criminal or civil, without bringing sufficient evidence to support the case. If this were not so, then we would see the worst abuses of justice that may be found in certain other countries, where trials are mere farces and accusations of guilt are enough to put innocent people behind bars, and even worse.

Innocence is presumed, guilt must be proved

In this country, we believe a person to be innocent until proved guilty. Furthermore, it is for the plaintiff or the prosecution to substantiate his case. The defendant does not have to prove his innocence. This duty is called the **burden of proof**.

Standard of proof

The **standard of proof** refers to the amount of evidence needed to support the plaintiff's or prosecution's case.

In *civil cases*, the standard is sufficient to prove liability **on a preponderance or balance of probability**. This means that, after reviewing **all** the evidence, the judge must decide that in all probability the defendant did the act complained of.

In *criminal cases* the standard of proof is much greater. The prosecution must prove to the jury's or magistrates' satisfaction that the accused committed the crime **beyond all reasonable doubt**.

The burden of proof

In English criminal law it is up to the prosecution to prove the case against the accused. In other words the **burden of proof** is on the prosecution.

Alternative methods of settling disputes

Of course, court action should only be the last resort, after other ways of settling your problems have failed. Also, it must be remembered that only **legal** problems can be solved in court. If there is no relevant law or legal rule, then the courts are no answer.

In some legal situations, there are alternatives or additional methods (see Figure 2.4 and text).

Do-it-yourself

The law has often encouraged self-help and in most situations will prefer that the parties have tried to resolve their problems relatively amicably, *before* resorting to court. Indeed, occasionally, judges have been known to tell the parties to go away and to try to sort things out for themselves, as they feel that the courts are not the proper theatre for neighbour disputes and the like.

Some legal problems can be solved cheaply by one or more of the following methods.

- Reporting to the police, if a crime appears to have been committed.

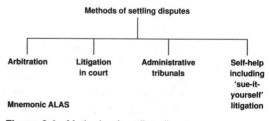

Figure 2.4 Methods of settling disputes

- Reporting to local authorities for breaches of statute, e.g. pollution, offensive trades, noise, unhygienic eating places, short weight, breach of byelaws, building regulations and town and country planning law. The appropriate department must be used (see the telephone directory).
- Seeking advice from Citizens Advice Bureaux.
- Writing to consumer programmes on television and radio, e.g. 'Watch-dog'.
- Reporting to the Director-General of Fair Trading, if an unfair trading practice appears to have been operated.
- Writing to the appropriate disciplinary bodies or governing bodies for trade, profession or industry, e.g. the Law Society, the Chartered Institute of Building, the Association of British Travel Agents.
- Writing to the Member of Parliament for your constituency.
- Seeking advice from your trade union.

In some situations, appropriate action can be taken by yourself to resolve a difficulty. This may be, for example, a problem concerning your tax affairs or a dispute over a small building project. A number of points should be noted.

(a) Avoid legal problems by thinking ahead

The crunch is always going to come at some time or another. So, it is far better to predict problems and, by doing so, they may then be avoided. Whilst the following does not strictly relate to standard form building contracts, the principles are the same.

Important details of contracts/accounts should be committed to paper and copies kept and filed properly. Anything agreed by you and the other party should be evidenced in writing and copies kept by both parties. (Carbon copy books are useful for this purpose.) An accurate record of expenditure and receipts relating thereto should be maintained and a note of any moneys received from clients. Money should be paid into your account as soon as possible and into a deposit account for preference. Accounts should be

rendered as quickly as you are able and followed up if unpaid. Maintaining friendly relations with others is essential. Pleasantness costs nothing and can often solve problems like magic.

(b) If a problem does occur

Either visit or telephone the other side (be it customer, electricity board or taxman) and try to resolve the matter before it gets worse and without undue delay. Keep a note of the date, time, the person to whom you spoke and what was said, for future reference (evidence if it goes to court).

If that fails, follow up soon after with a letter, keeping a copy referring to your previous conversation.

If there is still no reply or solution, send a recorded delivery letter referring to your previous visit, telephone call and letter and setting a time limit for a reply.

If there is still no reply or settlement then the usual courses are open to you in addition to the ones mentioned above. In some cases financial help with legal expenses is available. This is called legal aid and advice on whether or not you are entitled to it can be obtained from any solicitor's office with the legal aid symbol displayed. A list of legal aid solicitors can be supplied at your local Citizens Advice Bureau.

Administrative tribunals

If you feel that you have been unfairly dismissed, your first reaction is not to take your employers to court, but to take them to an industrial tribunal (see p. 237).

Such a tribunal is one of many types which has been set up by Parliament to deal with such diverse matters as the discipline of the professions, e.g. the General Medical Council, labour law matters by the industrial tribunals, rent disputes by the rent tribunals and land matters by the Lands Tribunal.

A tribunal as its name implies is composed usually of three people, one of whom is often a lawyer, the others representing certain aspects of the type of work/matter being dealt with.

The tribunal's procedure is less formal than that of the courts and a problem can be dealt with more quickly than in the over-burdened court system. The powers and remedies of the tribunal vary depending on the subject-matter and the authority given to it by Parliament. Reference should thus be made to the appropriate Act.

Arbitration

Definition

This is the formal settlement of a dispute by a third person (or persons) chosen by the disputing parties. It is an alternative method to court action. But, it is impossible to avoid recourse to the courts by agreement, as any such agreement or contractual clause to that effect is **void** at common law for being contrary to public policy. Thus, arbitration decisions, known as **awards**, may be appealed against in the High Court, but only on a question of **law** (not fact) and the court may confirm, vary or set aside such an award or send the award for reconsideration back to the arbitrator.

Relevant law

The law governing arbitrations is found in the Arbitration Act 1996 which came into effect on 31 January 1997, in any particular Act relating to a special type of arbitration and in ordinary contract law. Should there be any ambiguity or uncertainty regarding any part of the Act, it specifically states the principles behind it in S.1 and that the Act shall be construed accordingly. Thus by S.1(a) the object of arbitration is to obtain the fair resolution of disputes by an impartial tribunal without unnecessary delay or expense. S.1(b) states that the parties should be free to agree how their disputes are resolved, subject only to such safeguards as are necessary in the public interest. Finally, by S.1(c) in matters governed by Part 1 of the Act the court should not intervene except as provided in Part 1. Such a statement of principles is an unusual step to take. It may have been introduced as a nod towards *Pepper* v. *Hart* 1995, a case which for the first time allowed the courts to refer to Parliamentary material such as Hansard, where there was uncertainty about a legislative term.

Contract law applies because the essence of an arbitration is that the parties agree to the dispute being decided by arbitration and agree to being bound by the arbitrator's award. Thus, agreement is essential at all stages of the proceedings.

When can you arbitrate?

Disputes may be settled by arbitration in a number of different situations.

Under a special Act of Parliament

Some Acts specifically provide for disputes to be referred to arbitration. Reference must therefore be made to the relevant Act for the scope of the arbitrator's powers and the conduct of the proceedings. If these matters are not dealt with then the Arbitration Act applies. S.94(1) of the Act specifically states that the provisions of Part I of the Act (the provisions relating to arbitration pursuant to an arbitration agreement) will apply to all arbitrations under any Act whether or not passed before or after the Arbitration Act. Furthermore, the Arbitration Act only applies in so far as it is not inconsistent with the rules and procedures contained in the special Act and also provided that it has not been excluded by the special Act (S.94(2)).

In any legal dispute

Anyone with a legal dispute may decide to opt for arbitration instead of going to court. Both parties must agree to the arbitration either orally or in writing, and must decide who is to be the arbitrator or who can choose if they cannot agree. They must also decide on the scope of the arbitrator's powers and that they will both abide by any decision he makes. The provisions of the Arbitration Act apply to written agreements only (S.5).

(Note: the **Small Claims** system in the county courts is an arbitration procedure with the district judge acting as an arbitrator. S.92 of the Act specifically states that the Act does not apply to those types of arbitrations.)

Where court proceedings have already started

Sometimes a judge may decide that arbitration is better suited to the type of case being heard and he may order the case to be transferred for arbitration before a circuit judge who has the relevant expertise and who has been empowered to deal with Official Referee's business. Such cases usually require a detailed examination of documents or involve complex scientific or technological matters. Also under S.9 a party to existing legal proceedings may request the court to stay such proceedings in order to allow the matter to be decided by arbitration as had been agreed in an arbitration agreement by the parties. In such a case the court should grant a stay unless the agreement itself is null and void, inoperative or incapable of being performed.

By prior agreement to arbitrate in the event of a dispute

Many contracting parties recognise that they may very well disagree at some stage of the contract. To avoid the necessity of going to court to settle their differences, a clause is inserted into the contract to submit the dispute to arbitration. Article 5 of the JCT Standard Form of Building Contract is such a clause. Such clauses do *not* oust the jurisdiction of the courts. Such an agreement is an agreement in writing under S.5 of the Act.

Advantages and disadvantages of arbitration

Arbitration is a very popular method of settling disputes in the commercial world and is frequently used in relation to building contracts.

Reasons for popularity

1. Privacy. Arbitrations are held in private, whereas court proceedings are open to the general public. Few people wish to wash their dirty linen in public, least of all commercial companies whose share prices may be adversely affected by bad publicity.
2. Expertise of the arbitrator. The arbitrator is chosen for his expertise on the subject-matter of the dispute, be it ship-building or construction work. Often, valuable time is wasted in court proceedings whilst lengthy technical explanations are made for the benefit of the judge. However, as disputes get more

complicated, involving many different disciplines, this may prove to be less of an advantage than it used to be. Also most disputes involve many legal issues and, whilst an arbitrator may have technical expertise, he often does not have sufficient legal expertise to cope with the complexities involved. Those arbitrators with the most expertise tend to be the most booked up and may not be available when needed.

3. Own choice of arbitrator. There can be no choice of judge in court actions. But, in arbitrations, the parties have the opportunity of choosing someone they both trust and respect. If they cannot agree on the appointment, then they can refer the choice to someone else. Article 5, Part 4, Cl. 41, refers to the settlement of disputes by arbitration.
4. Convenience of parties. Arbitrations can be conducted at any place and at any time, for the convenience of the parties and arbitrator. Whilst a court may be convened somewhere other than a court house for special reasons, this is fairly unusual. (Such reasons have included holding the court in the court yard as the defendant had a contagious disease and at the house of a defendant who was a recluse and never went out.)

   There is very little say in the timing of court cases, unless the matter is urgent, e.g. when an injunction is required.
5. Procedure more flexible than court proceedings. S.34 now states that it is for the tribunal to decide all procedural and evidential matters, subject to the parties' agreement.
6. Right of appeal on a point of law (S.69). Unless otherwise agreed a party to the proceedings may appeal on a point of law arising out of the proceedings. An appeal may not be brought without all parties' agreement or leave of the court and the court will only give its leave if it is satisfied on a number of matters, e.g. that the question is one of general public importance and that the decision of the arbitration tribunal is open to serious doubt.
7. Remedies. The remedies that may be awarded are by agreement or, in the absence of such

an agreement, the tribunal has the same remedies as a court. Thus, it has the power to award injunctions, specific performance or rectification of a contract or make a declaration on any matter (S.48).

8. Provisional awards. The parties may agree that a provisional award be made to be taken into account in the final award (S.39). This is different from interim awards.

Disadvantages

1. Length of arbitration. Once a court case has actually been listed for trial, the proceedings are quite swift and reluctant litigants can be brought to heel by the wide powers of the judge. For example, if the plaintiff does not proceed with any of the requirements of the case, the defendant can ask for it to be struck out for want of prosecution.

Arbitrations on the other hand have been known to 'drag on for ever' to quote Lord Denning (former Master of the Rolls). This is because delaying tactics can be used to a party's advantage and the arbitrator has fewer powers than judges to force the delaying party to comply with his orders. By S.33 of the Act the arbitrator is under a duty to adopt procedures suitable to the circumstances of the particular case avoiding any unnecessary delay or expense. Fast track arbitration procedures introduced by the JCT is available and, whilst not widely used, could be useful where the parties are willing that the procedure should work.

However, under S.41 Arbitration Act 1996, extra powers may now be applied for. The parties are free to agree on the powers of the tribunal, in the case of a party's failure to do something necessary for the proper and expeditious conduct of the arbitration. If the arbitrator is satisfied that there has been inexcusable delay on the part of the claimant, thereby jeopardising the fair outcome of the proceedings or such as has caused serious prejudice to the respondent, then the claim may be dismissed (S.41(3)). If such an order

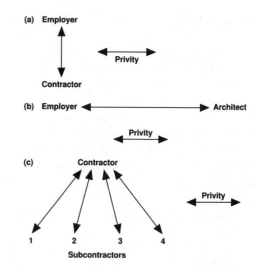

Figure 2.5 Parties involved in a building contract situation illustrated separately

is made, the arbitrator may continue the proceedings **in default** of the attendance or otherwise of the other party (S.41(4)).

2. Only parties to the agreement may arbitrate. Only parties to a contract have rights and duties in relation to that contract. This is called **privity of contract** (see p. 73).

It follows that only parties to an arbitration agreement can arbitrate unless there is some clause contained in such an agreement whereby other parties involved in the dispute, but not party to the agreement in question, may have their disputes settled at the same time.

For example, a building contract is between a main contractor and an employer; the employer may also have a contract with an architect and in turn the contractor may also have contracts with subcontractors, nominated or otherwise (see Figure 2.5).

When one looks at the total building project, the contractual relationships are as shown in Figure 2.6. Parties who have no contract with each other, such as the architect and subcontractors, have no privity of contract. Any dispute between the employer and contractor or employer and architect or contractor and subcontractor must be arbitrated separately,

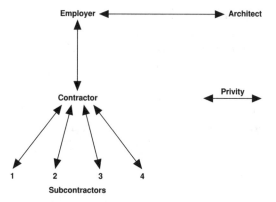

Figure 2.6 Parties involved in a building contract situation illustrated together

unless their contracts are connected in some way and all parties have agreed that related disputes should be heard by the arbitrator at the same time.

Thus, the JCT form provides for the joining of separate disputes to be heard by the same arbitrator at the same time, provided the issues are substantially the same as or connected with issues raised in a related dispute between the employer and a nominated subcontractor or the contractor and a nominated subcontractor or between the contractor and/or the employer and a nominated supplier: Article 5. There are stringent conditions attached to this right.

The parties are free to agree that the arbitration proceedings be consolidated with other arbitrations or that concurrent hearings shall be held on such terms as they agree (S.35). In the absence of such an agreement the arbitrator has no powers equivalent to those of a judge to consolidate actions. In court actions, however, even if the disputes are contractual, the judge has the power to have the separate proceedings consolidated and heard at the same time. If the disputes are tortious then any number of people may be joined as co-defendants, if they appear to share the blame.

3. Costs of a court hearing will usually be cheaper. Although it is well known that court proceedings are neither cheap nor necessarily quick, arbitration may sometimes be more

expensive if there has been undue delay. Legal aid is never available for such proceedings and all the costs, including the cost of a venue, must be borne by the parties, whereas some element will already have been borne by the state in a court case, e.g. the cost of the court house and judge.

4. Reasons are not necessarily given for the arbitrator's award. However, under the 1996 Act, if an appeal is made and one of the parties asks for reasons to be given, then the High Court may require the arbitrator to give his reasons for making his decision. Arbitrators now must anticipate this happening and remember this when making an award.

In court cases, the judge will give full reasons for his decision in the **ratio decidendi** (see p. 23).

5. Arbitration only takes place if the parties agreed at the outset. Problems often occur as to the interpretation of the arbitration clause in the contract and, if disputed, this will have to be decided on by the courts. This is before the actual problem can be dealt with.

Alternative dispute resolution

There has been an increase in non-litigious proceedings in recent years. Arbitration as described above is obviously one of the most commonly used methods in the construction industry and with the new inproved Arbitration Act 1996 may find increasing favour among detractors of arbitration.

But even this has been considered too slow or cumbersome or too weighed down with complicated rules. What is often required is some fast dispute resolution which may be used during the contractual period, allowing the problem to be solved and the work carried on as soon as possible or even without a hiatus. Alternative dispute resolution (ADR) has been successful in the United States and Australia and is being increasingly used in this country. A Centre for Dispute Resolution (CEDR) was set up in 1991. Both the CEDR and the Institute of Arbitrators publish model forms and guidelines.

There are a number of different types of ADR, mediation or conciliation, evaluation and mini-trials and now adjudication (see below). They are operating in an increasing number of ways. Even the court system is encouraging the use of ADR. A Practice Direction from the Commercial Court (see p. 32) in 1996 now requires the parties to a case which is being listed for more than a 10-day hearing to state whether they have considered using ADR and if they have not done so to consider it now. Also a report by Lord Woolf MR in 1996 called 'Access to justice' has recommended out of court settlements and a better organisation of the litigation machinery so that those cases that have to go to court are better managed, cheaper and quicker.

The following are some examples of ADR.

Mediation

Both sides are brought together, often in neutral surroundings, and are encouraged by a trained mediator to settle their dispute amicably. This is often used in neighbour disputes and in family disagreements. Some of these schemes are operated by charities such as Mediation UK.

Sometimes a construction contract will contain a mediation clause requiring mediation first prior to any litigation, in the hope of avoiding going to court. The contract terms may refer to mediation rules of a particular body, may state that in the absence of agreement on the appointment that the mediator will be appointed by some other mutually agreed body such as the CEDR and may give an agreement as to bearing the costs of the mediation and venue. Should one of the parties go directly to court without first trying mediation as agreed, the court may stay the proceedings and refer the dispute back to mediation. See *Northern Regional Health Authority* v. *Derek Crouch Construction Company Ltd* 1984.

Even the courts are trying to increase the use of mediation. In 1996 a one-year pilot scheme was started within the court administration system whereby prospective litigants who are going to sue for between £3,000 and £10,000 in the county court may, by ticking an appropriate box on a form sent to them, agree to opt for mediation in order to settle their dispute. Should this fail, they retain the right to go back to court in the usual way. Both parties pay a fee of £25 and a mediator from certain recognised organisations supervises a three-hour mediation session. This is all arranged within 28 days from application and takes place after court hours. If the parties cannot agree by the end of the three-hour session then the case will proceed in the normal way. (This is *not* the same as the Small Claims procedure in the county court (see p. 31).)

Evaluation or expert determination

This may be provided for in contracts. Here the dispute is to be settled by an expert who it is expressly stated is not acting as an arbitrator under the Arbitration Act 1996 and thus is not subject to all that entails. In such cases, the courts would not interfere with such a decision unless the parties agreed, so that the expert's 'judgment' would be final on that issue.

Mini-trials

These are similar to arbitrations and are heard by an executive tribunal usually consisting of non-legal representatives of the parties to the contract who with their chairperson will try to determine the dispute.

Adjudication

The new Housing Grants, Construction and Regeneration Act 1996 has introduced a procedure called adjudication. As the statutory provisions will only apply unless the contrary is agreed, many contractors will probably incorporate their own adjudication clause to avoid the Act. The relevant part of the Act is Part II, Ss. 104–117.

S.108 states that parties to a construction contract have the right to refer a dispute arising under the contract for adjudication under a procedure which complies with this section.

The word 'dispute' means any 'difference'. Construction contracts must now include clauses to cover the following matters. The contract shall enable a party to give notice at any time of his intention to refer a dispute to adjudication. It must provide a timetable with the object of securing the appointment of the adjudicator and referral to him of the dispute within seven days of the notice. It must also require the adjudicator to reach a decision within 28 days of referral or such longer time as is agreed between the parties. The contract must allow the adjudicator to extend the period of 28 days by up to 14 days with the consent of the party who referred the dispute. The contract must impose a duty to act impartially and enable the adjudicator to take the initiative in ascertaining the facts and law (S.108(2)(a–f)).

By S.107, Part II of the Act only applies where the contract is in writing and any other agreement, on any matter, is only effective if in writing, for the purposes of the Act. 'Writing' is defined broadly. So, it covers where the contract itself is in writing, whether or not it has been signed by the parties; where the agreement has been made by an exchange of communications in writing; and lastly, if it is evidenced in writing. In other words there does not have to be a standard form contract.

By S.108(3) the contract shall provide that the adjudicator's decision is binding until (and presumably if) the dispute is finally determined by legal proceedings, by arbitration (if the contract provided for this or the parties agreed by arbitration) or by agreement. Sensibly, the parties may agree that the adjudicator's decision finally decides the dispute (S.108(3)).

If the contract does not comply with the requirements of S.108(1)–(4) then the Scheme for Construction Contracts Adjudication provisions apply. (The Scheme is provided for in S.114 where the Minister responsible, by regulations, may make such a Scheme containing provisions referring to this part of the Act.)

S.104 defines a 'construction contract' as an agreement with a person for any of the following:

(a) the carrying out of construction operations;
(b) arranging for the carrying out of construction operations by others, whether under subcontract to him or otherwise;
(c) providing his own labour, or the labour of others, for the carrying out of construction operations (S.104(1)).

It also covers agreements to do architectural designs or surveying work or to provide advice in building, engineering, interior or exterior decoration or on the laying-out of landscape in relation to construction operations (S.104(2)). It does not include employment contracts (S.104(3)).

S.105 defines, in a lengthy and wide ranging definition, the words 'construction operations'. They cover the construction, alteration, repair, maintenance, extension, demolition or dismantling of:

(a) buildings or structures (forming or going to form part of the land) whether permanent or not and
(b) works (forming or going to form part of the land) including walls, roadworks, powerlines, telecommunication apparatus, aircraft runways, docks and harbours, railways, inland waterways, pipelines, reservoirs, water mains, wells, sewers, industrial plant and installations for purposes of land drainage, coast protection or defence (S.105(1)(a–b)).

The definition also includes the installations in a building or structure of fittings including heating, lighting, air-conditioning, ventilation, power supply, drainage, sanitation, water supply, fire protection, security or communication systems (S.105(1)(c)).

Cleaning either internally or externally during the course of construction, alteration, repair, extension or restoration is included in the definition (S.105(1)(d)).

Preparatory work or work integral to the main work in also covered. Thus, site clearing, earthmoving, excavating, tunnelling and boring, the laying of foundations, erecting, maintaining or dismantling scaffolding, restoration of the site, landscaping, providing roadways and other access are all included (S.105(1)(e)).

Finally, even painting and decorating is included (S.105(1)(f)).

What are not included in the definition of construction operations are the following: drilling for oil or natural gas; extracting minerals, such as coal, whether by open-cast, tunnelling or boring or constructing underground works; assembling, installing or demolishing plant or machinery or steelwork on a site whose main activity is to do with either nuclear processing, power generation, water or effluent treatment or the production, transmission, processing or bulk storage (other than warehousing) of chemicals, pharmaceuticals, oil, gas, steel, food and drink; the delivery or manufacture of building or engineering components, materials, plant or machinery or components for heating, lighting, air-conditioning, ventilation, power supply, drainage, sanitation, water supply, fire protection or security or communication systems, unless the contract also includes their installation; and the making, altering and repair of artistic works such as sculptures and murals (S.105(2)).

It should be particularly noted that by S.106 this part of the Act does not apply to construction contracts with a **residential occupier** (or any other contract the Secretary of State designates). Such a contract means one where the work to be done is in one of the parties' dwellings, whether occupied or to be occupied. 'Dwelling' means a dwelling-house or flat. So, obviously, the intention is to cover business and business type contracts only and not 'business to residential occupier' contracts. (This is similar to the Sale of Goods Act 1979 approach to the conditions implied into consumer contracts: see p. 113.)

3 Contract

Introduction – common misapprehensions

Many people think that a contract must be some complicated 'legal looking' document covered with red seals and full of complex clauses in small print. To some extent, those who deal with the standard forms of building contract may be forgiven for thinking this. But a standard form of any sort, be it a building contract or credit or hire agreement, is in fact in the minority. The majority of contracts are not even in writing. They are entered into in a seemingly casual way every day by millions of people when they buy things in shops, or petrol from a garage, use child minders, go to the cinema or buy a drink in a public house.

A contract is thus used in all transactions in which one person bargains with another, even if the subject-matter seems to be trivial.

Why then are people under the misapprehension that contracts *must* be in writing? This appears to have arisen for three reasons.

First, most important contracts such as buying a house, renting a television or car, buying goods on credit or hire purchase, take the form of complex printed documents. As the printed standard forms are prepared by central bodies or credit companies or even the company selling or hiring the goods or services, they are intended to cover all eventualities, and thus will contain a large number of terms and, regrettably and inexcusably, many which are not easily understood by the general public. Whilst a person in theory could ask to have a special one-off contract drawn up between the parties, in practice, the seller etc. will merely refuse to do business with the 'individual'. Because these lengthy printed contracts are contracted at an important stage in the average person's life, the trouble and harassment caused by trying to understand the terms will probably be vividly remembered and they assume that all contracts must be in writing. This is not so.

Second, time and time again, people are unable to take legal action against someone for breach of contract because they cannot prove that what they are alleging is true. This is not simply because they had nothing in writing. This is because they cannot provide the evidence to support their case, be it written or oral or any other sort. **Good sense** tells the businessman, therefore, that he should commit his contracts into writing and preferably get the other side to sign the agreement. Failing this, he must have witnesses to support any oral dealings.

Third, a small number of contracts have to be in writing or at least there must be a note or memorandum to record what was agreed. For example, a contract to transfer an estate or interest in land must be in writing (S.2 Law of Property (Miscellaneous Provisions) Act 1989). Others have to be in a **deed** (a document is a deed when it is validly executed, i.e. signed and witnessed). Such a contract is called a **specialty** contract and it gives the parties the advantage of being able to sue within **twelve** years of the breach of contract. This is one of the reasons why they are used in the building industry. Ordinary contracts called **simple** contracts allow the parties only **six** years in which to sue.

As these written contracts are concerned with important and expensive matters such as buying a house, transferring shares in registered companies, assigning copyrights and bills of exchange, they are extremely memorable.

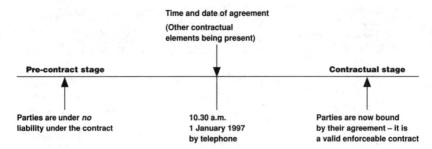

Figure 3.1 The importance of the timing of the agreement in contract

The second major misapprehension is a development of the first. This is that parties are not bound by what they have agreed unless the agreement is in writing. For some reason, many people appear to think that there are degrees of 'bindingness' and that oral agreements are less binding than written ones. Once again this is absolutely fallacious unless the contract happens to be one of the small proportion that must be in writing. Of course this misapprehension is again to do with lack of evidence. If there is no evidence, then the party wishing to enforce the contract may be in difficulties. Because so many people avoid their contractual liabilities through lack of evidence, the misapprehension grows daily. But this is wrong, even in situations where there is going to be some formal written contract at a later date (which should merely confirm what was orally agreed at an earlier stage).

Parties become contractually liable **immediately** there is an **agreement** between them, and the actual time of day may be of vital importance.

This is one of the most important elements of a contract (see pp. 51–63). **No** agreement, **no** contract and **no** contractual liability (see Figure 3.1).

Contract law

The law of contract is found mainly at common law and equity and in some legislation.

Definition

A contract is a bargain or agreement between persons giving them rights and duties in law. It does not usually have to be in writing but it must be the sort of agreement which would give rise to legal duties. The majority of contracts are oral (see the Introduction).

Examples

Buying a cup of tea in a canteen or a beer in a public house; travelling on a train having paid a fare; hiring a band to play at a rock concert; putting in central heating for a friend at less than the normal charge; building a garage based on a written estimate and a telephone conversation agreeing the rest of the terms; agreeing to build a multimillion pound development using a JCT standard form of contract; a contract of employment; a loan from a building society.

In each of the examples, there are a number of obvious common aspects. There is an agreement between the parties to give or do something in return for something given or done by the other side. There is an exchange of goods or services or something else of value between the parties. This is known as consideration. It would also seem that in the most trivial of situations, such as purchasing a cup of tea, the parties expect to be bound by their promises, and if they fail in some way, such as putting salt in the tea by mistake, that they will have to rectify the situation. This third element is absent for example in social arrangements which usually are not contracts as there is no intention to be legally bound by the terms and thus there is no legal redress if one party breaks a promise. So if you 'stand your girlfriend up' she can take no legal action against you.

Elements of a contract

Thus, the main elements for a valid binding contract are that there must be:

1. An agreement (see p. 51).
2. Consideration (see p. 50).
3. An intention to be bound by the agreement (see above).

The main ingredients presuppose the existence of certain other requirements. These are:

4. That there is no fundamental mistake over the subject matter (see p. 64).
5. That the parties are capable of entering into that particular type of contract (see p. 66).
6. That the contract is for a legal purpose (see p. 67).
7. That if the contract is supposed to be in a particular form, e.g. writing or in a deed, that it is (see p. 70).
8. That the parties consented to the agreement of their own free will (see p. 71).

Validity of contracts

Validity means that the agreement takes effect as a contract, giving the parties rights and duties which are enforceable in the courts.

If any of the elements which are essential to the contract are missing, then the purported contract is **void** (not valid). This means that the agreement has no effect in law, gives no rights and duties to the parties and is unenforceable in the courts. **It never was a contract** right from its inception. This is called being **void ab initio**, void from the beginning. For example, consideration is essential on both sides. Thus, if I give £100 to X and ask for nothing in return, this is a gift and not a contract. I cannot later ask X to do something in consideration for giving him the £100 earlier.

Some matters make an otherwise valid contract **voidable**. This means that if a party wishes to avoid the contract, because of some important defect in the contract, the contract then becomes void (see Figure 3.2).

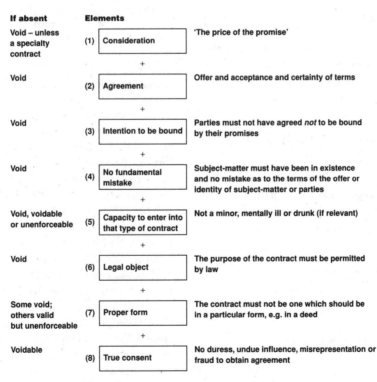

Figure 3.2 Requirements of a valid contract

Consideration

Definition

All simple contracts must have consideration given by each party to the other as the price of each other's promises. It may take the form of money, goods, services, promises not to sue, work etc. It may be positive or negative, but it must represent the **price of the promise** (see Figure 3.3). If there is no consideration or consideration given only on one side (as in a gift) there is **no** valid contract; it is **void** and thus no duties are imposed on the recipient.

Executed and executory consideration

Consideration, however, does not have to be given immediately on making the contract, as for example when buying a newspaper. It is quite usual for the consideration to change hands at a later date. For example, one may enter a contract to hire a tent for a date two months away. The total cost of hire may be paid immediately or a deposit paid and the rest paid when the tent is actually collected. A promise to perform in the future is called **executory** consideration. Consideration given at the time of making the contract is called **executed** consideration.

Consideration does not have to be equally balanced in value

The consideration on one side does **not** have to match the value of the consideration on the other. So, it is perfectly valid to sell houses, land and other valuable property for a nominal sum

(there may of course be tax complications, but that is not a concern of contract law). One hears of peppercorn rents, or even more romantically of a single red rose acting as ground rent. These are perfectly valid as they have some value, even if it is not equal to the value of the other party's consideration.

Where amount of consideration is not agreed, a reasonable sum can be implied

If there has been no agreement on the consideration, but it is implicit that consideration is going to be exchanged, then the court can order that a reasonable sum should be paid. If it is clear that the work was to be done gratuitously, then it is a gift and not a contract: *Lampleigh* v. *Braithwaite* 1615.

Consideration must be legal

Any consideration given must be legal; otherwise the contract is **void** (see Illegal contracts on p. 67).

Consideration must not be 'past'

You cannot make a contract in which one person promises to do something for another in exchange for a consideration which has already been performed (past consideration). So, Bill cannot say to Arthur 'Remember that fence I mended for you last week while you were on holiday, I'd like £50 for that'. When Bill had mended the fence, he had done so freely, or if he had intended to be paid for it, he had no way of contracting with Arthur because Arthur was away. If, on the other hand, Arthur had been present at the time Bill mended the fence, it is possible that it could be implied that Arthur should pay a reasonable sum. It is clear therefore in contract that, before work is started, the consideration must be clearly defined by reference to a fixed sum, or to rates, or some other agreed method: *Roscorla* v. *Thomas* 1842; *Re McArdle* 1951.

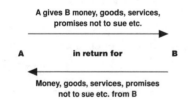

Figure 3.3 Diagram to illustrate the concept of consideration

Consideration must relate to the contract in hand

When one is making a bargain, the parties agree on the relevant consideration. If, however, they are already under an obligation to do something and then expect extra payment for performing what they have already promised to do, they will usually be unsuccessful in any action to claim this extra sum, even if it has been promised. There are various examples of this principle. Thus, for example, if one is already under a duty in law subject to a compulsory witness order to appear in court, then one cannot claim a sum offered to compensate for loss of time: *Collins* v. *Godefroy* 1831. Similarly, if one has already contracted with the other party to do something and is then promised an extra sum to finish the job, the promise is unenforceable because there is no consideration behind the extra promise: *Stilk* v. *Myrick* 1809. However, in *Williams* v. *Roffey Bros and Nicholls (Contractors) Ltd* 1991 a promise by a main contractor to a subcontractor of an extra sum if he finished the work on time, in order to avoid the payment of liquidated damages to the client, was held to be enforceable. This was because the avoidance of the 'penalty' was a big advantage to the main contractor and could be seen to be additional to the agreed consideration in the contract.

The agreement

The second important element of a contract is that there must be an **agreement** between the parties on the subject-matter and terms of the contract.

Most agreements can be broken down into two parts – an **offer** and an **acceptance** of that offer.

The offer

Anyone with legal capacity can make an offer, orally, in writing or by implication from conduct, such as gestures. A combination of any form is perfectly valid, such as a telephone conversation, which is oral, plus a written estimate. Sellers or buyers may make offers, as may contractors or employers or indeed any type of contracting party.

Who makes the offer has to be assessed in each set of circumstances in order to determine who may accept the offer.

Examples
'Will you buy my car for £2,000?'; 'Can I buy your car for £2,000?'; offering to pay for goods at the cash desk of the supermarket; 'Collect 5 tokens from side of packet and we will give you a free seed dibber'; 'Lost – Black kitten called Tiger, Reward £5'.

The person making the offer is called the **offeror**, and the person to whom the offer is made is called the **offeree**.

Invitations to treat are not offers
Displays of goods in a shop or shop window or a 'for sale' board outside a house are **not** offers. They are called **invitation to treat** and are mere invitations to others to make an offer. This is why no-one can force shopkeepers to sell goods in their windows as they are not making an offer: *Fisher* v. *Bell* 1961.

Intentions only are not offers
An announcement of an intention to sell or merely giving someone a piece of information concerning something is not an offer. So, if a person asked you, 'If you were going to sell your car, how much do you think you will get for it?', your reply '£1,000' merely amounts to the giving of information, and there is no offer. Therefore he cannot 'accept' your 'offer' as it is not an offer in law. If he then said, 'Alright, I'll give you £1,000 for it', you are then in the position of being able to accept or reject the offer, as he is now making the offer to you: *Harris* v. *Nickerson* 1873; *Harvey* v. *Facey* 1893.

Invitations for tenders
One often sees advertisements in trade papers containing the words 'invitation for tenders'. In these, a prospective employer invites contractors to put forward tenders for a particular job. Invitations for tenders are thus merely invitations to treat and, on their own, should not create contractual relations. Occasionally, the terms of such an invitation could give rise to contractual obligations in special circumstances. This happened

in *Blackpool & Fylde Aero Club* v. *Blackpool BC* 1990. Here the Council owned and managed the local airport. It decided that, when the Club's concession to run pleasure trips came up for renewal, it would invite tenders from seven parties, including the Club. In the invitation document, the Council stated that 'the Council do not bind themselves to accept all or any part of any tender. No tender which is received after the last date and time specified shall be admitted for consideration.' Because of a Council employee's failure to open the post-box until after the deadline, the Club's tender was not considered, being too late. The Club sued the Council for breach of contract and negligence. The Court of Appeal agreed with the Queen's Bench decision that the Council was liable for both and was liable in damages. In other words there *was* a contract in relation to the dealing with the tenders and there was negligence also.

Offers to individuals or the general public

Offers can be made to a particular person as in the example above, or to a particular group, e.g. 'Members of the Sporran Football Club can have 20 per cent discount on football gear', or the offer can be to the general public as in the reward example: *Carlill* v. *Carbolic Smoke Ball Co.* 1893 (see Figure 3.4).

Communications of offers essential

Offers must be **communicated** to the offerees. Otherwise, how can a person know of the existence of the offer and be capable of accepting it. Any method of communication is acceptable, from a letter to semaphore, telephone to telex, sign language to gestures: *Taylor* v. *Laird* 1856.

The offer must not be vague or couched in an indefinite way

Otherwise, even if there is an apparent acceptance, the courts may have difficulty in finding a binding contract (see p. 49): *Guthing* v. *Lynn* 1831.

Tenders

An invitation for tenders is an invitation to treat (see above). However, the tender itself is an **offer**, which, if accepted, forms the contract.

Preparing the tender can be a costly business. Obviously the cost of preparation should be taken into account when making the tender and be absorbed into the contract price. The unsuccessful builder must write off the cost. If there is a request to the builder to prepare additional designs, working drawings or plans etc. and if no payment for doing so is mentioned, it may be implied that a reasonable sum should be paid. In *Marston Construction Co. Ltd* v. *Kigass Ltd* 1990, following a fire, a contractor was invited to tender to design and build a new factory. At a meeting between the parties, the contractor was informed that he would be awarded the contract but that no contract could be entered into until and after the insurance monies were forthcoming. Additional design work was required which the contractor prepared. No indication was given to the contractor that the client would meet the costs. Subsequently no insurance monies were paid, so no contract was entered into. The court held that there was an implied request for the additional work and thus a reasonable sum should be paid. In practice, the contractor would be advised to state in writing that he would have to charge for the additional work should no contract be forthcoming and get this agreed before the work was done.

Conditional offers

Often an offer is made subject to a condition, e.g. as in Marston's case above where the contract was only going to be awarded if insurance monies were received. In the building industry conditions could be on receipt of planning permission, funding etc. The offer thus does not become effective until the condition is fulfilled.

Auctions

Each bid at an auction is an offer which may or may not be accepted.

Termination of offers

An offer, once made, does not go on for ever. It may be brought to an end in a number of ways.

Figure 3.4 An advertisement for the carbolic smoke ball (by kind permission of the Illustrated London News Picture Library)

Revocation

Here the offeror retracts his offer and communicates this to the offeree **before** a valid acceptance can be made. If the offeree hears from another source that the offer no longer stands because, for example, the goods have been sold, he cannot then accept the offer: *Byrne* v. *Van Tienhoven* 1880; *Dickinson* v. *Dodds* 1876.

Lapse

An offer will lapse if not accepted within a reasonable time or if a time is specified within the contract. Reasonableness depends on the subject-matter of the contract. For example, offers concerning perishable goods such as ready mixed concrete must be accepted without delay: *Ramsgate Hotel* v. *Montefiore* 1866.

Rejection

This may be a flat rejection or it may be a **counter offer**, e.g. 'Yes, but I'll only pay £950'.

A **conditional acceptance** also amounts to a rejection, e.g. 'Yes, but you'll have to put new tyres on the car'.

Death

If either party dies **before** acceptance, the offer ends. This is obvious. Occasionally, however, it is possible for there to be a valid acceptance, despite the death of the offeror. For this to be valid, the contract must obviously not be of a personal nature. Nor must the offeree have known of the other party's death. Selling a car is not a contract of a personal nature, so it is possible for a contract to be valid and binding on the deceased's personal representatives, who must then pass the car to the purchaser. The purchase price will then form part of the deceased's estate. In the case of a customer who wanted a decorator to decorate her house, this is a contract of a personal nature, and her death would terminate the offer.

Acceptance of the offer

Once again this can be in many forms, oral, in writing, or implied from conduct.

In the five examples of offers given previously, what sort of acceptances can be made? In the fourth example, merely collecting the tokens and sending them to the company would be enough. There is no need to tell them that you are now going to start collecting and will contact them in due course. Similarly, finding the kitten (in the fifth example) and returning it to its owner is sufficient to claim the reward (provided the finder has notice of the offer).

Who can accept?

If the offer is 'particular', i.e. it is made to one person or a particular number of persons, then only that person or one from the group may accept. If the offer is made to the general public, then anyone who has notice of the offer may accept the offer.

Communication is essential

The offeree must take some positive step to accept the offer. He cannot do nothing and then claim that there is a contract. So an offeror must not say 'If I don't hear from you, then I will assume the deal is on' and cannot rely on the lack of response amounting to an acceptance. The offeree might be on holiday, or even dead: *Felthouse* v. *Bindley* 1862.

If the offeror specified a particular method of acceptance such as first-class post or by hand, then that method must be complied with. It is a condition of the offer.

There are a number of exceptions to the rule that the acceptance must be communicated.

First, communication may be **waived** either expressly or impliedly: *Carlill* v. *Carbolic Smoke Ball Co.* 1893. Also acceptance can be implied from conduct. The token and reward contracts are examples of this: *Williams* v. *Carwardine* 1833. Acceptance, and thus the contract, is made at the time of communication. In the case of telephone or telex it is instantaneous.

Third, where acceptance is by **post** (and no particular method of acceptance has been specified) a valid acceptance is made immediately the letter is stamped, addressed and posted as the Post Office is the agent of the recipient.

As the rule strictly applied could lead to ridiculous situations, the courts may find that there was no agreement until the acceptance was actually communicated, depending on the facts of the case: *Household Fire Insurance Co.* v. *Grant* 1879; *Entores* v. *Miles Far East Corporation* 1955. Most sensible business people specify which method they want to use in order to avoid the postal exception.

Acceptance must be of all the terms offered

(See conditional acceptance and counter offers above.)

Auctions

An offer (bid) is accepted at the time the auctioneer brings his hammer down or in some other customary manner (S.57(2) Sale of Goods Act 1979 as amended). At that time he can sign the contract on behalf of both parties, despite originally only being the agent of the vendor.

If there is an offer and an acceptance of that offer, is there always an agreement which is binding on the parties?

Generally, the answer is yes. However, as we have already seen, all the rules must have been adhered to; otherwise there will probably be no agreement. Four further points need to be discussed.

Agreements made 'subject to contract'

It is usual in conveyancing (see pp. 104–10) to agree to buy and sell land 'subject to contract'. Thus, the solicitors' letters on behalf of vendor and purchaser will contain those words, so that the parties will not be bound by their agreement until there has been a proper formal contract agreed, signed and exchanged between them. However, those words are not magic, and if the court finds that there was a definite agreement to buy and sell between the parties, and that they had regarded the formal contract merely as written confirmation of what had already been agreed, then they will be liable under that contract. The court thus strips away all the words and writing to find the parties' true intention at the time of the alleged 'contract'. Because of S.2 Law of Property (Miscellaneous Provisions) Act 1989 (see p. 70) it may now be that the use of these words is unnecessary. Nevertheless, we will probably continue to see them in solicitors' letters as a failsafe device.

Lack of certainty of terms

If there appears to be such a lack of definition over what had been agreed by the parties, the court may take the view that there is no contract, either because there is a lack of consensus ad idem (meeting of minds) or because the parties were still at the negotiating stage, whereby they had agreed that a contract of a certain type was going to be entered into but had not defined the terms.

Nevertheless, the court may still find that the parties had agreed in contract, even if they had not agreed on all minor terms, which could perhaps be safely ignored.

Also, where the parties have not agreed on all terms, it may be implied that a **previous course of dealings** will control some (though not all) of the matters not expressly agreed in the present case, e.g. the amount of pay which could be the same as that paid on a prior occasion.

Letters of intent

These are being used increasingly in the construction industry. They may be no more than they imply by their name, i.e. that the parties hope that at some time in the future they will enter into a contract, usually if some pre-condition is fulfilled, such as obtaining a main contract, and these are often called 'comfort letters'. They are designed to have the effect of giving advance notice, often to subcontractors, that there is work in the pipeline and that specialist materials will need to be ordered if they are to be available at the start of the building work. Problems occur if work begins and there is no written contractual basis other than the letter of intent.

Unfortunately in recent years there have been a series of cases in which the courts have found that, taking into account all the circumstances and looking beneath the words on paper, there is indeed a contract in being and thus all that that implies in law. The problem then remains of what terms can be found in the contract. As often all that is in writing is the letter of intent, that will be looked at as the prime source. So if builders are on site doing the job mentioned in the letter and if problems occur, obviously, in the absence of any other writing, the letter of intent will be examined minutely if there is a dispute between the parties. It must not be thought in these situations that the letter **is** the contract. It will merely be evidence that the contract exists (unless, of course, no other interpretation could be made). The terms of the contract will then be made up of what was said in the letter, what has actually been done or orally agreed and of course as implied by the law.

It is imperative, therefore, if a letter of intent is used to 'warn' of the impending placing of a contract, that on being given the work to do the parties are placed on a proper contractual footing with a written contract entered into as soon as possible, preferably before the work starts and problems arise.

Preliminary work costs may be recovered on a quantum meruit basis if the letter of intent does not state that the work is being done at the contractor's own risk.

Standard forms

In simple contract situations it is usually easy to determine when the agreement has been made, with the acceptance of the offer. 'Would you like to buy my camcorder for £100?' 'Yes!' It is less so when the parties are using their own standard forms, which they amend and pass back and forth. Possibly to complicate matters even further, the parties, whilst appearing to be still in the negotiating stages, start work on a project. Obviously, in such situations there must be a contract, but on what basis are they working, on whose terms and when did their contractual obligations begin? This situation was described as the 'battle of the forms' in *Butler Machine Tool Co. Ltd* v. *Ex-Cell-O Corporation (England) Ltd* 1979. Very often, the courts take the view that the last form will be the basis of the contract. But this is no hard and fast rule. In *Chichester Joinery Ltd* v. *John Mowlem & Co. PLC* 1987, the main contractors JM were sent a quotation by CJ. Later JM sent a printed enquiry form which referred to conditions printed on the reverse. There were no such conditions printed there. After further meetings JM then sent a purchase order which referred to terms and conditions set out on the reverse and which required an acceptance by signature within seven days. 'Any delivery made and accepted will constitute an acceptance of this order.' CJ failed to sign and return the order and, before delivery was made, sent a printed form with 'Acknowledgement of Order' written at its head. This stated that it accepted JM's order 'subject to the conditions overleaf'. In a dispute the Queen's Bench Division held that the contract had been concluded on CJ's terms, which was,

in law, a counter offer which had been accepted by conduct when JM accepted later delivery of material goods by CJ on site. One can see from this example how complex such situations can be.

Terms of the agreement – express or implied and innominate

Of course, most contracts are not like the simple examples used above. Take for example a building contract. In addition to the basic elements of the bargain, i.e. building work for money, there are a myriad other details which must be settled as part of the agreement. For example, when work is to be started, what materials must be used, allowances for delay etc. It is far easier at the outset to agree to all these terms **expressly**, either orally or in writing, to cover all eventualities. But often terms will be agreed in a slightly different way. For example, standard forms of contract such as the JCT building contract may be used. This contains standard clauses with which the building industry is familiar. Parts of the contract may be altered by agreement. The terms of such an agreement are still **express**, however.

Contractual terms may also be **implied**, i.e. not specifically stated. They may be implied by the courts as being present in the contract, by virtue of past dealings between the parties, trade usage or custom, or because of an Act of Parliament such as the Sale of Goods Acts 1979 or the Supply of Goods and Services Act 1982 (see below), both amended by the Sale and Supply of Goods Act 1994.

Also, the courts will imply a term if it is obvious that the parties must have intended it to be present in their agreement. The leading case is *The Moorcock* 1889. To quote Scrutton L J in *Reigate* v. *Union Manufacturing Co. (Ramsbottom)* 1918, such a term will be implied in situations where, if the parties had been asked whether such and such a term should be implied, they would have answered 'Of course, so and so will happen; we did not trouble to say that; it is too clear'.

The standard form contracts used in the building industry, such as the JCT contracts, are designed to avoid some of the problems mentioned. They are drafted after lengthy negotiations between all the interested parties representing various aspects of the construction industry which make up the Joint Contracts Tribunal (JCT). The detailed terms of, say, the JCT 80 Standard Form of Building Contract should, by consistent use, arm the builder with a ready source of well drafted, practical contract terms. Coupled with the multitude of cases decided on such terms, this should give more certainty to the contractual process in building. By altering or amending terms, the parties go into uncharted waters and it is then that disputes often occur, with interpretation having to be done by the courts. Choosing the appropriate standard form of contract and becoming familiar with it **before** breaches occur would help all parties considerably and avoid much litigation.

Terms implied into contracts for work and materials (e.g. building contracts) by the Supply of Goods and Services Act 1982

Part II of the Supply of Goods and Services Act 1982 deals with the supply of services. Part II covers contracts relating to pure services where no goods are transferred, such as for hairdressing, dry-cleaning, professional services of accountants, solicitors etc. and carriage of goods, as well as contracts where goods are transferred in conjunction with a service such as building, repairs etc. This second category is thus covered not only by this part of the Act but also by Part I of the Act covering the transfer of goods (see p. 122). Also, by Part II of the Act contracts of hire are also covered if the contract provides a service in addition to the hire of goods such as hiring a television set (see p. 124).

Definition

By S.12 a contract for the supply of a service is 'a contract under which a person ("the supplier") agrees to carry out a service ... whether or not goods are also –

(i)   transferred or to be transferred; or
(ii)   bailed or to be bailed by way of hire

under the contract, and whatever is the nature of the consideration for which the service is to be carried out'.

Thus, Part II only applies if there is a **contract** between the parties. In such cases, where there is no contract, the rules of common law have to be relied on. If there is a contract then the party suing can choose to sue either in contract or in tort (see p. 160).

It should be noted that the Act specifically excludes contracts of employment (of service) or apprenticeships.

Section 13 – care and skill

This section imposes an implied term on suppliers of services in the course of a business that the supplier will carry out the service with **reasonable care and skill**.

The standard of care required would, it seems, be that of the average, competent builder, drycleaner, plumber, architect etc. performing his usual job and would probably be approached in the same way as the standard of care in negligence (see p. 175). However, it would appear that no guarantee of success at something would ever be implied as a result of this term (such a term would have to be expressed). However, as a result of certain express terms in a contract, a term may be implied by common law that the result of the contract should be, for example, fit for a particular use, e.g. in *Greaves & Co. (Contractors) Ltd v. Baynham Meikle & Partners* 1975. In this case a firm of consultant structural engineers (BM) designed a warehouse under a subcontract for G Ltd, the main contractors employed by Duckhams Oil. BM were made aware of the use to which the warehouse would be put including the fact that forklift trucks would be crossing the floors carrying and stacking oil drums. After the building had been completed, the floors began to crack due to vibrations from the trucks. G Ltd accepted liability under the main contract and paid damages but then claimed an indemnity from BM. The Court of Appeal held that they were liable because the 'package deal' contract had specified the type of building and the purpose to which it would be put and thus there could be an implied term in the subcontract that the engineers would design a building that would be reasonably fit for that purpose.

Section 14 – time for performance
By S.14(1) a supplier acting in the course of a business must carry out the service within a reasonable time, if the time has not been fixed by the contract or has not been fixed by prior dealings between the parties or in some other manner agreed by the parties. What is reasonable is a question of fact (S.14(2)). If the supplier does not perform his contract within a prescribed time, then this is a breach of contract even if time was **not** made of the essence and the party suing may claim damages as it is a breach of warranty. If time is of the essence and the supplier has failed to complete within the specified period, the other party may repudiate the contract. If time is not of the essence (i.e. time for performance is not a condition) the customer can then give reasonable notice making it so, e.g. 'If you do not repair my window within seven days, I shall go elsewhere'. If the repair has still not been done within that period, the customer can regard the contract as at an end.

It would appear that if a mercantile contract contains a clause relating to time, then time for performance is of the essence and thus such clauses are conditions. For time of payment to be of the essence, however, it would have to be the subject of an express term. Of course, in most building contracts, time is always of the essence as dates for completion are expressly stipulated.

Are all statements referring to the agreement actually part of the contract?

Not necessarily. One has to differentiate between the **terms**, which are part of the contract, and **representations** which are not. A representation is a statement of fact made before the contract is made, and which is not intended to become part of the contract. Some representations become terms by undergoing a metamorphosis. Those that do not remain in their original state and, if they are incorrect, give the misled party no rights to sue for breach **of the contract** itself. This is because the representation did not become part of the contract. However, there are alternative remedies which are available for **misrepresentation** (see p. 72).

It is obviously difficult to differentiate between representations and terms, especially if the contract is not in writing. The following tests may be applied, which whilst not wholly conclusive may indicate whether a statement made regarding a contract is a term of that contract or a mere representation.

- Statements made by an expert, relied on by the other party, who is a layman, are more likely to be contractual terms, e.g. 'This car has got a new engine', stated the garage proprietor.
- Statements made some time before the contract is agreed upon are less likely to be terms, unless reaffirmed at the time of making the contract.
- Oral statements which are not incorporated into later written contracts are less likely to be contractual terms, unless the court views the contract as consisting of both the document and a prior oral statement.

What happens if the parties later disagree over what was agreed in the contract?

Where the contract is in writing, then the general rule is that **only** the written contract should be evidence of what was agreed (the parol evidence rule). However, as we have seen before, many contracts are in writing **and** in conversation, e.g. a builder's estimate and telephoned instructions orally agreed. Thus, evidence of what was agreed orally must, of course, be admitted in a dispute, e.g. a letter confirming the contents of the telephone conversation which was not challenged at the time of receipt.

Where the written document appears to cover all the matters in dispute, then it is unlikely that any orally agreed terms will override the written agreement, unless exceptions to the rule can be applied.

Exceptions
Evidence that a word in a written document was used in a customary or trade way will be admissible.

Evidence may be admitted to show that there was an orally agreed pre-condition that the contract was not to come into being until that pre-condition had been met.

If the written document was executed in a mistaken way by the parties, then an application in equity may be made for **rectification** of the document. Rectification can be ordered only of the *written* agreement and not of the **actual** agreement between the parties. Thus, if a building contractor agreed to build a house and swimming pool for a client, then if the formal contract in a deed has been mistakenly executed, omitting to mention the pool, the contract may be rectified at a later date. Obviously, evidence of a cogent nature must be put forward by the party wishing to have the deed rectified. (Plans and drawings of the building and pool and estimates would probably be good evidence.) Also, it must be proved that at the time the written contract was executed the parties were **still in agreement** over all the terms, and it was only afterwards that one of the parties tried to stick to the letter of the written agreement.

(Note: had the parties been in agreement **after** the mistake was discovered, they could remake the contract by a new deed (see termination of contracts, p. 76).)

Categories of terms

Terms, whether express or implied, may be either **conditions** or **warranties**.

A condition is usually a term of the utmost importance to the contract, which, if breached, allows the other party to repudiate (end) the contract.

A warranty is a term of less importance to the contract concerning some subsidiary matter, breach of which allows the other party to merely claim damages.

Unfortunately, the words 'condition' and 'warranty' are used haphazardly in practice and often the court has to decide whether a term is one or the other. If it is found that a term, labelled warranty, is in law a condition, then the right to repudiate the contract will arise.

Some lawyers also talk of innominate terms. These terms have not been described as either conditions or warranties in the contract, nor are there any statutes or any other references which can put them into one of the two categories. The courts now take the view that one must look at the breach of the term and examine what was the result of the breach. If this was extremely serious, then the term must be treated as a condition. If it was not serious, then it should be treated as a warranty: *Cehave NV v. Bremer Handelsgesellschaft mBH The Hansa Nord* 1975.

Exclusion terms

As a person can virtually agree to anything (within the law) in a contract, can a party exclude or exempt his liability under a contract? In other words, can he restrict his contractual duties in some way, so that he will only be liable for breach of contract in a limited number of situations? The answer is yes, but subject to certain restrictions imposed at common law and by statute.

Statutory restrictions on limiting contractual liability

- The Sale of Goods Act 1979 (see p. 115) as amended.
- The Supply of Goods (Implied Terms) Act 1973 (see p. 121) as amended.
- The Supply of Goods and Services Act 1982 as amended.
- The Unfair Contract Terms Act 1977. 'A person cannot by reference to any contract term ... exclude or restrict his liability for **death or personal injury** resulting from **negligence**': S.2(1) of the Unfair Contract Terms Act 1977.

By S.2(2) 'In the case of other loss or damage, a person cannot so exclude or restrict his liability for negligence, except in so far as the term ... satisfies the requirement of reasonableness.'

The above sections only apply to persons who may be liable in the course of a business, or from the occupation of premises used for business purposes of the occupier.

Where one contracting party is a **consumer**, or one contracts by reference to the other's **written** standard terms of business, then by S.3, as against that party, the other cannot by reference to any contract terms, exclude or restrict any liability of his in respect of his breach of contract. Nor can he claim to be entitled to render a contractual performance substantially different from that which was reasonably expected of him or to render no performance at all, unless such clauses satisfy the test of reasonableness. By S.12 a consumer is one who is not doing it in the course of a business.

(Note: by Sched. 1 of the Act, the above sections do not apply to insurance contracts, contracts for the transfer of land, patents, trademarks, copyright or design rights, registered designs, technical or commercial information or other intellectual property, contracts concerning the formation or dissolution of a company or its constitution or rights or duties of its members or securities. Section 2 does not apply to contracts of employment except in favour of the employee.)

By S.13 the Act prevents making the liability or its enforcement subject to restrictive or onerous conditions. It also prevents excluding or restricting any right or remedy in respect of the liability or subjecting a person to any prejudice in consequence of his pursuing any such right or remedy.

(Note: by S.13(2), an agreement in writing to submit present or future differences to arbitration does not amount to excluding or restricting liability.)

Section 3 Misrepresentation Act 1967 incorporated by Section 8 Unfair Contract Terms Act 1977

If a contract contains a term purporting to exclude or restrict liability for a misrepresentation or a remedy available as a result of a misrepresentation, then that term is **void** unless it is reasonable in the circumstances.

Unfair Contract Terms in Consumer Contracts Regulations 1994

These Regulations were made by authority of the European Communities Act 1972 in response to EC Directive 93/13 and came into effect in July 1995. They are wider reaching than the Unfair Contract

Terms Act 1977 and should **not** be ignored. They will affect builders' standard contracts with non-business customers as the Regulations state that they apply to 'any term in a contract concluded between a seller or supplier and a consumer where the said term has not been **individually** negotiated' (R.3(1)). Thus, where the parties have agreed to all the terms themselves, these rules will not apply. If, however, the builder uses a prepared standard form of whatever type, he is bound to observe these rules.

The builder will fall within the category of 'supplier' as this is defined as a 'person who supplies goods or services and ... is acting for purposes relating to his business'. A consumer is a 'natural person who ... is acting for purposes which are outside his business'. It must be noted that a registered company is not a natural person (see p. 148).

Unfair terms

By R.5 if an **unfair** term is included in the contract between a supplier or seller and a consumer, then that term is **not** binding on the consumer. An 'unfair term' is defined as 'any term which contrary to the requirement of good faith causes a significant imbalance in the parties' rights and obligations under the contract to the detriment of the consumer' (R.4(1)). When deciding what is unfair many factors are taken into account. These include the nature of the goods or services, all the circumstances surrounding the conclusion of the contract at **that** time and all other terms of that contract or any other contract on which it is dependent (R.4(2)). A builder's subcontract and main contract would be such interrelated contracts.

The requirement of good faith is spelt out in Schedule 2. When assessing whether there has been good faith, regard must be taken of four factors: the **strength** of the bargaining position of the parties, whether the consumer had been **induced** to agree the term, whether the goods or services were a **special order** from the consumer and whether the seller or supplier had dealt **fairly and equitably** with the consumer.

By R.5(2), if the contract is capable of existing without the unfair term, then the rest of the contract can continue. Thus, if it cannot exist without the unfair term, presumably, as a corollary, the whole contract is invalid.

Schedule 3 helpfully gives an illustrative list of 17 types of terms which may be regarded as unfair. A few examples of interest to the builder follow. Terms which attempt to exclude the liability of the seller or supplier for the death or personal injury of the consumer resulting from an act or omission of the seller or supplier are unfair (Para. 1(a)). This is virtually the same as S.2 Unfair Contract Terms Act 1977 (see p. 59). Paragraph 1(e) states that a term which requires a consumer who fails to fulfil his obligation to pay a disproportionately high sum in compensation is unfair. Finally Para. 1(o) provides that obliging the consumer to fulfil all his obligations where the seller or supplier does not perform his is unfair.

Construction of written contracts

Regulation 6 requires that the seller or supplier shall ensure that any written term of the contract is expressed in plain, intelligible language. If there is any doubt about the meaning of a written term, the interpretation most favourable to the consumer shall prevail. The latter rule is known as the **contra proferentum** rule and confirms the existing rule used to construe ambiguous terms in documents or legislation by the courts.

Choice of law clauses

Some unscrupulous people may try to avoid these rules by stating in the contract that the law of a non-EU country applies. Regulation 7 states that in such circumstances these Regulations will still apply if the contract has a **close connection** with the territory of the EU. It will be interesting to see how this is going to be applied in practice.

Duties and powers of the Director-General of Fair Trading

One significant difference from the Unfair Contract Terms Act 1977 is the provisions contained in R.8 which are designed to prevent the continued use of unfair terms. If someone complains to the Director-General that a contract term drawn up for general use is **unfair** then the Director-General must consider this complaint unless it

appears to be frivolous or vexatious. The Director-General may then obtain an injunction against anyone using or recommending such a term in contracts with consumers. The court may grant an injunction on such terms as it thinks fit.

Finally, it should be noted that not all types of contracts are covered. By Schedule 1 the regulations do not apply to contracts of employment, contracts relating to rights of succession or rights under family law, contracts relating to the incorporation and organisation of companies or partnerships nor terms which are statutory or regulatory requirements in the United Kingdom or requirements relating to international conventions.

The Regulations were drafted hastily and have been much criticised. It will be interesting to see how they are interpreted by the European Court of Justice and domestic courts.

Exclusion clauses in standard form contracts

The building industry is particularly conversant with standard form contracts. Unfortunately, familiarity breeds contempt and many a person when trying to pursue a remedy for breach of a contract is caught by an exclusion term which they had not bothered to read or had merely overlooked at the time of making the contract.

Standard form contracts are subject to the statutory restrictions above.

In addition, the common law has developed a systematic approach to the treatment of exclusion clauses in written (and standard form) contracts. The courts and Parliament recognise that exclusion clauses between parties who are in a relatively equal bargaining position, such as between parties in business, are in need of less protection than the consumer contracting with a businessman. This is illustrated by the different levels of protection under the Acts above. The following rules must also be noted.

Notice of the term

Reasonable notice of the exclusion terms must be given **at the same time as** or **before** a contract is made. This is obvious, as otherwise the party adversely affected may possible not have otherwise entered into the contract.

In *Olley* v. *Marlborough Court Ltd* 1949, guests at a hotel had their property stolen, unaware, until they had entered their hotel bedroom where a notice was displayed, that the hotel only took responsibility for property handed to the manageress for safe custody. The court held that the notice was not valid. The notice should have been exhibited at the reception desk, where it ought to have been specifically brought to their attention.

Notice of such terms could, however, be inferred from previous dealings between the parties. This is particularly so in business, where valuable time would be lost drawing a party's attention to contractual terms of which he should already have knowledge.

Is the term really part of the contract if it is only displayed in a receipt or ticket?
Many of us have received tickets or receipts which to our surprise contain reference to 'company conditions' or an exclusion clause, of which we were not aware before we had entered into the contract. The courts take the view that if such receipts or tickets are merely receipts and no more, then the party adversely affected by the term is not bound by the term, as the ticket is not part of the contract. The exclusion should have been brought to their notice when or before the contract was made: *Chapelton* v. *Barry UDC* 1940 (man injured by collapsed deckchair).

What if the document containing the exclusion term has been signed by the affected party?
Persons sign documents at their own peril, for the courts always believe that they must have intended to be bound by agreements to which they had affixed their signatures. The golden rule is thus, never sign something which you have not read and understood. Only in exceptional circumstances will someone be able to avoid the consequences of signing such an agreement.

These are:

1. Fraud. Where he was induced to enter into the contract by the other party acting fraudulently.
2. Misrepresentation. Where the other party innocently or negligently (or fraudulently)

made a statement of fact, which caused the other party to enter into the contract, e.g. *Curtis* v. *Chemical Cleaning & Dyeing Co.* 1951, where a dry-cleaning shop assistant misled a customer by telling her that the receipt which had to be signed excluded liability for damage to sequins and beads only. This was untrue.

3. Pleading non est factum (it is not my deed). Very occasionally, there occurs a situation in which the party signing a document is under a total misapprehension as to the nature of the document. If it turns out that the document is totally different in nature to what he intended to sign, e.g. an IOU and not a will, then when someone attempts to enforce it against him, he may plead non est factum – it was not my deed. A party who is careless in signing cannot use this defence.

Contra proferentum rule
Where there is an ambiguity concerning the exclusion clause, then the courts will construe the clause against the party who inserted it into the contract, e.g. 'Cars are parked in this car park at the owner's risk'. This is ambiguous; therefore the car park owner cannot rely on such an exclusion clause in contract. It may thus happen that he is liable in both contract and tort.

Can an exclusion clause avoid all liability under the contract?
This is possible, but each case must be viewed in the light of the Unfair Contract Terms Act 1977 (see above), the Unfair Terms in Contracts Regulations 1994 and *Photo Production Ltd* v. *Securicor Transport* 1980 (HL). In this case (a pre-Unfair Contract Terms Act case), the plaintiff company employed the security company to patrol their property for a relatively small fee. One of the security company's employees set alight to the plaintiff's premises. The plaintiff company sued Securicor for damages, who then relied on an exemption clause which stated that under no circumstances would the defendant company be liable for **any injurious act or default** by **any employee** of Securicor.

The House of Lords held that the parties were in an equal bargaining position, and therefore the exemption clause was valid and reasonable in the circumstances. If the security company had not inserted the exclusion clause and thus laid themselves open to liability, the amount they would have had to charge would have been much greater. The plaintiffs would be covered by insurance and their original bargaining positions would have taken these factors into account. The 'reasonableness' tests of the Unfair Contract Terms Act 1977 would thus be applied to any future cases tried since the Act came into effect.

Intention to be bound in law by the agreement

Generally, where there has been an agreement between parties with consideration on both sides, a contract will automatically arise, the presence of consideration indicating that it is not a gratuitous arrangement. Thus, in the majority of cases, parties do not have to set out to create contracts; they arise naturally from their behaviour.

Nevertheless, as we have already seen, in certain situations a contract will not be found, either because of the nature of the agreement or because the parties specifically state that the agreement is not legally binding. Thus, mere social arrangements are generally not contractual.

Domestic agreements, such as an arrangement to pay a regular sum for housekeeping to a wife, while the husband is abroad, may not be held to be contractual, but then again the opposite may also be shown.

Commercial agreements are generally presumed to be contractual. However, the parties may agree that their arrangements should not be binding at law, in which case the courts may very well refuse to go against the wishes of the parties. The words used to achieve this, however, must be clear and unambiguous. The burden of proving the meaning of ambiguous words intended to exclude legal relations falls onto the shoulders of the party who wishes to rely on them.

Summary

The three vital elements of a contract are thus: consideration; intention to create legal relations; agreement (mnemonic CIA) (see Figure 3.5).

If any one is not present, then the contract is void, and therefore unenforceable in the courts.

If they are all present, the parties can seek to enforce the contract in the courts.

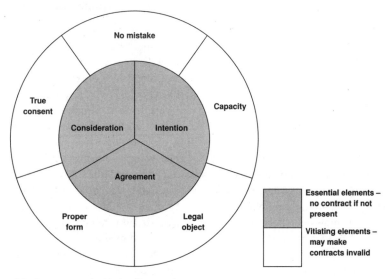

Figure 3.5 Essential elements and vitiating factors in a contract

Nevertheless, in certain circumstances, the courts will not enforce the seemingly valid agreement.

What follows is an outline of the factors which may make a contract unenforceable.

Mistake – can a mistake affect the validity of a contract?

In most situations, a mistake concerning some aspect of a contract will have **no** affect on its validity and the parties will be bound by the terms of their agreement. Thus, parties must make sure that they have understood, in every sense, the terms of their agreement; otherwise, they may be forced to perform a contract which they would prefer to avoid.

The following are typical examples of such mistakes.

1. *Mistakes over price.* If both parties agree to buy and sell an article for a particular price, neither party can later avoid his contractual responsibilities if it turns out that the price is incorrect and that the article is too cheap or too expensive.
2. *Mistakes over quality.* Similarly, agreements over an article based on the mistaken belief that it is an antique or a reproduction or has some special quality cannot be avoided (unless the seller had deliberately misled the buyer to believe that it was genuine or there was a misrepresentation) (see p. 72).

Types of mistakes

Mistakes can thus be divided into three categories: unilateral; mutual; common.

A unilateral mistake is where only **one** party is mistaken over some aspect of the contract, e.g. a purchaser thinks he is buying a guitar which he mistakenly believes belonged to John Lennon and the other party may or may not know this.

(Note: if the seller has told the purchaser that this is not so, but the purchaser thinks he knows better, the contract could not be avoided on finding out the truth.)

If on the other hand the seller has misled the purchaser, then the purchaser may have redress in fraud or misrepresentation.

A mutual mistake is where the parties are in agreement but are at cross purposes, e.g. Sam agrees to sell his push bike for £200, whereas Bill believes that Sam is selling his motor bike.

A common mistake is where the parties are in agreement but have both made the same mistake over some aspect of the contract. For example, Simon agrees to sell his holiday home in the West Indies to Barry; neither of them are aware that it has recently burnt down. Alternatively, if both buyer and seller of a chair think it is genuine Chippendale, then if this is untrue there is a common mistake.

Can a mistake ever make a contract void?

Yes. If a mistake is of such immensity that it robs the contract of any effectiveness then the contract will be void at common law. There are few hard and fast rules and each type of contract and mistake is treated in a different manner by the courts depending on whether the mistake is unilateral, mutual or common. If a mistake makes a contract void, it is said to be **operative**.

Unilateral mistake – at common law

An error of judgement, as illustrated above in the guitar example, would amount to a unilateral mistake and would generally not be a ground for avoiding the contract. The contract is viewed objectively. For example in *Tamplin* v. *James* 1880, J had made the highest bid at an auction for the sale of a public house, mistakenly believing that a piece of land was included in the sale. There had been no misdescription or ambiguity in the auction particulars. J was sued for breach of contract and was resisting a claim for specific performance of the contract (see p. 79). The court awarded the remedy against him. It was not an operative mistake.

Occasionally, unilateral mistakes may be operative if the mistake is of fundamental importance to the contract and one party knew of the other party's mistake, e.g. where a con-man had tricked someone. Such mistakes have occurred in relation to someone's identity.

Mistakes over the identity of the other contracting party

If someone believes that he is making a contract with a **particular** person but they make the contract in each other's company, as for example in a shop, then the contract generally cannot be avoided on finding out the truth of the matter.

Occasionally, however, the mistaken identity of someone will be of such importance that on finding out the truth the contract could be declared void for mistake. Contracts in which dealing with a particular person is important are such a rarity that the courts apply stringent pre-conditions before declaring such a contract to be void. The court would have to be satisfied that:

- The mistaken party intended to contract with another person and not the person with whom he made the contract (who was aware of the other's intention).
- He considered the identity of the other party to be of fundamental importance **at the time he entered into the contract**. In other words had he known of the true identity of that person he would not have entered into the contract.
- He checked the identity of the person, e.g by asking to see his bank cards or other means of identification. See *Ingram* v. *Little* 1961; *Lewis* v. *Averay* 1972. See also unilateral mistakes as to the nature of the document (see p. 62).

Unilateral mistake in equity

Strictly speaking, if a contract is void ab initio, no declaration of this is necessary and in most cases the fact that the contract was void ab initio would be pleaded as a defence in an action for damages or specific performance.

Equity, however, may assist further by: refusing to award a decree of specific performance; or formally setting aside the contract by declaring it to be void; or allowing rectification of a valid contract provided the mistaken party can prove that the document should have contained a term

of benefit to him and the other party knew of this omission. For example, in *A Roberts & Co. Ltd* v. *Leicestershire CC* 1961, the plaintiffs tendered to build a school over a period of 18 months. The defendant council drafted a contract for execution by the plaintiffs in which the work would be completed over 30 months. The plaintiffs failed to notice the mistake and the defendant council officers, knowing of the builders' mistake, failed to have the agreement altered before execution by the council. The court allowed rectification for a unilateral mistake and the clause was altered to the shorter period (see p. 59).

Mutual mistakes at common law

A mutual mistake will only operate to avoid the contract if the mistake is such as to remove the real basis of the contract, i.e. the mistake is of fundamental importance as for example was the mistake over the sale of the bikes above. Obviously there is no agreement between them as there has been a mistake over the offer or the identity of the subject-matter. Thus, in *Raffles* v. *Wichelhaus* 1864, a seller contracted to sell a ship's cargo aboard the *Peerless* sailing out of Bombay. The buyer was under the misapprehension that he was buying the cargo of another ship which was also called the *Peerless* and also in Bombay. Neither knew of the existence of the other ship. The contract was held to be void for **mutual** mistake.

If, on the other hand, a seller agrees to sell facing bricks by sample at a special bargain price, when he later discovers that the sample shown was of a brick of much higher quality he will be forced to sell at the bargain price. Here both parties are mistaken. The seller thinks he is selling 'X' bricks at 'Y' price. In fact he is selling 'Z' bricks at 'Y' price. The purchaser thinks he is buying 'Z' bricks at 'Y' price. Therefore the parties are at cross-purposes over the type of bricks. There is thus a mutual mistake but not one which is so fundamental as to avoid the contract. The parties would be bound. A reasonable man viewing the contract would have inferred that there was an agreement between the parties, albeit based on a mistake.

Mutual mistake in equity

Once again equity follows the law and, in the exceptional circumstances in which a contract is void, it will be void in equity and no equitable relief may be given.

If it is valid, however, equity also follows the law and will generally refuse equitable relief and the mistaken party will be forced to perform the contract as agreed. Occasionally, however, a court will use its equitable discretion and may refuse a decree of specific performance.

Common mistake at common law

A common mistake will only make a contract void at common law if the common mistake is concerned with the existence of the subject-matter or other fundamental basis of the contract **at the time the agreement** was entered into. Thus, for example, if the goods to be bought already belong to the purchaser, the contract is void. Also, if the property to be sold has been destroyed or never existed, then the contract is void.

So in *Scott* v. *Coulson* 1903, an insurance agreement was entered into on the life of someone who was already dead. Obviously, the contract was void for **common** mistake, as both parties were under the same misapprehension over the existence of the subject-matter.

Thus, where parties contract on the basis of a common mistake, which is not concerned with the above matters, the mistake is not operative, e.g. if Simon agrees to sell what he thinks to be a genuine antique to Bill who is also under the same misapprehension, despite the existence of a common mistake, the contract is not void at common law.

Common mistake in equity

Where the contract is void at common law, equity will follow the law and equitable remedies would therefore be refused as no rights and duties would flow from a void contract.

However, if the contract is valid at common law the court may use its equitable discretion and may award the following remedies

1. *Rectification of written contracts.* See p. 59.
2. *Set aside agreements on terms fair and just to both parties.* Whilst the court will admit that the contract is valid at common law, the judge may nevertheless be unwilling to grant a decree of specific performance and may even go so far as to set the contract aside on terms favourable to both parties. The parties may even be sent away to remake their contract.

Mistakes over type of document signed

See p. 62.

Who can make contracts? Capacity

Generally anyone, i.e. any legal person, can enter into a contract. This is called **capacity to contract**.

However, there are certain people who because of their age or mental state are restricted in the types of contracts that they may enter into.

Minors

Fortunately, the law has set out to protect minors from unscrupulous people trying to take advantage of their youth, by forcing them to perform contracts which perhaps they should never have entered into in the first place.

Unfortunately, the degree of protection varies depending on the type of contract.

Voidable contracts

These contracts are valid, unless avoided by the **minor** before or within a reasonable time of becoming 18 years of age. Thus, a minor is bound by its terms, unless he avoids the contract.

Examples

1. Interests in land. By S.1(6) of the Law of Property Act 1925 no minor may hold a legal estate in land. Nevertheless if someone attempts to grant a lease to a minor he will then hold an equitable interest in the land and will have to pay the rent and comply with the covenants contained in the lease.

2. Shares in a company. Similarly he is bound by any obligations imposed on owners of shares unless he repudiates.
3. Shares in a partnership. Whilst no creditor of the partnership may sue the minor partner personally for a debt, he can still pursue his claim against the joint assets of the partnership.

Valid contracts

Contracts for necessaries

Necessaries include **services**, such as education and medical treatment, and **goods**, which are defined by S.3(3) of the Sale of Goods Act 1979 as being 'goods suitable to the condition in life of the minor . . . and to his actual requirements at the time of the sale and delivery'. Such necessary goods should therefore be things which one would expect a minor to require having regard to his 'condition in life'. This would take into account his status, profession and income. So basic food, clothing, bicycle and watch would probably be regarded as necessaries, but not antique goblets or very expensive watches. Furthermore, such goods would have to fit in with 'his actual requirements at the time of sale and delivery'. Each case is decided on its merits.

By S.3(2) Sale of Goods Act, in a contract for necessaries, the minor must pay a reasonable sum for the goods. So if no price was set or if the minor refused to pay because the sum was extortionate he would still have to pay but only a reasonable sum.

Usually problems only occur where the supplier of the goods or services has not been paid. If he insists on being paid in cash or making sure that cheques have been cleared before delivery there is no need for recourse to action. If it is needed, the supplier is protected provided the contract is for necessaries or the following applies.

Contracts under the Minors' Contracts Act 1987

This Act repealed, for the most part, the Infants Relief Act 1874 which now only applies to pre-1987 contracts. Thus, the old common law rules apply in the absence of statute law, i.e. most minors' contracts are unenforceable unless they are ratified after their 18th birthday, unless the rules relating to necessaries above applies.

The Minors' Contracts Act introduced two new important provisions.

First, by S.2 if an adult **guarantees** a minor's contract then even if the contract is not enforceable against the minor for some reason, e.g. because it was not for necessaries, the contract **is** enforceable against the guarantor. So, beware of fresh faced youths asking you to act as their guarantor!

Second, by S.3 a court can order the restitution, i.e. the return or transfer, of property acquired under a contract which is unenforceable against the minor. So, for example, if a minor contracts to buy an expensive motor bike, as the bike is not a necessary the contract is unenforceable against the minor. But if he refuses to pay for it, the dealer can ask the court for a restitution order and ask for the bike's return or for the proceeds of sale if he has sold it.

Contracts of service

Provided a contract of apprenticeship or service is beneficial to the minor, the courts will regard the contract as valid. Thus, in *Dunk* v. *George Waller & Son Ltd* 1970, where a master broke an apprenticeship agreement, the apprentice could sue on the contract.

Corporations

See p. 148.

Mentally disordered persons

These have varying rights depending on whether they are subject to the Mental Health Act and whether the other contracting party knew of the disability. If the contract is for necessaries then they must pay for the goods.

The contract must be legal or if legal not void

Generally, the law of contract allows parties to make contracts for whatever purpose they wish even if they are not to their best advantage.

However, the contract must be of a type that is permitted by law nor must it be void, otherwise the parties will be unable to enforce the contract in the courts.

Illegal contracts

Contracts may be illegal at common law and by virtue of statute law. In both cases, such contracts are generally **void** ab initio. Thus, no rights and duties will flow from the contract at all. Money paid over under such a contract will usually be irrecoverable except in certain situations. These effects are due to the fact that the law never allows someone to be able to benefit from a wrong to which he was a party. This is encapsulated in the latin maxim *Ex turpi causa non oritur actio* – there can be no action arising from an evil cause. Also, because of this principle, any collateral contract or contract arising out of or as a result of the illegal contract will also be illegal and therefore void.

Contracts illegal in common law on the grounds of public policy

- Sexually immoral contracts.
- Contracts to commit crimes or torts, e.g. Rentakilla contract killing.
- Contracts detrimental to the state, e.g. trading with the enemy.
- Contracts for the sale of public honours such as knighthoods.
- Contracts to impede the administration of justice.
- Contracts to avoid paying taxes; a clause can make the whole contract unenforceable.

Contracts illegal by virtue of statute law

It is unusual for legislation to expressly forbid contracts. If it does so, then it is usually for the protection of the public. Most illegal contracts, however, would fall under the previous heading as being contrary to public policy in common law.

Nevertheless, certain pieces of legislation do expressly make some types of contracts illegal.

Thus, contracts designed to collectively enforce the maintenance of prices within a trade or industry are prohibited under the Resale Prices Act 1976, so that in general sellers can sell goods at whatever price they wish. There are two exceptions to this, e.g. ethical and proprietary drugs and books which must normally be sold at fixed prices throughout the country.

More often, legislation makes only certain terms of contracts illegal or imposes conditions on the way in which contracts should be carried out, e.g. by requiring a licence to be obtained before operating. In such cases, the contracts will be unenforceable on the part of the person behaving illegally but will be enforceable by the innocent party. If the illegality is only a peripheral matter, then the courts may allow the person who committed the wrong to be able to enforce it (this shall not be presumed in every situation).

Void (but not illegal) contracts

At common law on grounds of public policy

The following are examples of such contracts:

- contracts restricting the freedom of marriage or otherwise contrary to the estate of marriage;
- contracts intended to oust the jurisdiction of the courts;
- contracts in restraint of trade (see post).

As such contracts are void, they are unenforceable in the courts. However, if a term can be severed from the contract so as to leave the rest of the contract untainted, the courts will allow the contract to be enforced minus the offensive term. Of course, sometimes it is impossible to sever the objectionable term, and if this is not possible the whole contract will be void.

As such contracts are only void, and not illegal, it is probable that money or property paid over under the void contract could be recovered. (Cf. illegal contracts above.)

Finally, unlike illegal contracts, subsequent or collateral contracts arising as a result of the void contract will not necessarily be void.

Special mention must be made of the third type of void contract – contracts in restraint of trade.

Contracts in restraint of trade

Such contracts attempt to restrict parties from carrying on their business or profession in any way they wish. Whilst such contracts are prima facie void, they may be regarded as valid, provided they are **reasonable** as between the parties and are in the **interest** of the **general public**.

Builders may encounter restrictive terms in contracts of employment. For example, a builder may employ a person to work for him under a contract containing a term that, if the employee should leave that employment, he would not attempt to compete with his former employer. Other terms in contracts of employment which have been held to be reasonable are that the employee should not divulge trade secrets. As the courts do not like such terms, they will only allow them to stand if the following requirements are met:

- The employee must be in a position whereby the customer/client would prefer to deal with him rather than his employer **or**
- the employee must have access to trade secrets *and*
- the area of restraint must be reasonable, e.g. a builder could possibly impose a limit on his employee carrying on a business within five miles of his former employment. Each case must be judged on its merits however. Restrictions have been allowed to cover the whole of the United Kingdom.
- The duration of the restraint must be reasonable. A lifetime ban would probably be unreasonable.
- The term must not merely restrict competition as this is unreasonable.
- The term will probably be invalid if the person imposing it was in a superior bargaining position to that of the other party (this is usually the position in the case of employers).

If a builder sold his business with the **goodwill**, then the courts are very ready to allow the validity of a term in the contract prohibiting the seller from soliciting his former customers'

business or using the firm's name (if the name was sold as part of the goodwill). A restriction on carrying on a similar business in a particular area will be subject to the same test of reasonableness as before. It should be noted, however, that such restrictive terms must be of benefit to the business only.

Similarly, the courts are less likely to interfere with terms in restraint of trade between parties on an equal footing, such as businessmen.

Contracts void under statute law

Certain specified contracts may be made void under statute law. This means that the legislation, direct or indirect, has not made the contract illegal, but the contract will be void ab initio. Thus, whilst we are permitted to place bets with bookmakers, by virtue of various gaming and wagering acts, the contracts are legal but unenforceable.

Certain contracts which offend against the Restrictive Trade Practices Act 1976 are held to be void as being contrary to public interest, unless on investigation by the Restrictive Practices Court they are found not to be so. A restrictive trade agreement is a contract by which manufacturers, suppliers or exporters limit the supply or distribution of goods in some way. This may be so as to keep prices high or orders constant, for example, and thus may be detrimental to the interest of the public. Such contracts have to be registered with the Director-General of Fair Trading. These contracts deal with such matters as recommended prices, terms on which goods are to be supplied, quantities in which they are available or the classes of people to whom or area in which they may be sold.

A customer protection scheme proposed by the Building Employers' Confederation has recently been approved by the Restrictive Practices Court and is now operational. It provides a guarantee that unsatisfactory work will be remedied in exchange for a charge of 1 per cent of the contract price. The scheme is compulsory for all members of the Confederation.

Finally, by virtue of the European Communities Act 1972 which ratified the Treaty of Accession by which the United Kingdom went into the EEC,

Art. 85(1) of the Treaty of Rome is now part of English law. This makes certain contracts void if they attempt to impede competition within the European Communities. It only applies to trading agreements between the nationals of different EU states. It does not affect domestic UK agreements.

Form of contract

Simple contracts

These may be in any form. They may be in writing, they may be oral or implied from conduct or they may be a combination of any form, e.g. partly in writing and/or partly oral and/or partly implied from conduct. However, there must be consideration given by each party to the other.

Contracts which must be in writing

There are a certain number of contracts which must be in writing. (Note: 'in writing' does *not* mean being in a deed. See below.) Thus the contract itself must be in a written form although it can be on any type of paper.

The following are examples: bills of exchange (see p. 132); consumer credit agreements to which the Consumer Credit Act 1974 applies; transfers of shares in registered companies; transfers of debts (see p. 127); contracts of marine insurance. These contracts are **void** if not in writing.

Additionally by S.2 Law of Property (Miscellaneous Provisions) Act 1989, a contract for the sale or other disposition of an interest in land (such as a lease, option to purchase or a mortgage of land) can now only be made in writing and only by incorporating all the terms which the parties have expressly agreed, either in one document or (as in conveyancing where contracts are exchanged) in both (S.1(1)). (Note: this is the contract part of the sale of land. The actual conveyance, i.e. the transfer of the estate or interest, which is the performance part of the contract, must be by deed. See below.)

By S.2(2) the terms may be incorporated in a document, either by being set out **in** it or by reference to some other document. Such 'some other

document' could be a standard form of conditions of some organisation such as the Law Society. Lastly, by S.2(3) the document **actually incorporating** the terms or, if contracts are **exchanged, one** of the documents must be signed by or on behalf of each party to the contract. However, they do not both have to sign the same one, as is usual in conveyancing.

Builders will be interested that S.2 does not apply to public auction contracts which still become binding at the drop of the auctioneer's hammer. Also contracts for leases for less than three years and contracts regulated by the Financial Services Act 1986 are excluded (S.2(5)).

If S.2 is not complied with then the contract is invalid in law.

Contracts which must be evidenced in writing

This does **not** mean that the contract must be in writing. There must, however, be some written evidence of the existence of the contract in order to be able to sue under the agreement.

Guarantees

A guarantee is where one person agrees to pay a debt for another if the debtor should default. Such contracts must be evidenced in writing (S.4 Statute of Frauds 1677); otherwise the contract is unenforceable, i.e. it is not void but is unenforceable against the guarantor. It could, however, be used as a defence.

Contracts of employment
See p. 227.

Contracts which must be in a deed

'Contracts' which have consideration on only one side must be contained in a deed. So, if your great aunt promises to give you £5,000 on your 21st birthday and asks you to do nothing in return (love and affection is not good consideration), then if you are going to be able to enforce her promise to you, the promise must be contained in a deed; otherwise her promise is unenforceable. If she gives you the money it is a mere gift.

Certain other matters must be contained in deeds. Strictly speaking, they cannot be called true contracts although they are usually the result of contracts. These are transfers of British ships, transfers of estates or interests in land (see p. 92).

A contract which is in a deed is said to be a **specialty** contract and it has the advantage of extending the limitation period during which one may sue from the normal six years to twelve years. This is one reason why many building contracts are in a deed.

By S.1 Law of Property (Miscellaneous Provisions) Act 1989, a deed is a document which makes it clear on the face of it that it is intended to be a deed, either by describing itself as such or by stating in the document that it is being executed and signed as a deed (S.1(1)). Execution of the deed is done either by the maker signing in front of one witness, who then signs (attests) the deed, or by another person signing on behalf of the maker at the maker's direction and in his presence and the presence of **two** witnesses, who after each attest (S.2(2)). This last method is used where the maker cannot sign for some reason, because he is physically incapable. The deed must then be 'delivered', i.e. treated by the maker as binding as a deed. A seal therefore is no longer a requirement of a valid deed.

Occasionally, a person will execute a deed but will deliver it **in escrow**. This means that despite execution and delivery to the other party they do not regard it as binding until some pre-condition is fulfilled. This is very common in conveyancing (see pp. 104–10).

True consent

The parties must enter into the contract freely, without force or undue influence. Nor must they have been duped into entering into the agreement by lies or misled by untrue statements.

Duress – common law

Threatening someone with violence or actually harming them will render a contract **voidable** at the option of the threatened party. It may even be void but there are few cases on this point.

Undue influence – equity

Undue influence is where someone exerts his influence over another in order to obtain a benefit from that person. Such influence may be presumed where a **fiduciary** (of good faith) relationship exists between the parties, e.g. as between a doctor and patient, solicitor and client, trustee and beneficiary, parent and child, minister of religion and disciple etc. This presumption is of course rebuttable in certain circumstances, e.g. where the giver of the benefit had independent advice in relation to the intended contract or where there is other cogent evidence of that party's free will. Also in *National Westminster Bank plc* v. *Morgan* 1983, it was made clear that the relationship between bank manager and customer is not presumed to be fiduciary – although in the facts of that case, there was such a relationship.

Undue influence may also arise where there is **no** fiduciary relationship but there is evidence that undue pressure was brought to bear on someone to gain advantage. This has occurred where that person is not very clever or else has delusions.

If there has been undue influence then the person on whom undue pressure has been brought may set aside the contract and recover any property transferred. If there is no fiduciary relationship between the parties then undue influence must be proved to have been brought on the party wishing to rescind the contract.

Once a contract has been affirmed, rescission is impossible. Thus, if a person in a position to rescind has acted in such a way as to lead one to believe that he is not going to rescind, then the courts may refuse to set the contract aside. An action to recover money or property should be taken therefore as soon as possible after the contract has been made; otherwise the courts may view a delay in pursuing a claim as amounting to implied affirmation.

Where third parties have bought the subject-matter of the contract, having had no notice of the undue influence, rescission of the original contract is then impossible. However, rescission may be allowed where purchasers have had notice of the undue influence or where the transferee gave no consideration for the gift.

Misrepresentation – common law, equity and statute

For a definition of misrepresentation and a comparison with terms of contracts see p. 58.

It must be remembered that a misrepresentation must

- be a statement of fact, not of opinion or of law;
- be untrue;
- have induced the other party to enter into that particular contract with the maker of the statement (representor) (or his principal).
- The statement must have been relied on by the other party (representee), e.g. in *Attwood* v. *Small* 1838 a purchaser of a mine chose to have independent mining experts to check whether the claims of the vendor of the mines' production capacity were true. The experts confirmed the statement and the contract went ahead. Later the purchaser went to court for an order to rescind, claiming that the statement was a misrepresentation. The House of Lords refused the order as it was clear the purchaser had not relied on the vendor's statement.

However, the courts would allow rescission if the misrepresentation was part of the reason for entering the contract. A plaintiff who knew the misrepresentation was untrue cannot then claim rescission. However, the burden of proving that he did know the truth is placed firmly on the shoulders of the defendant.

There are three types of misrepresentation.

1. Fraudulent

Here the false statement is made 'knowingly, or without belief in its truth, or recklessly, careless whether it be true or false' per Lord Herschell in *Derry* v. *Peek* 1889. It must therefore have been made dishonestly.

Remedies

1. A plaintiff may sue under the tort of deceit (also called fraud) under which he can claim damages for the loss he has suffered as a result of the tort (whether or not he also claims rescission of the contract).
2. **And/or** rescission by the representee at common law (see below). It should be noted that the representee could, if he wished, claim damages in deceit but still affirm the contract, i.e. go through with the contract but claim damages.

2. Negligent misrepresentation

The usual rules of negligence apply and an action in tort for negligence could possibly be brought where a person makes a statement, negligently hoping to induce the other to enter into a contract.

Remedies

1. A possible action for negligence (see above).
2. **And/or** rescission by the representee (see below).
3. Under the Misrepresentation Act 1967. By S.2(1) (expressed in an extremely roundabout way), where someone makes a misrepresentation in a negligent manner, he will be liable to pay damages '**unless** he proves that he has reasonable ground to believe and did believe up to the time the contract was made that the facts represented were true'.

   Thus, unlike in fraudulent misrepresentation where the plaintiff has to prove the existence of fraud, the burden of proof is shifted to the shoulders of the defendant and he has to show that he reasonably assumed the statement to be true.

   Whilst this section gives a slightly easier method of claiming damages under the Act, this method can only be used where the misrepresentation was made prior to a contract. Thus, where no contract was entered into, for some reason, a common law action for negligence is the only method that can be used.

3. Innocent misrepresentation

Here the misrepresentation is made innocently without any element of negligence.

Remedies

1. Rescission – see below.
2. Under the Misrepresentation Act 1967. Damages in lieu of rescission under S.2(2). For damages to be awarded, S.2(2) requires that the opportunity to rescind must be still available. If it has been lost then the right to damages will be lost also (under this head). The damages will be awarded at the discretion of the court.

Rescission generally

Contracts brought about by misrepresentations are voidable. This means that they are valid until the representee rescinds the contract on discovering the truth of the matter. If he does not rescind when he realises he has been misled, he is said to affirm the contract.

Rescission is possible at common law when the representee's action of, for example, informing the police or returning the property or otherwise informing the representor of the rescission makes the contract void. At common law, however, this is purely a practical matter; there is no action for rescission as in equity. Such court action may be advantageous if the representor refuses to admit that the contract is void and tries to enforce it in some way.

Rescission is no longer possible if

- the contract has been affirmed after finding out the truth (see p. 72);
- restitution in *integrum* is impossible to achieve – this means that it is impossible to put the parties back into the positions they were in prior to the contract, e.g. if the goods have been consumed;
- if an innocent third party has obtained property rights over the subject-matter of the contract to be rescinded (see p. 71).

Summary

If the basic elements of a contract are present and none of the factors which are grounds for making the contract void or voidable can be applied, the contract is valid (see Figure 3.2).

The doctrine of privity

The doctrine of privity is that only the parties to the contract have rights and duties thereunder. This has been illustrated by reference to building main contracts and subcontracts (see p. 42), under which employers and subcontractors cannot generally sue each other.

However, it has always been possible to obtain a collateral warranty from the subcontractor. This is a promise by the subcontractor that his work will be of the quality required by the main contract. As this is a contractual promise, it must either be contained in a deed (unlikely) or there must be consideration. This may take the form of a promise in consideration for being made the nominated subcontractor or it could be a separate fee for some element not found in the main contract, e.g. for design work.

It should not be forgotten that the doctrine only applies to contract. It is usually irrelevant in tort. See *Junior Books* v. *Veitchi* 1982 (see p. 170).

Some aspects which are exceptions to the doctrine are found in the doctrine of the undisclosed principal in agency (see p. 84); the transfer of contractual rights such as debts (see p. 127); third party rights in motor insurance (see p. 273); restrictive covenants (see p. 102); a beneficiary's right to sue under the trust created by other people (see p. 86); and the transfer of rights of action to the trustee in bankruptcy.

How do contractual rights and duties end?

A contract does not go on for ever, with the parties owing each other contractual duties ad infinitum. Contracts end in a number of different ways. These are by performance, agreement, frustration and by breach of contract.

Performance

Most contracts end by the parties performing their part of the bargain. This might have occurred quickly, e.g buying a cup of tea for 80p, or it may

Contract

take some time as in a building contract. Once there has been performance the parties do not have to perform any more duties, although there is a residual right to sue, if it turns out at a later date that there has been a breach of the contract during the contractual stage, even if the damage shows up later.

This is of course subject to the limitation periods (see p. 80) and to anything agreed by the parties in the contract itself. For example, the final certificate under a JCT contract does not prevent the employer from suing at a later date, as such an action will probably relate to hidden defects which would not have been obvious during the defects liability period. Even though the contract states that the certificate is conclusive as far as the contractor's claims are concerned, this will be no bar to a claim for breach of contract within the limitation periods.

The major rule relating to performance

Performance has to be complete; anything less than what was agreed will not do. Thus, if a builder enters into a lump-sum contract, e.g. to build X a house for £50,000, and for some reason the builder fails to complete the house, abandoning it before substantial performance, he cannot claim anything under the contract, nor can he claim on a **quantum meruit** (see below) as it is his fault that the contract has not been performed. A number of exceptions to this rule have been developed, however.

Exceptions

Substantial performance

If there has been substantial performance of the contract then this will discharge the liability of the parties. What is substantial performance? This amounts to performance as close as a reasonable man would expect even if it has been performed badly. Obviously, in building contracts even with variations the final building cannot be exactly as specified down to the last nut and bolt. Whether it is substantial will depend on the circumstances of the case. Of course, if the work has been done

badly and the client refuses to pay, then provided there has been substantial performance the court would order payment but subject to netting off the cost of putting it right.

Where one party accepts partial performance by the other

This must have occurred voluntarily and, if this is so, the other party can then claim to be paid on a **quantum meruit** basis, i.e. be paid a reasonable sum for the work he has completed.

(Note: this is only so where the acceptance of the part performance was freely undertaken.)

Divisible contracts

These are difficult to define. Suffice it to say that where a contract is made up of separate instalments of work to be done at different times, such as a maintenance contract, then payment is not conditional on completion of the whole contract. Whether this is true or not, however, will depend on what was agreed between the parties. If payment (or other consideration) cannot be recovered without complete performance on the other side then such a contract is said to be an **entire** contract and will not be divisible.

Complete performance prevented by the other party

If one party who is ready and willing to perform his part of the contract is prevented from completion by the other from doing so, then he cannot sue on the contract because it has not been completed, but he can claim a **quantum meruit**.

When can performance be expected?

This is normally up to the parties to agree. In building contracts this is usually strictly laid down because of the importance of getting the job done as soon as possible to avoid additional expense or liquidated damages. Under S.14 of the Supply of Goods and Services Act 1982, however, 'where, under a contract for the supply of a service by a supplier acting in the course of a business, the time for the service to be carried out is not fixed by the contract, left to be fixed in a manner agreed by the contract or determined by the course of dealings between the parties, there is an implied term that the supplier will carry out the service within a reasonable time'. By S.14(2) 'what is a reasonable time is a question of fact'.

In all other contracts if no date is agreed then usually none will be implied by the law. Time is **not of the essence**. Many people are caught by this rule and if performance is necessary by a certain date then they should make this a condition of the contract, preferably by setting a date in writing, e.g. when hiring a wedding suit or getting household insurance before leaving on holiday. Such a clause makes time of the essence.

If performance does not seem to be likely in the near future, and no date was set for performance in the contract, a party can later make it a condition by giving notice that if the work is not complied within X days (a reasonable time) he will regard the contract as at an end. This is useful when dealing with delivery of goods (see p. 115).

The same rules apply if performance amounts to being paid. Time is usually not of the essence unless it is made a condition of the contract or there are statutory rules.

What if a party is willing to perform his duties but this is refused?

If performance amounts to paying for something, if this is refused he need not offer to pay again. However, he is still liable to pay under the contract. He should pay the correct amount of money in legal tender if in cash. The other party is under no obligation to give change (see Figure 3.6).

If performance is not to do with payment but is concerned with goods or services, then offering to perform, which is then refused, cancels out any further liability under the contract and indeed it gives the offeror the right to sue for breach of contract. Thus, for example, if Jim takes a week's holiday to put up new guttering on Mrs Smith's house, when he arrives on the Monday as agreed and she turns him away, he can then sue her for breach of contract as he is ready and willing to perform his part of the contract; if he wishes he can of course waive the contract (see p. 76).

	Maximum
Gold coins	Any amount
Bank of England notes	Any amount
Silver and cupro-nickel coins of more than 10p	£10
Silver and cupro-nickel coins up to 10p and less	£5
Bronze coins	20p

Figure 3.6 Legal tender

After a contract has been agreed, can one side satisfy the contract requirements by giving less consideration than was agreed?

Generally the answer at common law is no, unless the contract is brought to an end in some way, thereby discharging the parties from their contractual liabilities. The basic rule is known as the rule in Pinnel's Case (1602).

Thus, unless equity helps out, someone must always pay the price agreed, even if a creditor agrees to accept a smaller sum in satisfaction of the contract terms. This is obvious because the original contract has not been performed and any subsequent agreement is not contractual as it is unsupported by consideration and therefore **not a valid contract**. Thus, in *D & C Builders Ltd* v. *Rees* 1966, Mrs Rees knowing of the builders' financial difficulties offered them a lesser sum than was agreed for building work done. Knowing they might get nothing if they waited, they accepted the offer. At a later date they sued for the balance. The court took the view that she could not rely on an old exception to the rule in Pinnel's Case that payment in a different form to that originally intended is good consideration for the promise to accept a smaller sum in satisfaction of the original contract.

Exceptions

Accepting a substitute consideration, accord and satisfaction – common law

Mrs Rees tried to rely on the exception to the rule in Pinnel's Case that a creditor may, if he wishes, accept something entirely different in the place of the original debt, even if the substitute, such as a car, is worth less than the debt. This is perfectly valid in common law and will bind the creditor,

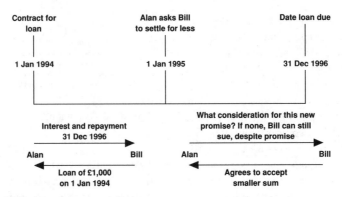

Figure 3.7 Diagram to illustrate the example below

who may **not** then go back on his promise. Even an early payment of a smaller sum will fall within this category, presumably because it may be of considerable value to a creditor with cash flow problems. Mrs Rees failed in her defence as the court took the view that a cheque was not sufficiently different. Nowadays it is virtually the same as cash.

Promissory estoppel – equity

Occasionally equity will help out by stopping (estopping) a creditor from going back on his promise to accept a smaller sum in satisfaction of a debt, provided the debtor has relied on that promise and has changed his circumstances accordingly. This is only allowed if it is used as a defence, 'a shield and not a sword': *Combe v. Combe* 1951.

Example (see Figure 3.7)

Alan borrows £10,000 from Bill repayable on 31 December 1996 with interest at 10 per cent per annum. Alan realises he will not have the money to pay the debt and asks Bill to settle for less. Bill will not be bound by any promise he makes unless (a) he agreed to accept a smaller sum at an earlier date; (b) he agreed to accept a ring in satisfaction of the debt; (c) equitable estoppel bars him from suing Alan.

(Note: the parties could of course agree to bring the contract completely to an end but it would have to comply with the requirements below.)

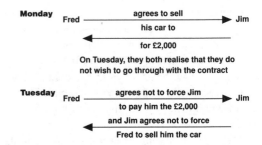

Figure 3.8 Waiver

Agreement

If neither of the parties have performed their part of the bargain, then they can of course agree by another exchange of promises to end the contract. (See Figure 3.8.) They will have then made a new contract to end the old. This is called a **waiver**.

Contracts ended by deed are of course always valid as consideration is not necessary on both sides.

It is also common for parties to provide in the contract itself for a date for the contract to end automatically, e.g. as in a fixed term contract of employment (see p. 233), or for a procedure for either party to terminate the contract, usually by notice, again as in contracts of employment. (These are, of course, subject to statutory rules concerning compensation and periods of notice. See pp. 234–5.)

If one party has completely performed his part of the bargain, the parties cannot waive the agreement as in Figure 3.8. This is because the promise

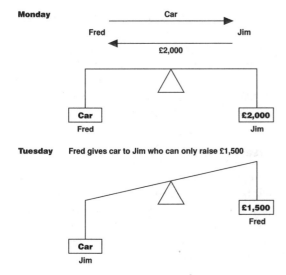

Monday

Tuesday Fred gives car to Jim who can only raise £1,500

Figure 3.9 Diagram to illustrate the example given below

not to sue for complete performance would not correspond with having already performed a part of the agreement. See Figure 3.9.

As we can see from the second balance in Figure 3.9, Fred's consideration, the car, far outweighs Jim's £1,500. Does he have to accept the sum? The answer is NO, because Jim is already bound by the contract. (See p. 75.) But, if he wishes for some reason to be kind, then he must make a new contract in one of two ways.

He can execute a **deed of release**. Remember a deed does not have to have consideration from the recipient of a gift. Or, he can make a new contract with some new consideration element which is different from that already agreed. This is called **accord and satisfaction**. So, Jim could offer him £1,500 and a ring (even if only worth a small sum). Had the situation involved the repayment of a loan, Jim could offer to pay the sum at an earlier date (which is considered to be new consideration). (Note: equitable estoppel may also apply (see p. 76).)

Frustration

A party cannot avoid his contractual obligations merely because costs have risen or circumstances have changed to make the contract difficult to perform. At the pre-contractual stage, the parties should have thought through the consequences of their contract and taken into account the possibility of changes in circumstances. They should insert, if necessary, terms in the contract to cover such eventualities, e.g. fluctuation clauses in a building contract. If the parties agree to end contractual obligations as shown previously, all well and good. Otherwise, these continue unless the contract can be shown to have been **frustrated**.

Frustration will only terminate the contract where some outside happening occurs, which is the fault of neither party and which has had such a dramatic and fundamental effect on the performance of the contract that it is either impossible to perform or its performance is substantially different from that originally envisaged.

The following are examples where frustration has occurred.

Illegality after the contract is entered into
Trading with an enemy after war has broken out would make a previously legal contract frustrated.

Death or severe illness in personal contracts
Whether the illness frustrates the contract will depend on the facts of the case and take into account the position of the employee in the organisation. For example a drummer with the Barron Knights pop group was held to have been lawfully dismissed after his illness made it impossible to work on more than four nights a week: *Condor* v. *Barron Knights Ltd* 1966.

State intervention
The compulsory purchase of land for example or requisitioning of property during a war frustrates a contract in relation thereto.

Destruction of subject matter of contract
See sale of goods p. 120 and mistake p. 64.

Fundamental change in surrounding circumstances uncontrolled by either party
In *Metropolitan Water Board* v. *Dick Kerr & Co Ltd* 1918, D K Ltd contracted in 1914 with the board

to build a reservoir over six years. The contract contained an extension of time clause, but owing to the war the board, after two years' work on the reservoir, ordered D K Ltd to stop work.

D K Ltd claimed that this order frustrated and therefore terminated any further obligations to build the reservoir. The court agreed as at the end of the war the contract to build the reservoir would have been entirely different from what was originally envisaged. It should be noted that the outbreak of war does not frustrate a contract automatically, unless the contract amounts to trading with the enemy, which is illegal. The surrounding events must be viewed to determine the effect of the war and the date on which frustration may or may not occur. See *Finelvet A G* v. *Vinaya Shipping Co Ltd, the Chrysalis* 1983 which was concerned with the outbreak of the Iran/Iraq war.

Effect of frustration

Under the Law Reform (Frustrated Contracts) Act 1943 a contract is automatically terminated by frustration from the date of the frustrating event.

By S.1(2), money paid before the frustrating event is recoverable and money due to be paid will no longer have to be paid. However, if the party who had been paid or was due to be paid had incurred contractual expenses prior to the date of discharge, then the court may, if it considers it fair, allow that party to keep or recover the whole or part of the money paid or payable. The amount cannot exceed the amount of expenses actually incurred. Any expenses after the date of discharge cannot be reimbursed.

In addition to the protection given by S.1(2), S.1(3) protects a party who has already performed part of his contract before the discharge by frustration.

Under S.1(3) if one party has obtained a valuable benefit before the date of discharge as a result of anything done by the other party to the contract, then he must pay a sum, being no more than the value of the benefit he has received. Such a sum will be as much as the court thinks just. Thus, if a building contract is frustrated, then if the builder has already built a proportion of the works he should be paid for the work done. The

wording of the section, however, is a little difficult to explain and there are few cases on the effect of frustration in such situations.

The Act does not apply to contracts of insurance and certain contracts for sales of goods (see p. 120) concerned with agreements for the sale of specific goods which have perished.

Breach

If one party refuses to perform the contract at all, i.e. he repudiates the contract without good reason, he commits an anticipatory breach. Alternatively, during performance of the contract if he fails to comply with all his duties he is also in breach of contract.

Remedies for breach of contract

If there has been a breach of contract, the breach may be such as to allow the innocent party to be able to treat the contract as discharged as where there has been a breach of a condition or it may be such as to allow the party only to claim damages as where there has been a breach of warranty.

Obviously there may be situations in which, despite there having been a substantial breach, the injured party will still choose to continue with the contract but will be allowed to claim damages. If the other party does not wish to perform his part of the contract at all, then a decree of specific performance may be asked for in the Chancery Division of the High Court.

Finally, if the innocent party wishes, he may claim to be paid on a quantum meruit for the work he has actually done, if the breach is such as to prevent him finishing the contract or if the contract is invalid (and thus a claim under the contract is impossible).

Treating the contract as discharged

The innocent party will not have to do anything further under the contract. If he chooses he may ask the court for a declaration to this effect. He will also be able to resist a claim for specific performance by the other party by raising the breach as a defence.

Damages

The rule in Hadley v. Baxendale 1854 – remoteness of damage

The actual damage for which the court awards compensation (damages) in contract must be 'either arising naturally, i.e. according to the actual course of things, from such breach of contract itself or such as may reasonably be supposed to have been in the contemplation of both parties, **at the time they made the contract**, as the probable result of the breach of it', per Alderson B in *Hadley v. Baxendale* 1854.

This means that the courts will only award damages for damage which is not too remote from the breach in fact and in law. (Note the use of the two different words. Damage means what one usually means by damage. Damages means compensation.) *Hadley v. Baxendale* concerned a millowner who sent a broken crankshaft for repair. He arranged for the defendant carrier to deliver it to the repairers. Due to the defendant's negligence, there was considerable delay in transporting the shaft and the mill had to close down for much longer than was expected by the millowner. The millowner therefore sued the carrier for damages. As he had failed to inform the carrier that he had no spare crankshaft and that he would be forced to close for the period of repair, the court only awarded damages for the delay. It did not give damages for the loss of profits occasioned while the mill was closed. This would have been too 'remote' from the actual breach. The carrier would have realised that he was inconveniencing the millowner and would have to pay damages for this breach of contract. However, as he did not know the mill was closed, there was no way he could have reasonably contemplated the additional loss.

In *Victoria Laundry (Windsor) Ltd v. Newman Industries Ltd* 1949, a claim for loss of profits was allowed by the court as it would have been reasonable to contemplate such loss. The laundry had ordered a new boiler which arrived much later than expected. The laundry company sued for damages for normal loss of profit plus damages for the loss of new dyeing contracts which they

had had to turn away. The court awarded damages for the first loss but not the second, using the principles of *Hadley v. Baxendale*.

Plaintiff must mitigate his loss

If there has been a breach of contract the plaintiff must not make the most of it. He should attempt to alleviate his position by reducing any financial loss he may suffer. If he fails to make a reasonable attempt the court will not award all the damages he was expecting. Thus, if a purchaser of goods has refused to take delivery, the seller must try to sell those goods elsewhere.

Prior agreement on quantum of damages

It is very common for businessmen to include a term in their contracts so that, if certain breaches occur, a specified amount of damages will be paid by the party in breach. This is designed to avoid (but not prevent) litigation. Such damages are called **liquidated** damages as the amount is **fixed** by the contract. Normal unknown damages are unliquidated and are only determined by the court at the end of the action.

It should be noted that such clauses are sometimes referred to as penalty clauses. However, a penalty clause is a liquidated damage clause which is designed to punish the party in breach. Such clauses are prima facie void. Whatever words are used to describe such clauses, the courts will strip their description away and decide whether they are valid liquidated damage clauses or void as penalties. If the clause is invalid because the sum specified is so huge that it is obviously intended to punish the party in breach, then damage will be awarded but only to compensate for the true loss. Where a valid liquidated damage clause is inserted in a contract, then even if the actual loss is greater than allowed for by the clause, the court will not award greater damages than agreed in the contract.

Specific performance – in equity

A decree of specific performance will be awarded at the discretion of the court if it considers that the legal remedies are inadequate. Such a decree requires the party in breach to perform his part of the contract.

Decrees will thus be awarded in situations where the subject-matter is unique, e.g. a Ming vase or a particular piece of land, as in such cases damages will hardly compensate the plaintiff.

As the remedy is equitable, equitable principles are applied. Thus, if a plaintiff has behaved inequitably or he has not performed his part of the contract, then a decree will not be awarded. He who comes to equity must come with clean hands; he who seeks equity must do equity.

Specific performance will not be awarded in contracts for personal services, e.g. contracts of employment, as it would be inequitable to force someone to work for another.

Nor will a decree be granted if the court would have to provide a constant monitor to see that it was carried out, e.g. as in a building contract.

Damages in lieu of specific performance

If the High Court is of the opinion that the plaintiff has a case in law and is entitled to a decree of specific performance in equity but for some reason refuses to award a decree, then they can award damages in lieu. Damages may also be awarded in addition to a decree.

Injunctions – in equity

Some contracts require performance in a negative way, e.g. a promise not to do something in exchange for money. Thus, a request for a prohibitory injunction might be a more suitable method of enforcing such a contract.

Mandatory injunctions are also available but they are less usual. Mandatory injunctions order a party to do something in order to remedy a breach. Thus, for example, in *Charrington* v. *Simons & Co.* 1970 the defendants were ordered by injunction to remove a road they had built in contravention of a contract.

Damages in lieu of an injunction

Once again, the court may, if it prefers, order damages in lieu of or in addition to an injunction.

Limitation of actions

It is unfair on a defendant and unreasonable of a plaintiff if he (the plaintiff) fails to take action within a reasonable period of his right to do so. The Limitation Act 1980 (a consolidating Act) lays down strict periods after which the plaintiff will be unable to sue in court.

In *contract*, the limitation periods are: for simple contract, six years after the date the contract should have been completed; for specialty contracts, twelve years.

If there has been an express defects liability clause inserted into the contract as is usual in building contracts, then time is said to run from the end of that period. If a defect has been deliberately concealed by fraud then the limitation period may be extended to take account of such fraud: S.32.

(Note: for limitation periods and date of accrual of action in tort see p. 162.)

It must be noted that limitation is a defence and, if not pleaded, an action which is statute-barred can still be taken.

It is possible for time to run afresh, if there has been an acknowledgement in writing by one party that he still owes a fixed sum or liquidated amount under a contract or if there has been a payment made in relation thereto. The Limitation Act periods apply only to claims for money and do not apply to requests for equitable remedies such as a decree of specific performance or an injunction. Nevertheless, as we have already seen, 'delay defeats equity' and the equitable doctrine of laches, upon which the Limitation Acts have been founded, would apply. Thus, if there was any serious and unexplained delay in taking the matter to court, no equitable relief would be given.

In conclusion, the stages of a contract are shown in Figure 3.10.

Agents

Definition and examples

An agent is someone who is appointed to act on behalf of another person called a **principal**.

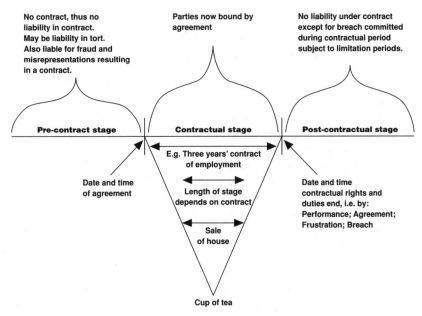

Figure 3.10 Stages in a contract

Agents represent their principals in many ways. See Figure 3.11. For example, a solicitor undertakes legal proceedings on our behalf. An estate agent undertakes to find prospective purchasers for someone selling his house. Sales representatives sell goods on behalf of their companies. Architects and consulting engineers act on the employer's behalf in relation to building contract work undertaken by a contractor. Site agents act as their employer's representative on site.

For our purposes, however, we are most interested in a common aspect of most types of agency. This is the agent's ability to enter into a contract with a third party on behalf of his principal. A number of points emerge from this.

First, the principal must have capacity to enter into the contract, even if the agent does not. Thus, it is quite possible that an agent could be a minor.

Second, once the contract has been set up between the third party and the principal, the agent drops out of the contractual situation altogether, so that there is privity of contract between the other two parties only. The agent is therefore not liable at all under the contract (but see p. 83).

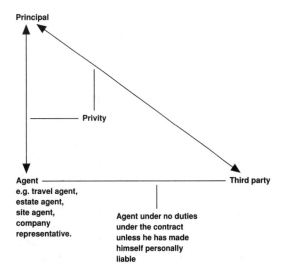

Figure 3.11 The relationship between the principal, agent and third party in agency

Third, if the relationship between principal and agent is of a contractual nature, then in consideration for arranging the contract between the third party and the principal, the agent will be paid a fee known as 'commission'.

How does one become an agent?

An agent may be appointed in a number of different ways, be it implied from behaviour, by ratification, express appointment, out of necessity or by estoppel (mnemonic IRENE).

Occasionally, agents may be appointed by deed called a **power of attorney** which gives the agent the right to do anything on behalf of the principal, e.g. sign a conveyance, collect the principal's rents and even take out a grant of letters of administration to a person's estate. Powers of attorney are often used where the principal is going to be out of the country on business or is severely incapacitated. (Note: a power of attorney is essential if the agent is to sign a deed.)

Implied appointment

Becoming an agent by implied appointment is very common and particularly occurs in employment situations, where it is implicitly part of the employee's job that they may, for example, buy things on behalf of their employer. This sort of agency is essential; otherwise the employee would be liable on such contracts if the employer failed to pay.

Ratification

This type of retrospective appointment can occur in two situations.

First, where someone who is not an agent acts as if he is, his 'principal' can, if he so chooses, ratify (confirm) the contract. He will then take over the responsibility for the contract. If he does not ratify, however, and he is under no compunction to do so, then the 'agent' remains personally liable on that contract. This situation could occur where an employee, who had never acted as agent before, suddenly attempts to contract on his behalf, perhaps because he was offered a bargain while his employer was absent, which he thought was too good to miss. He is of course taking the chance that his employer will ratify the contract.

Second, ratification can occur where the agent acts outside the scope of his known authority. For example, a buyer may have authority to buy goods on his principal's behalf up to the value of £5,000. If he wishes to purchase goods for more than that

amount, prior permission must be given. The third party must know of this restriction. If he fails to ask for permission and agrees to purchase goods for £8,000, he will be personally liable on that contract if the third party was not aware of the limitation to the agent's authority. This is so unless the principal ratifies the contract.

(Note: ratification is only permitted if the agent told the third party that he was acting as an agent for a named principal or one who could be identified. The principal must have had capacity to enter into the contract at the time of making the contract.)

Third, the contract must be capable of ratification. Thus, a void contract cannot be ratified.

Express

Express appointments may be made in any way, e.g. orally or in writing. If the agent has to enter into a contract under seal, however, he must be appointed by deed also (a power of attorney).

Necessity

There are a few unusual situations in which someone may act as an agent and bind a 'principal' even though that person has no authority to do so. For the acts of the agent to be binding;

- There must be a genuine emergency.
- It must be impracticable to get the principal's instructions before the act is done. This has made the agency of necessity less important nowadays with easy forms of communication.
- The act must be done in good faith.
- The act must be done for the benefit of the principal and not merely for the convenience of the agent.
- The agent must already be acting on behalf of the principal as an agent.

It would be unusual for a builder to find himself acting as an agent of necessity. It is possible to imagine that such a situation could occur in relation to the supply of ready-mixed concrete. If the driver was sent to the incorrect address, through no fault of his own, he could attempt to sell it elsewhere in order to prevent it solidifying. Although he may not have authority to do this, he may come within the ambit of the agency of necessity if the above pre-conditions are met.

Agency by estoppel

This occurs where a principal allows third parties to believe that someone is acting as his authorised agent. In such cases he will be estopped from denying the agency (even if in fact that person had no authority at all), if such third parties rely on such a representation to their detriment. This rule applies even if the principal had no intention of creating an agency.

Thus, if husbands, partners or employers allow their wives, co-partners or employees to act as their agents, they cannot later avoid responsibility for the agents' acts, unless they had previously expressly notified third parties with whom the agent has been dealing of their refusal to be bound by the agent's acts.

How wide is the agent's authority? Is the principal always bound by the acts of his agent?

The courts regard an agent's authority to be **actual** or **apparent** (or ostensible).

Actual authority means the authority which the agent actually has by agreement (express) or by implication of the job or as implied by the law or what is usual in a particular business.

A principal will always be bound by the contracts made by his agent which he authorised.

Apparent (or ostensible) authority is the authority which outsiders such as third parties assume the agent to have. This authority may be implied or what is usual in those particular circumstances or the authority which occurs in situations where the principal represented someone to be his agent – as in agency by estoppel.

Thus a person will be responsible for contracts which have been entered into by their agents in accordance with what is **usual** for a particular job. Thus, if it is normal for a site agent to make certain types of purchases, it is immaterial if that particular site agent has no such authority (unless the third party knew of such a restriction). The principal will therefore be liable on that contract.

Therefore an agent's authority depends on the type of agency, what has been agreed between the agent and principal, and on what the law implies into the agency situation.

Exceptions to a principal's liability for an agent's contracts

An agent will be personally liable on contracts made on behalf of his principal in the following situations.

Agreement to be personally liable

If for some reason the agent agrees to be personally liable, then he can sue or be sued on the contract, whether or not the principal is joined with him. This could occur where the third party thinks the principal may not pay and asks for the agent's assurance that he (the agent) will be responsible.

Contracts signed in the agent's own name

An agent signs contracts in his own name, without any qualification, at his own peril. This is because the courts will need strong evidence to the contrary that there was no intention that the agent should be bound by the contract. Thus, agents should always sign documents or letters with their name only if accompanied by words such as 'for and on behalf of'; 'pp Joe Bloggs' (pp – *per pro*, Latin for 'for and on behalf'); 'as agent for Joe Bloggs': *Plant Engineers Ltd* v. *Davies* 1969.

Similarly, if an agent signs a cheque in his own name and not on behalf of his principal, he will make himself personally liable for that cheque. Thus, personal cheques should always be avoided in business.

What is the extent of an agent's liability if he appears to be acting on his own?

In such cases, if he has not revealed that he is an agent, then a third party, who is of course under the misapprehension that the agent is a principal, can enforce the contract against the agent. If he finds out about the agency, he can if he wishes claim against the principal (provided he knows his identity). He must elect to sue one or the other. He cannot sue both. Conversely, the principal can take over the contract and claim any rights under it.

If the agent has revealed that he *is* an agent, but not the identity of the principal, e.g. as at international art auctions, then the agent is not liable under the contract unless it falls within one of the exceptions above.

If a person pretends to be an agent, then he is obviously personally liable on any contracts he makes.

Duties of an agent

During the course of an agency, an agent owes his principal a number of duties, many similar to those owed by an employee to his employer. These are:

- A duty to use such care and skill as he possesses in performing his functions as an agent.
- A duty of obedience. He should do what is required of the agent, unless the act is illegal. If he fails in a contractual duty he will be in breach of contract.
- A duty to perform his contract in person, unless it is usual in a particular kind of agency to delegate certain duties. The agency itself cannot be delegated.
- A duty not to deny that his principal has title to the principal's property.
- A duty to account for any money received by him as an agent on behalf of his principal and must hand this over to the principal.

- Finally and probably most importantly he must act in good faith, i.e. he must not compete with his principal, nor make secret profits such as a discount allowed by the third party (see p. 231).

Termination of agency

This occurs by operation of law or as a result of the parties' actions.

Operation of law

Agencies are automatically terminated on

- death of the principal or agent,
- insanity of the principal or agent,
- bankruptcy of the principal or agent.

Also, agencies may be terminated by

- frustration (see p. 77),
- performance of the contract (see p. 73),
- effluxion of time (see p. 76).

Acts of the parties

The parties may terminate their agency by mutual agreement either under the terms of their contract or by subsequent agreement. The principal may revoke the agency at any time for a good reason. If the revocation was unjustified, then the agent may sue the principal.

4 Law of property

Introduction

What does property mean?

What sort of property do you own? Make a list. The word property comes from the French word 'propre', meaning one's own, and thus property in law is anything capable of **ownership**.

Your list should therefore include 'things' such as motor cars, clothes, jewellery, furniture, household goods, animals, food, cheques, copyright, patents, money and debts, as well as your house and land. You cannot own people or corporations.

You will notice that out of this list some property is **tangible**, i.e. it can be touched or is physical. Other types of property are **intangible** and cannot be touched, such as copyright, patents, shares in a company, or goodwill.

Possession and ownership

Property can be possessed or owned. What is the legal difference?

Possession

This was recognised much earlier in history, as it is relatively easy to determine who is the possessor of something. It is a question of **fact** and needs no law to determine it. Because of the ease of proving possession, it is often said that 'possession is nine tenths of the law'. Indeed, the person in possession of property has a better right to it than anyone else, **except for the true owner**. Thus, a squatter **has** rights, but only **possessory** rights, and they can never be better than those of the owner. If he remains on the land long enough, however, he may acquire ownership of the land by adverse possession. Similarly, if a person finds a lost ring he may have it returned to him by the police, but, if the true owner seeks its return within the limitation period, then the finder may be sued for the ring itself or its value. 'Finders, keepers' is only partly true.

Ownership

This concept encompasses all the legal rights that can attach to property, so that the owner can do with the property whatever he wishes, including giving up possession. Ownership is therefore better than mere possession as the owner is entitled to possession as well, and can claim any property back that has been wrongly taken out of his possession.

Can ownership be separated from possession?

It is usual for the owner of something to keep it in his possession. Sometimes, however, the possessor may not be the owner of the property. The building contractor, for example, takes possession of the employer's site. He often hires (leases) equipment, such as cars, lorries or compressors. He sends his equipment for repair, and the repairer takes possession of the equipment temporarily. The thief has unlawful possession of any goods he steals from the site.

Legal categories of property in English law

English law categorises property in a number of different ways, such as tangible and intangible property. The most important legal classification, however, is historical and rather outdated – real and personal property (see Figure 4.1).

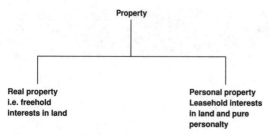

Figure 4.1 Categories of property in English law

Real property

This consists of **freehold** interests in land (see below).

Personal property

This consists of **leasehold** interests in land, which are called **chattels real**, as well as **pure personalty**, which consists of **chattels personal**, e.g. things such as jewellery, bricks, lorries etc. as well as intangible things such as shares, patents, debts etc. (see p. 111).

For our purposes, however, it is far easier to divide property into two areas, **land** and **all other property**.

Land

Introduction

You will remember, from the history at the beginning of the book, that under the Norman feudal system only the king could actually **own** land. Although the feudal system gradually died out, this principle in theory still holds true today. Thus, only the Crown can own land. All other persons merely own **estates** or **interests** in the land. The words estates and interests are difficult to explain and reference should be made to the leading textbooks on land law for a full explanation. Suffice it to say that, for our purposes, 'estates' are interests in land, which continue either indefinitely or for fixed or determinable periods and which are capable of being owned. Interests are, as their name implies, interests which can be owned in relation to *other* legal estates in land.

Estates and interests

Estates

Before 1925 there were many different types of estates that could be owned in land. However, under S.1 of the Law of Property Act 1925 only **two** estates are now permitted at common law. All others must now exist in **equity** as *interests* behind a trust (see Figure 4.2).

Uncle Fred leaves his house 'Dunworkin' in his will to Biggles, the banker, and Seth, the solicitor, to hold on trust for the benefit of his sister, Phoebe, for her life, after which it is to be sold and the proceeds given to Uncle Fred's favourite charity, Brian's Home for Sick Parrots. Biggles and Seth, as trustees under Uncle Fred's will, are the owners of the legal estate and their names will appear on the title deeds as legal owners. In equity, however, Phoebe, as the beneficiary under the trust, actually owns the property.

On Phoebe's death, the purchaser is not concerned with the equitable interest in the house. He merely wishes to acquire the legal estate from Biggles and Seth, free from the equitable interest, with vacant possession (if there was a breach of trust, then the charity or Phoebe, if she was still alive, could sue the trustees).

Interests

Section 1 of the Law of Property Act 1925 also stated that in future only **five** legal interests could exist in someone else's land. All other interests from then on had to be equitable.

The reason why the number of estates and interests was reduced was to simplify the transfer of land. As there is now a smaller number of legal estates and interests, fewer enquiries need to be made during conveyancing (see pp. 104–10).

What is meant by land?

Subject to the restrictions in **use** of land, which we will examine later, a landowner will, in addition to owning the 'crust' or 'surface' of the land, also own the airspace above the land and the earth below.

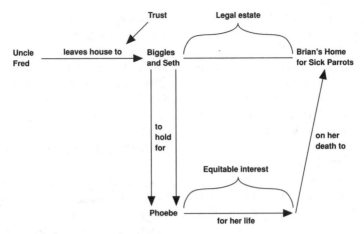

Figure 4.2 Diagram to illustrate an equitable interest behind a trust

Over the centuries this has been steadily whittled away. Gold and silver in a mine belong to the Crown. Articles of gold or silver found hidden also belong to the Crown as treasure trove, if the true owner of them is unknown. Any other chattels found on land may be judged to belong to the owner of the land rather than to the finder, as in *City of London Corporation* v. *Appleyard* 1963, in which a safe containing banknotes was discovered by two workmen in the basement wall of a house they were demolishing. The lessee of the house was held to have priority over the finders.

Coal and petroleum which is discovered on land now belongs either to the state or to public corporations, such as the British Coal Corporation, by virtue of Acts of Parliament.

In addition to the land any buildings or things which have become attached to the land then become part of the land. Therefore, when a purchaser wishes to buy a house, his solicitor is just as concerned with the land because the house and land represent the property purchased. If the house were pulled down, what would the householder own? Attachments which originally were chattels are known as **fixtures** and may include such things as fitted kitchens, toilet-roll holders, rose bushes and machines bolted to a factory floor.

However, machines standing by their own weight have been held *not* to be fixtures. Similarly, builder's materials, stacked for convenience to form a wall, were also found not to be fixtures.

Fitted carpets are probably not fixtures. Removal of such fixtures, on sale, without prior agreement, gives the purchaser the right to sue for breach of contract. So if you are moving and wish to remove some fixtures you must make sure that this is made clear in the contract.

If a river runs through land, then the bed belongs to the owner of the land. If his land merely adjoins the river, then he and the owner of the land on the opposite bank will each own up to the middle line of the river, unless the contrary is shown from the title deeds.

Similarly, an owner of land adjacent to the sea will own up to the highwater mark (from the highwater mark to the low water mark belongs to the Crown).

The landowner also owns the crops on the land and may catch and kill wild animals – but he cannot own them till they are dead, as wild animals are free creatures, incapable of ownership.

Additional rights relating to the land

At common law the landowner, in addition to owning the land, will own certain rights, some of which have been reduced by statute (see p. 88).

One such right which is of importance to builders is that of **support**. A landowner is entitled to have his land supported from underneath and from the sides. Thus, if someone quarries under land, with the owner's permission, he must make sure that the surface does not collapse. If subsidence occurs he may be sued. Should there be a building on the land which collapses as a result of removing the support, damages may only be awarded in relation to the building if the land would have collapsed whether or not there had been something built on it. Thus, there is **no** natural right of support for **buildings** (but see in relation to nuisance).

A right of support for a building may, however, be acquired as an **easement** (see p. 97).

If the damage was caused by natural processes such as wind or rain, there can be no right of action.

Limitations of the use of land

Unfortunately for the landowner, whatever type of estate he owns, the idea that the Englishman's home is his castle has been considerably eroded in the twentieth century.

For the good of society, restrictions have been placed on the landowner's use of his land. Some are common law creations but most have been created by statute.

The following are merely some examples which should be familiar to the builder.

Development – statute

No 'development' of land as defined by S.55(1) Town and Country Planning Act 1990, as amended by the Planning and Compensation Act 1991, can be undertaken unless planning permission has been expressly granted by the local planning authority, or is permitted under a development order. At the worst a local planning authority could get a court order requiring the demolition of unauthorised work.

Nuisance – a tort (see Ch. 7)

A landowner should not commit a nuisance on his land which affects another occupier of land; otherwise, he may be sued for damages or an injunction. Actionable nuisances include creating dust, noise, smoke, smells and vibrations and permitting tree roots to undermine neighbours' walls.

The commission of a public nuisance could lead to a criminal prosecution.

Dumping of rubbish and toxic waste – statute and tort

Not only could this amount to a nuisance, but also indiscriminate dumping of rubbish is controlled by statute, principally the Environmental Protection Act 1990. Under this Act there are regulations concerning toxic or dangerous waste which must be complied with. Infringement of these is punishable in the criminal courts.

Compulsory purchase – statute

There are numerous Acts under which local authorities, government departments, public corporations and now, worryingly, even privatised companies are authorised to compulsorily purchase private land for such varied uses as motorways, road widening schemes, slum clearance and public buildings. The Acquisition of Land Act 1981, Land Compensation Act 1973 and Planning and Compensation Act 1991 govern the procedure for compulsory purchase.

Compensation is paid at open market value, with extra for removal expenses, and where the valuation is disputed, appeals may be made to the Lands Tribunal.

Preservation orders – statute

Under town and country planning legislation, trees may be subject to tree preservation orders, and buildings of architectural or historic interest may become listed buildings. If this is so, the trees may not be cut down or lopped nor the buildings demolished or altered without consent from the local planning authority, i.e. the district council.

The rule in Rylands v. Fletcher – tort

Where a landowner brings on his land something dangerous which he allows to escape, he will be strictly liable for damage caused. Such dangerous things include water in a reservoir, fire, explosives and animals (see p. 190).

Negligence – a tort

Similarly, a landowner should not use his land in a negligent manner which may result in damage to another person or property (see p. 167).

Occupier's liability – tort and statute

An occupier should see that his visitors are safe when entering his property. Thus he should make sure that it is not in disrepair and warn the visitors of potential dangers (see p. 184).

Rights of entry of public officials – statute

There are many public officials who have statutory rights to enter another's land. Normally, that entry would amount to trespass, unless the official is a police officer who is trying to prevent a breach of the peace or a serious crime being committed. Such officials include inland revenue officials, building inspectors, planning, health and safety inspectors, environmental health officers, gas, electricity and water authority officials, bailiffs and police. Usually written authority to enter should be produced for the landowner's inspection.

Withdrawal of natural rights – tort

If the landowner interferes with certain natural rights enjoyed by his neighbour, e.g. he undermines the support to his land, he can be sued in tort (see p. 87).

Water – statute

The Water Resources Act 1991 restricts the abstraction of water without first obtaining a licence, subject to exceptions, and makes it an offence to pollute river water, springs or lakes.

Restrictive covenants – common law

These are terms written into deeds of transfer of land which restrict the use of the land **being sold** in some way for the benefit of the land **being retained**. Typical examples are the prohibition of businesses or having caravans or livestock in the gardens (see p. 102).

Easements – common law

These are private rights over the land of another, e.g. a right of way. Where an easement exists, the owner of the land subject to the easement must allow the owner of the easement to exercise it (see p. 97).

Trespass – tort

A trespass to land takes the form of an unauthorised interference with the possession of another person's land in a direct way, e.g. walking on the land, swinging a crane through the airspace, or dumping rubbish (see p. 205).

Civil Aviation Act 1982 – statute

By S.76 no-one can sue for trespass or nuisance caused by aircraft merely flying over one's land, provided the appropriate regulations are complied with.

The two legal estates: S.1 Law of Property Act 1925

These are (i) the fee simple absolute in possession and (ii) the term of years absolute.

The fee simple absolute in possession

This estate is the only *legal freehold* estate and is the nearest to true ownership that ordinary people can get. No-one has a better right to the land than the owner of the fee simple. There can only be one fee simple estate in a piece of land at a time. This is the estate that you would have if you bought your house 'freehold'.

This estate

- lasts indefinitely;
- can be sold, given away, split up, leased and inherited by will or on intestacy;
- is in possession; therefore the owner has exclusive rights to remain on the land.

Other freehold estates

(Note: these are **not** legal **estates**; now only equitable **interests**.)

Before 1926 there were other freehold estates. e.g. Phoebe's estate on p. 86 was a freehold estate for life and lasted until her death. This could have been sold, but as it would have ended on Phoebe's death, any purchaser would have merely bought an interest **pur autre vie** (during the life of another).

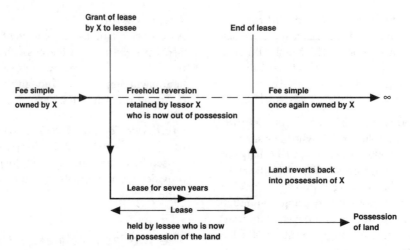

Figure 4.3 Diagram to illustrate a lease

Similarly there were freehold estates called fee tails, i.e. the estate was inherited by a person within a limited category or class of people consisting of the donee's descendants. It was designed to ensure that the land was transferred in one piece from generation to generation.

These estates since 1925 are now only **equitable** and must exist as **interests** behind a trust.

Terms of years absolute (leases)

There can only be one fee simple owner of a piece of land at a time, but there may be many leases.

The owner of a lease is therefore an estate owner just as much as the fee simple owner. But his estate is limited in time to fixed or determinable period, e.g. 99 years or a weekly tenancy. He must be given exclusive possession of the property (even if he later gives exclusive possession to a sub-lessee).

So, if the person owning the fee simple grants a lease the lessee gets exclusive possession and the fee simple owner loses it. The part of the fee simple that he retains is called the **freehold reversion**.

It is the same legal freehold estate, but minus his right to possession (see Figure 4.3). He is then called the **freeholder** or **lessor** and the person holding the lease the **leaseholder** or **lessee**.

Leases are sometimes called **tenancies** especially where the period of leasing is quite short. In that case, the lessor is called the **landlord** and the lessee the **tenant**. This is once again an indication of how history has influenced English land law, as it embodies the concept of land holding or tenure.

A lessee can **assign** his lease, i.e. sell it or transfer it by gift or will, in which case he transfers **all** that is left to run of the lease. Alternatively, he can **sub-let** by creating a lease for a period shorter than his own lease. This is very common in business, where there may be a chain of leases.

The freeholder may also transfer his freehold reversion, and the person to whom it is transferred will then have all the reversionary rights, as well as the right to receive **ground rent** (see Figure 4.4). The transfer of freehold reversions to unscrupulous freeholders has become a problem in recent years.

The essential elements of a lease

(i) The lessee/tenant must have *exclusive right to possession of the property. Thus, lodgers and persons in hotels, bedsits or service apartments are not tenants but mere* **licensees** (see p. 103).

The description used in the lease or agreement to describe the relationship is not conclusive, and the court may strip away the description and examine the true relationship beneath.

(ii) The duration of the lease must be **fixed** or it must be capable of being **determined**.

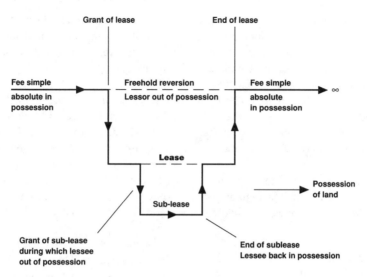

Figure 4.4 Diagram to illustrate a sub-lease

The intention of the parties is very important. If a relative comes to live in your holiday cottage until he buys somewhere, he is probably not a tenant. He is more likely to have a **licence** to occupy the land (see p. 103).

If there is a dispute, however, the court would look at all the relevant circumstances, and even if you swore that there had been no intention to create a lease, only a licence, they may still say a lease has been created.

Types of leases

Fixed period leases

These are common in business and domestic land ownership, e.g. 1, 3, 7, 99 or 999 years. At the end of the period the right to exclusive possession reverts back to the lessor. The lease cannot be ended by mere **notice to quit**, but if the lessee is in breach of a covenant then the lessor can take steps to remove the tenant.

Similarly, the lessee cannot give notice to end the lease unless a clause is written in, giving the parties the opportunity to terminate at certain specified times.

The lessee can surrender the lease back to the lessor but only if this is agreed between the parties.

Periodic tenancies

(e.g. weekly, monthly, quarterly and yearly tenancies).

These continue indefinitely until ended by notice to quit from either party. Whilst they do not seem to be for a fixed period, the law regards them to be so, e.g. for a week. If notice to quit is not given, then a further week's tenancy comes into being, and so on.

The length of notice depends on what sort of tenancy it is, e.g. a week's notice for a weekly tenancy, a month's notice for a monthly tenancy, six months' notice for a yearly tenancy. If the premises are let as dwelling, then four weeks' notice must be given.

The duration of the tenancy depends either on the terms of the tenancy or on the period for which rent is to be paid. So a person paying weekly rent would have a weekly tenancy.

A person who **holds over** after the expiration of a fixed tenancy and pays rent which is accepted as rent will have a new **periodic** tenancy created.

The following are not leases

Tenancies-at-will

It is uncertain whether a tenancy-at-will is within the definition of a term of years absolute, though it seems to be more than a licence.

Here, the tenant occupies the land **with the landlord's permission**, but the tenancy may be ended at **any time** by either party. Thus, there is **no** certainty over the duration of the tenancy.

It may occur on holding over after a fixed lease has ended. If the landlord starts accepting rent in relation to a fixed period, e.g. a week or month, however, then a periodic tenancy will be created by implication. If either party dies, the tenancy-at-will ends.

Payment should be made by the occupier for his period of occupation. If it is a regular payment at periodic intervals, once again it becomes a periodic tenancy.

Tenancies-at-sufferance
If a lease ends, and the tenant holds over **without the landlord's permission** and **without the landlord accepting rent**, then the tenant is a tenant-at-sufferance.

It is *not* a true tenancy and the landlord may take legal steps to remove the tenant at any time.

(Note: it is important to determine whether a tenancy/lease exists, because owners of such a legal estate are given many statutory rights. For example, a leaseholder with a long lease at a low rent may seek the compulsory purchase of his freehold reversion from the landlord or the grant of a new lease at a limited rent under the Leasehold Reform Act 1967 as amended by the Leasehold Reform (Housing and Urban Development) Act 1993. Tenants may also be protected, under the Rent Acts, the Protection from Eviction Act 1977 and the Housing Act 1988, from eviction or unreasonable increase in rent.)

Creation of leases as legal estates
Leases for less than three years
These may be created orally or in writing or by deed. If the lease or tenancy agreement is not by **deed**, (see p. 71 for Definition), then, although no formalities are required, S.54(2) of the Law of Property Act 1925 must be complied with, i.e. the lease must take 'effect in possession for a term not exceeding three years . . . at the best rent which can be reasonably obtained without taking a fine'.

A lease takes effect in possession if the lessee can take possession immediately. A fine is a lump sum or premium, which is sometimes paid in return for a lower rent.

Leases for more than three years
These must be created by **deed** in order to create a legal estate (S.52 Law of Property Act 1925).

Leases which do not come within the above categories, and are not legal estates, may nevertheless be treated as leases, by virtue of either equity or statute, and acquire many of the rights and duties associated with leases held as legal estates.

A contract to create a lease which has not been made by deed will be enforceable if it is in writing. If not, it is *void*, i.e. wholly without effect (Law of Property (Miscellaneous Provisions) Act 1989). This rule does not apply to leases for less than three years provided the requirements of S.54(2) of the Law of Property Act 1925 (above) are complied with.

Rights and duties of lessors and lessees under a lease
As we have already seen, the lessee has virtually the same rights as the fee simple owner except that the lease does not go on indefinitely. He is also subject to the same restrictions on the use of the leased property.

What is different, however, is that the lease itself, whether written or oral, will contain provisions regulating the behaviour of both the lessor and the lessee. These regulations of behaviour are called **covenants** (promises). They may be **express**, i.e. actually written or spoken, or they may be **implied** by the common law or by statute, or the lease may talk of the **usual** covenants applying. Usual covenants are those which have become accepted over the centuries as being the '**usual**' ones found in a lease.

Thus, express covenants are those spoken about or written into the lease, which have been expressly drafted and tailor-made for that particular lease. Implied covenants do not appear at all in the lease, and the usual covenants will only apply if that expression is used in the lease.

Covenants normally found in leases

Whatever the type of lease and whether the covenants are express, implied, or usual, there will of course, always be covenants by the lessee to pay rent and rates, and by the lessor not to interfere with the lessee's 'quiet enjoyment' of the land, i.e. he will not physically prevent the lessee from enjoying the premises leased, e.g. by removing the doors or windows.

Also, the lessor impliedly promises not to '**derogate** from his grant'. This means, that if the premises were let for a particular purpose, then the lessor will not prevent its use for that purpose.

There is also an implied covenant that the property is fit for human habitation at the beginning of the tenancy, but this depends on whether the property is a house or furnished accommodation.

Covenants concerning repairs

The responsibility for doing repairs depends very much on the type of lease, and what sort of covenants are mentioned in the lease.

Lessor's implied obligations. If nothing is mentioned in the lease then at common law there is no duty on the part of the lessor to repair or maintain property he has leased. Under the Landlord and Tenant Act 1985 as amended, however, if a dwelling house is let for less than seven years the lessor must keep the exterior and structure of the premises in repair. He must also repair, and keep in working order, installations for the supply of water, gas and electricity, but not appliances using the supplies, such as cookers or fires, sanitation and water heating.

The lessor should be given notice of the disrepair by the lessee, unless the property to be repaired is within the control of the lessor, e.g. a common staircase, where the lessor shares the premises with the lessee.

The courts are quite willing to imply covenants relating to repairs, taking into account all the circumstances. Any lease which is silent on a point can be dealt with in this way.

Lessee's implied obligations. The extent of this implied duty to repair depends on the type of lease/tenancy he holds.

1. *Periodic tenancies.* A periodic tenant need not carry out repairs and maintenance, but he must take proper care of the property such as unblocking drains or turning off the water during winter holidays. This is called acting in a tenant-like manner. He is also liable for voluntary waste.

   (Note: **Waste** is the term used for something done, or not done, which alters the nature of land, for better or worse.

   In relation to a lease it is important that the lessor receives his property back in more or less the same state as he leased it.

   There are four types of waste: **permissive**, **voluntary**, **equitable** and **ameliorating**. Permissive waste amounts to allowing the property to fall into disrepair. Voluntary waste involves altering the land in a detrimental way, e.g. chopping down trees or demolishing buildings. Equitable waste consists of acts of wanton destruction, i.e. voluntary waste done maliciously, and ameliorating waste involves acts which improve the land.)

2. *Fixed term leases.* A lessee for a fixed term is liable for repairs and must keep the property in the same condition as he took it over at the beginning of the lease. He is liable for voluntary and permissive waste.

The 'usual' covenants relating to repairs. When the lease states that the usual covenants apply, the lessee must keep the property in repair and allow the lessor to enter and inspect the state of repair. If the lessor is under an express duty to do some repairs, then the lessee *must* allow him to enter the premises for that purpose.

Leases of special types of property may have slightly different 'usual' covenants.

Express covenants. These are specially drawn up for the particular lease, and can cover whatever the parties wish. However, express covenants tend to be similar, depending on the type of lease and property.

Generally, long leases normally impose duties on the lessee to repair. This duty may include replacing the whole building if it is destroyed.

This indicates the need to make sure the property is adequately insured to cover such an eventuality.

If the lease is short, then the lessor will usually agree to do all external and structural repairs.

At the expiration of a lease, with an express covenant to repair imposed on the tenant the lessor may inspect the property to view the state of repair and serve a **schedule of dilapidations** on the lessee itemising the work required to be done and the value of it.

By S.4 of the Defective Premises Act 1972, where the lessor has expressly agreed to do repairs then he owes a duty to those persons who might be affected by the state of the premises, such as visitors. This duty is to see that they are reasonably safe from personal injury or damage to their property, caused by a defect, of which he knows or ought to have known. Thus, although he is not in occupation, and therefore not within the scope of the Occupiers Liability Act 1957 he is under a similar duty to that owed by occupiers to visitors (see p. 184). The Act specifically restricts the lessor from excluding his liability.

Covenants against assigning and sub-letting

Assigning a lease is the transferring of all that is left of a lease. Sub-letting means that, whilst retaining the lease, the lessee sub-lets and allows into possession a tenant for a period less than his own lease.

If the lease says nothing against either of these, then the lessee is entitled to assign or sub-let his lease.

Most lessors expressly stipulate in the lease that there should be no assignment, sub-letting or mere parting with possession without consent. Statute has enacted that such a covenant is qualified by a proviso that the consent must **not** be unreasonably withheld.

Remedies for breaches of covenants

Both lessors and lessees can enforce the covenants by suing for damages at common law or seeking the equitable remedies of a decree of specific performance or an injunction.

The lessor may also seek forfeiture of the lease (see termination of leases below) or **distrain** against the goods of the lessee for rent arrears. This means that the lessee's goods may be seized.

If the tenant gives notice of disrepair to the landlord, the tenant may do the work himself and recoup the costs from netting off from future rents. This would be a good defence against a claim for non-payment of rent. This is another example of self-help (see p. 198). Under the Landlord and Tenant Act 1987 a manager may be appointed by the court to carry out the landlord's responsibilities and to receive the rents. This is additional to, and more satisfactory than, the right of the High Court and county court to appoint a receiver and takes precedence over those rights. Also, under the Housing Acts 1985, 1988 and the Local Government and Housing Act 1989 the local authority can serve a repair notice on the person having control of a dwelling house or house in multiple occupation, if the local authority considers the house to be unfit for human habitation. The notice will specify the necessary works. If the works are not carried out, a closing order or even a demolition order can be ordered.

Continued liability regarding covenants

Unitil the Landlord and Tenant (Covenants) Act 1995, the liability of parties to a lease, both the original and subsequent assignees, depended on the doctrine of privity of contract (see p. 73) and the doctrine of privity of estate. At its most basic, the latter doctrine meant that, if the covenants both touched and concerned the land and were not mere personal covenants, then subsequent assignees of both the freehold and the lease would be bound by the covenants. The combination of both doctrines meant that parties to a lease who had sold their interest could nevertheless be held responsible for a breach of covenant even though it was not that particular person's breach and though he no longer had an interest in that land. The harshness of those rules led to a Law Commission report recommending reform. However, the Act is not retrospective so the old rules still apply to pre-Act leases. As far as new leases are concerned, by S.5 the benefit and burden of the tenant's covenants end on assignment and also for the landlord provided either the tenant or the court gives consent (S.8).

Termination of leases

It should be noted that leases do not end on the death of either party and that the remainder of the lease or reversion will be transferred to the beneficiaries under the deceased's will, or on intestacy.

Leases may be terminated in the following ways

Expiry of a fixed term lease
This occurs automatically by effluxion of time.

Notice
Here the lessee or lessor gives notice to the other to terminate the lease. The length of notice and whether it is allowed depends on the type of lease.

Forfeiture, i.e. taking the lease away, 'eviction', repossession
If the lease contains an express forfeiture clause, or one giving the lessor the right to re-enter the land if the lessee commits a breach of covenant, or the lease is granted on certain conditions, then the landlord will have a right to forfeit on application to court.

The law no longer encourages forfeiture without good reason, and therefore there are many statutory requirements that must be complied with before forfeiture can be allowed.

Surrender
It is possible that the lessee may no longer be able to afford the rent, especially on a short lease, which tends to be expensive. He may ask to surrender the lease back to the lessor. If it is accepted, then the lease comes to an end. There is no obligation on the lessor to accept surrender.

Merger
This arises when the lessee acquires the freehold reversion. The two interests then merge (see Figure 4.5).

Disclaimer on bankruptcy
A trustee in bankruptcy can disclaim responsibility for a lease which is a burden to the bankrupt. This does not end the lease. It merely terminates the bankrupt's liability under it.

Frustration
If circumstances surrounding the lease change so dramatically as to destroy the effectiveness of the lease, then the lease may be ended by frustration: *National Carriers Ltd v. Panalpina Ltd* (1981).

(Note: there are many Acts of Parliament which protect lessees from eviction, or even give them

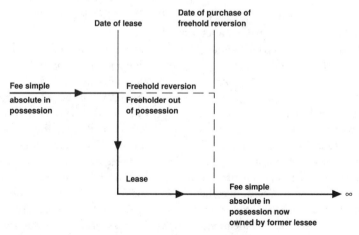

Figure 4.5 Diagram to illustrate the merger of freehold and leasehold

the right to claim new leases. One such example is that of business leases.)

Business leases

It is more usual for the businessman to lease his business premises. This is because freehold premises are very expensive, as they are usually in the most sought after parts of the country, e.g. in town centres.

Business lessees are given some statutory protection under the Landlord and Tenant Act 1954 Part II.

Compliance with the Act is essential

The lease can only be ended if the provisions of the Act are complied with.

This means that if there is a fixed period lease and the tenant holds over, having received only a normal notice to quit, then the lease will continue quite validly.

Notice to quit

A proper notice to quit must contain the following elements: (i) the lessor must give between 6 and 12 months' notice in a form laid down by the Act; (ii) the period of notice must not expire earlier than the lease would have expired normally, i.e. regardless of the Act.

On receipt of the notice the lessee has two months to inform the lessor that he will not give up the premises and may apply to court for it to grant a new lease.

If the tenant has a fixed period lease, he may request the landlord to grant a new tenancy under the Act and, having given such a request, must then apply to court for confirmation.

Granting a new lease

The court **must** grant such a lease, unless the landlord is able to prove that one of seven grounds apply to the lease. These include the tenant's failure to repair the premises as requested, failure to pay rent, substantial breaches of covenants, the provision of alternative accommodation, that the lessor wishes to demolish or reconstruct the premises, and that the lessor wishes to occupy the premises for his own domestic or business purposes.

The rent to be charged on the new tenancy is controlled by the court. Should the lessee not be granted a new lease, then, unless any of the three grounds which were for the benefit of the landlord applies, no compensation for loss of possession is paid.

Improvements

The tenant may have made improvements to the premises which have enhanced the letting value. Providing he gave the landlord notice of the carrying-out of the improvements, he may claim compensation under the Landlord and Tenant Act 1927.

Co-ownership of estates

It is perfectly usual for two or more people to own an estate in the same piece of land. Indeed, it is very common for husbands and wives to buy their matrimonial home and own it together. On the death of one of the spouses the whole of the estate then belongs to the survivor. This right of survivorship is the most important aspect of one type of co-ownership, a **joint tenancy**.

It would be inappropriate to have a right of survivorship in a business relationship, so there is another type of co-ownership in equity called a **tenancy-in-common**. Thus, if partners own property as tenants-in-common, if one partner dies, his share in the property goes to his next-of-kin or beneficiaries under his will.

There are some important differences between the two types of co-ownership which must be noted.

1. *Legal estates can only be co-owned by joint tenants, with a maximum of four.* As equitable interests can only exist behind a trust, if in our partnership example above there are 20 partners, four of them will have to own the legal estate as joint tenants, but they will hold it on trust for all 20 of them, as tenants-in-common. If someone wanted to buy the land, he will only be concerned with buying the legal estate from the four joint tenants. The trust is of no consequence to the purchaser.

2. *Joint tenants own the property as if they were one person.* Joint tenants own the whole estate as

if they were one for the same length of time, and are entitled to have possession of the whole of the property. It is rather like when a husband and wife buy a suite of furniture out of their pooled income and refer to it as 'our suite'. Both of them enjoy the furniture, and neither would point to a particular part of it and say that it was their own.

Interests in other people's land

Section 1 of the Law of Property Act not only restricted the number of legal estates to two, the fee simple absolute in possession and the terms of years absolute, it also reduced the number of legal interests or charges which can exist over someone else's land to five. We are concerned with only two, easements and a charge by way of legal mortgage. We are also concerned with *equitable* interests, such as restrictive covenants and licences.

Easements

An easement is a legal interest which one land-owner has over another's land, such as a right of light, a right of way, a right to take water, or a right of support, a right to have a sign on a

neighbour's building, or to use a clothes line on a neighbour's land.

These are very important, because without such easements, some pieces of land would be virtually useless and without value.

In theory, other sorts of easements may be recognised by the courts as the use of land changes over the years, but in practice they are reluctant to extend the types.

(Note: these are not public or private rights. They are particular rights attached to particular pieces of land. They only belong to the landowner, because of his ownership of that piece of land.)

There are certain essential requirements.

Dominant and servient tenements

There must be a **dominant and a servient tenement** of nearby land. In Figure 4.6, if the owner of Blackacre grants a right of way across his land to the owner of Whiteacre, then Blackacre is the servient tenement and Whiteacre the dominant. The servient tenement is the land burdened by the easement, and the dominant tenement is the land benefiting from the easement.

The position would be the opposite if the owner of Whiteacre granted a right of way over his land to the owner of Blackacre.

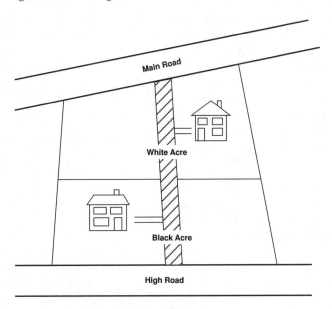

Figure 4.6 Dominant and servient tenements necessary for easements

When the land is sold or transferred in any way, e.g. by gift or on death, the easements will pass at the same time.

The easement must benefit the dominant tenement

The easement is for the benefit of the **land** and not merely the owner of that piece of land. Thus, in the example, there is no doubt that Blackacre and Whiteacre are more valuable if they each have a right of way to both Main Road and High Road.

In the example, dominant and servient tenements are close together. It is possible, however, for easements to exist where they are not so close but they should not be too far apart. So, a farmer down the High Road may have an easement across both Whiteacre and Blackacre, to get to Main Road. As this is a short cut for his animals, it will benefit his land. The dominant and servient tenements must be owned by different persons; otherwise the landowner has rights of possession and ownership over his own land.

The easement could still exist, however, if one person were the tenant of the other, as there are then two legal estates.

Types of easements

1. Rights of way

These may be limited to a particular time of day or a particular type of use, e.g. for cattle or pedestrians, or they may be totally unlimited.

There must be a definite path or route to be used, and the person who benefits from the easement must construct and maintain the right of way. The person exercising the easement must not then deviate from the defined route, unless he has been obstructed for some reason. (Note: this is actionable in nuisance, see p. 193.)

2. Rights of light

These can only relate to a particular window or aperture, and not to the building, as there is no normal or natural right to have light.

3. Rights of support

These may relate only to a particular building, such as a semi-detached house (for natural rights of support see p. 87).

4. Rights to water

These can relate to a pond, lake or river.

5. Miscellaneous

See above; also rights to use a letterbox or lavatory or a coal shed.

Acquisition of easements

These may be acquired in a number of ways.

(i) Statutes may indirectly create easements.

(ii) On the transfer of land an easement may be expressly granted by the transferor, or expressly reserved by him to benefit land he has retained.

(iii) Easements not expressly mentioned may be implied to give effect to the intentions of the parties if the land transferred is useless or less valuable without the easement. This would include easements of necessity. Thus if someone sells Pinkacre (Figure 4.7) without granting a right of way, then the shortest possible right of way will be implied between Pinkacre and Bottom Road.

(iv) By the rule in *Wheeldon* v. *Burrows* if **part** of a piece of land is transferred the transferee will receive all the quasi-easements which already exist over the retained land, providing they have been enjoyed continuously without interruption before transfer, were obvious on inspection, were necessary for the reasonable enjoyment of the land, and had been used by the transferor. The quasi-easements would then become proper

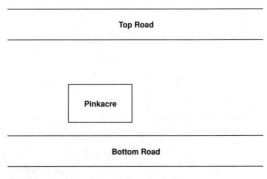

Figure 4.7 An easement of necessity

easements. (They were quasi-easements because, before transfer, the transferor owned **all** the land and the rights were merely normal rights of ownership.)

(v) By S.62 Law of Property Act 1925 a **conveyance** of land passes all **existing** easements to the transferee without the use of special words being used in the conveyance.

(vi) By presumed grant or prescription. This amounts to long use, i.e. exercising the 'easement' for a long period, as of right.

Prescriptive rights amounting to easements may be acquired in three ways. The oldest method is at common law. The claimant must prove that the easement has been in continuous use since time immemorial (1189) and has been exercised *nec vi, nec clam, nec precario*, i.e. without force, without secrecy and without permission from the landowner. Twenty years' continuous use raises a rebuttable presumption that it has been in use since 1189 (the beginning of Richard the Lionheart's reign).

The second method is by **lost modern grant** and has been described as 'that revolting fiction' by Lush L J in *Dalton* v. *Angus & Co* (1877). Here the court pretends that an easement was actually granted by a landowner but that the document has since been lost.

The third method is under the Prescription Act 1832 and the Rights of Light Act 1959. They are rather like Limitation Acts, but in reverse, as they lay down periods of use which will give rise to easements. An easement, other than that of light, enjoyed for 20 years as of right and without interruption is absolute and indefeasable (unless oral consent was given).

Where the right was enjoyed for 40 years, then if written permission is granted, this will defeat the right to an easement.

Rights to light must be enjoyed for 20 years to create an easement, and only written permission will defeat it.

A 'That's Life' programme on BBC television gave an account of a man who was charged a very nominal fee by British Rail for his light. British Rail owned the land next door. Whilst he never paid the bill, nor was it ever enforced, the purpose behind it was to amount to a written permission to enjoy the light. At any time, therefore, British Rail could build something to restrict this light to a particular window, and he would not then be able to claim an easement of light.

Termination of easements

By statute
An Act of Parliament may enact law which will at the same time extinguish certain easements, e.g. if an Act was passed to build a nuclear reactor on a piece of land which was subjected to easements, then these would be extinguished by the Act.

By release
Here the owner of the dominant tenement, either expressly or impliedly, releases the owner of the servient tenement from being burdened by the easement.

If this is done expressly it must be done by writing in a deed. This would be the appropriate method for a building developer who wishes to buy out the easements on his land, in which case the dominant owner must execute a deed of release.

An implied release occurs where the dominant owner indicated by his behaviour that he regarded the easement as at an end. Merely stopping using the easement would not be enough, unless it was for a long time, e.g. 20 years. Thus, the dominant owner should show some intention to release the servient owner, e.g. by altering his land, so as to make the exercise of the easement impossible, such as demolishing a house benefiting from a right of light.

By merging ownership and possession of the dominant and servient tenements
As indicated earlier, there can be no easement unless there is a dominant and a servient tenement. So, if one person acquires both pieces of land the easement must cease. It should be noted, however, that it is the fee simple that must be owned of both pieces, and not a mere lease.

Breach of an easement

If someone infringes an easement, this amounts to a nuisance and the dominant owner may sue for damages, or more importantly for an injunction.

Technically, he could also use the remedy of abatement (but see p. 198).

Mortgages

Introduction

People often remark, when they are going to buy a house, that the building society has given them a mortgage. This is not legally correct. What has actually occurred is that in order to pay for the house the building society has loaned money to the borrower, who, immediately after the purchase, grants a mortgage of the property to the society as security for the loan. Because it is the borrower who grants the mortgage it is he who is the **mortgagor** and the lender is the **mortgagee**, and not the other way round (see Figure 4.8).

Why?

Why does the building society, or indeed, any other lender of money, want a mortgage? Because it gives them security, which puts them in a better position than other creditors. If the debt is not paid the mortgagee can sell the mortgaged property and satisfy the debt. An ordinary unsecured creditor who has no security, like the milkman, can only claim in the borrower's bankruptcy. He may not even get the full amount owed, if there is not enough money to pay all the creditors.

Definition

A mortgage is therefore an interest in another's land, which secures a debt. It is like an invisible limpet which is not removed until the debt is paid off.

Mortgages may be either legal or equitable, which determines the rights and remedies of both parties.

Creation of legal mortgages

Since the Law of Property Act 1925 a freehold estate may only be legally mortgaged by either

1. a **demise** (transfer by lease) for a term of years absolute, subject to a provision for cesser on redemption, i.e. the land is leased for a fixed period, usually 3,000 years, which will come to an end when the debt is repaid, *or*
2. a **charge by deed**, expressed to be by way of legal mortgage. (This is a short document containing the above expression which, by virtue of the Act, is 'legal shorthand' for the concept in (1) above.) By S.87(1) of the Act, a mortgage created by a legal charge carries the same protection and remedies as if a lease for 3,000 years had been created.

Similarly, leases can be legally mortgaged either by sub-demise (sub-lease) subject to a provision for cesser on redemption, the sub-lease being at

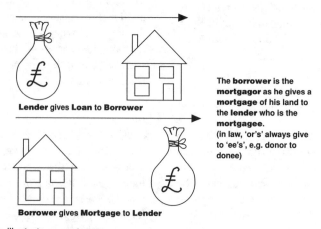

Lender gives **Loan to Borrower**

Borrower gives **Mortgage to Lender**

The **borrower** is the **mortgagor** as he gives a **mortgage** of his land to the **lender** who is the **mortgagee.** (in law, 'or's always give to 'ee's', e.g. donor to donee)

Figure 4.8 Diagram to illustrate a mortgage

least one day less than the mortgagor's lease, or by method (2) above.

Creation of equitable mortgages

1. Of equitable interests – *equitable mortgages*. If a person has only an equitable interest in land, such as an interest under a trust, then any mortgage he grants appertaining to that interest must be equitable. It should be in writing and be signed by the transferor or his agent (Law of Property Act 1925 S.53(1)(C)).
2. Equitable mortgages of legal estates. Where the requirements or the Law of Property Act 1925 are not met, as above, then any purported mortgage must be equitable.

    An informal mortgage, such as depositing title deeds with the lender to prevent the sale of the mortgaged property, creates an equitable mortgage. Although there need not be a contract in writing, under the Law of Property (Miscellaneous Provisions) Act 1989, common sense suggests that this is more sensible.

How does a mortgagee prevent the mortgagor selling the mortgaged property without the mortgagee's knowledge and therefore defeating the purpose of taking security

This is done in two ways:

1. By depositing the title deeds with the mortgagee. After the purchase of the property has been completed, the purchaser's solicitors will have to send the title deeds to the building society, or other mortgagee. The society will not release the title deeds until the debt has been paid, i.e. they have a lien over the deeds. If the property is going to be sold the solicitors will have to promise to hold the deeds for the society and not the mortgagor, and promise also to pay off the debt out of the proceeds of sale.
2. Registration. Where the property is the subject of a registered title, the fact that the property is mortgaged must be registered at the Land Registry (see p. 108).

Registration protects the lender's interest in the property, as anyone wishing to buy will search the registry records to discover the existence of any such interests.

Where the property is the subject of an unregistered title, mortgagees who do not have the title deeds, because their mortgage is a second or subsequent mortgage, must register their mortgage at the Land Charges department as their mortgage will otherwise be void against a purchaser.

The mortgagor's right to redeem the mortgage

The mortgagor must have the right to redeem the mortgage, otherwise the mortgagee's interest would be the more important. The mortgagee, therefore, is not allowed to impose any restrictions so as to prevent redemption. He may wish to do this, because the longer the loan continues the more interest he earns.

The mortgagee's rights if the mortgagor fails to pay the debt

1. Sue. He may sue for the money, as it is an ordinary debt. In practice, this is usually useless, as the debt would have been repaid if the borrower had any money.

2. Foreclosure. Once the date for redemption has been passed, or there has been a breach in the terms of the mortgage, which gives rise to full repayment being required, then the mortgagee may demand repayment in full. If, after a reasonable time has elapsed, no payment is forthcoming, the mortgagee can apply to the court for a foreclosure order.

If this is granted and the order is made absolute, the mortgagor's estate becomes vested in the mortgagee.

This is useful if the land is more valuable than the loan, and the lender would like to keep the land for some reason. In practice, however, foreclosure is not often used as the lender really wants his money back, and not land in lieu.

3. *Sale of the property*. If the mortgage is by deed and if there has been a breach of covenant *or* no interest has been paid for at least two months *or* notice has been served on the mortgagor who has still failed to pay after three months, then the mortgagee can put the land up for sale.

This is quite usual for building societies, as they need the money and not a transfer of land, as under foreclosure.

On the sale the purchaser gets the mortgagor's interest in the property, freed from the right to redeem.

The mortgagee must act in good faith, when selling the property. Unless it is a building society, however, the mortgagee does **not** have to get the best price for the land. Building societies are under a statutory duty to get the best price reasonably obtainable.

Any money left over after paying the debt and expenses should be paid to the mortgagor. If there are other mortgages on the property then they should receive the rest to cover their debts. The order in which they are paid depends on the order of registration.

The advantage of this remedy is that, unlike foreclosure, there are no preliminary court proceedings.

4. *Taking possession of the land*. Proceedings must be taken in the Chancery Division or county court first. The mortgagee's right to possession arises immediately a legal mortgage is created, as the mortgagee either has a lease or is treated as if he has one and is therefore entitled to possession. Thus, strictly speaking, the mortgagee can enter the mortgagor's land, even if there has been **no** breach of the mortgage.

In practice, this is unusual, unless the property is agricultural or commercial, where, by letting it or managing it better than the mortgagor, the mortgagee can raise more money on the property and thereby get paid the money owed.

The mortgagee must account strictly for the income which he has or should have received, e.g. rents from tenants. He must also keep the property in reasonable repair.

5. *Appointing a receiver under S.101 Law of Property Act 1925*. In any mortgage created by deed there is a statutory power to appoint a receiver. A receiver must be appointed in writing and can be appointed in the same circumstances as under a power of sale. The receiver should pay interest due under the mortgage to the mortgagee and pay any rents, rates, taxes and insurance premiums due on the land. He may, if the mortgagee tells him to, pay the cost of any repairs and pay off the principal sum outstanding.

In effect, the mortgagee, without taking possession, has appointed a manager to try to run the property better than the mortgagor or channel the funds in the proper direction.

Rights of mortgagees with equitable mortgages

These are very similar to the rights above, although in some respects they have been cut down or are limited because of the equitable nature of the interest or mortgage itself.

Restrictive covenants – equitable interests in land

Whereas an easement gives the owner of the dominant tenement the right to do something on the servient tenement, a restrictive covenant prevents the owner of the servient tenement from using his land in a particular way for the benefit of the dominant tenement, e.g. preventing building or carrying on a business there.

Creation

Restrictive covenants can only be created by deed (cf. easements).

Essential requirements

1. A dominant and servient tenement (see p. 97).
2. The covenant must be restrictive or negative in nature.
3. The covenant must 'touch and concern' the dominant tenement, e.g. a restrictive covenant to stop a business or trade being carried on 'touches and concerns' the dominant tenement, as it could affect its value.

Termination of restrictive covenants

1. Release (see p. 99).
2. Joining of ownership and possession of the dominant and servient tenements (see p. 99).
3. Statute.

Apart from individual statutes which may extinguish restrictive covenants, there is a general right under the Law of Property Acts 1925 and 1969 for the servient owner to be able to apply to the Lands Tribunal for the termination or modification on a number of grounds. In particular, it may be granted if the restrictive covenant is now of no practical use, as the surrounding neighbourhood has changed in character since the restrictive covenant was originally made.

As restrictive covenants can affect the value of land for development, building contractors either must be sure that there are none which affect their building work or try to get them terminated by one of the above methods.

In appropriate cases where a restrictive covenant was imposed a long time ago, it may be possible to take out a restrictive covenant indemnity policy with an insurance company.

Breach of a restrictive covenant

As restrictive covenants are only equitable interests, should a breach occur, an injunction may be sought. Damages, which are a legal and not an equitable remedy, may only be awarded if an injunction is an inadequate remedy under the Supreme Court Act 1981. As only equitable remedies are available, the judges use their discretion in awarding an injunction or damages in lieu and will proceed on an equitable basis, e.g. 'delay defeats equity'. Thus, if you have failed to seek an injunction immediately your next door neighbour opened his whippet kennels, or soon after, the courts may decline an injunction.

Licences

A licence is a right given by an occupier of land to permit a person to do something which would otherwise be a trespass on the occupier's land, e.g. the right to be a lodger, to watch a film at his cinema, or to swing a crane through his airspace.

The granting of a licence does not give the licensee any property rights, nor does it give him the right of exclusive possession of the property.

There are a number of different types of licences.

A bare licence

This occurs where no consideration is given by the grantee for the licence. So, if Albert allows Bertha to cross his land while Bertha's road is being dug up and Bertha gives Albert nothing for the right, when he withdraws his permission Bertha must stop using his land or she will become a trespasser. The grantor may withdraw his permission at any time.

Contractual licences

These are rather like bare licences, except that they have been created by a contract which stated they would not be revoked. So, when you buy a ticket for a football match or pop concert you are buying a contractual licence to be on the occupier's property for a particular purpose.

Equity regards such licences as irrevocable for that particular time and purpose.

Contractual licences are obtained by construction companies to operate cranes across another person's land.

Also, under clause 23 of the JCT standard form of contract a contractor is given a licence to occupy the site until the date of practical completion. It is implied that the employer will not revoke the licence unless there is good reason to do so under clause 27, when the employer may end the contract.

Licences coupled with an interest

In addition to being allowed to enter the occupier's land, the licensee has the right to take something away, such as fish. The right to fish or take turf are two types of a category of interests called **profits à prendre**, and are similar to easements, in that they must be created by deed or prescription.

These licences are therefore irrevocable, and, indeed, can be sold or given to another.

Licences arising out of estoppel

Where an owner of land behaves in such a way as to lead another person to believe that he has granted or is in the process of granting him a licence over that land, then the owner shall be estopped (stopped or prevented) from denying that there is such a licence in existence.

Conveyancing – transferring ownership of land

Introduction

This is the legal term used for the transfer of ownership in land from one person to another. For most people the conveyancing transaction will be the one important legal transaction that they will remember, as they will usually employ the services of a solicitor when they buy their home.

For the lawyer the most important aspect of the transaction is to make sure that the vendor transfers to the purchaser all that he owns in the property. This is called transferring the **title**. He may, of course, have agreed to sell less than he owns, e.g. where he leases the property and retains the freehold reversion.

Because there can be different estates and interests in the same piece of land, conveyancing is not as simple as, say, selling a book, which can be done by merely delivering it into the possession of the purchaser. A certain procedure must be gone through, with numerous checks and searches made, to make sure that the property belongs to the vendor, that all matters that affect the land are disclosed, such as mortgages, and that the transferee gets what he is paying for. Proving ownership (**proving title**) is done by producing documents of title and not producing the thing itself as in the case of the book, as this is impossible in relation to land.

Methods of conveyancing – is the land registered?

In England and Wales there are two systems of conveyancing – one where the title to the land is **unregistered**, the other where it is **registered**.

A national system of land registration was introduced by the Land Registration Act 1925. In this system, once title to the land has been proved to the satisfaction of the Chief Land Registrar, the Land Registry issues a document of title, which guarantees that the estate owner mentioned in it is the true owner. Subsequent purchasers of the land merely have to check the land registry records to discover the identity of the estate owner and whether he has encumbered the property with mortgages, etc.

Not all land is registered yet, but approximately three quarters of all conveyancing transactions concern registered land.

Where title to land is unregistered then proving title, i.e. proving the vendor is the estate owner, has to be done by examining the title deeds.

In unregistered conveyancing, title to land must be proved through a chain of ownership back to a 'good root' of title for at least 15 years. This means that if the vendor has owned the estate for over 15 years then all that is necessary is to prove back to his own conveyance, i.e. the deed which transferred all the estate to the vendor. Otherwise, one needs to go back further to the next conveyance which transferred the land. If there is a break in the title, then one must find out why; otherwise there could be a defect, which could mean, at worst, that the vendor does not own the property, and cannot pass on a good title to the purchaser.

The following outline will indicate the differences between the two systems as and where necessary.

Basically, the transaction falls into **three** stages – before contract, contract to completion and post-completion. To simplify matters we will take a normal domestic conveyancing transaction which the Law Society recommends using the National Protocol for the sale of freehold and leasehold property.

Before contract

1. What can he afford and which house will he buy? The purchaser usually sorts out his finances first and finds out how much a building society or bank will lend him, based on his salary. Once

he knows his price range he can then look for a suitable house. He may register himself with estate agents, whose job it is to find purchasers for vendors, or he may look at 'property for sale' columns in the newspapers.

2. Borrowing the purchase price. On finding a house he likes he will complete the building society or bank loan application forms and send them off with a survey fee.

3. Why is a survey needed? The survey fee is payable, as the society will send their own surveyor to look at the house, to see whether it is valuable enough to act as security for the loan. If what you are borrowing is close to the purchase price, e.g. a 100% or 95% mortgage loan, then the society's approval should indicate that all is well and that there are no serious defects. If you are borrowing a smaller amount, however, approval of the loan is no real indication of the state of the premises.

A recent case illustrates this point, and has led to changes in building society practice regarding surveys.

In *Yianni* v. *Edwin Evans & Sons* 1981, Mr Y wanted to buy a £15,000 house, borrowing £12,000 from his building society. Their surveyors, an independent firm, inspected the property and indicated to the society (whose agents they were) that it was good security for the loan. Mr Y did not get an independent survey, and went ahead with the purchase despite the building society's exclusion of liability attached in the loan offer. The exclusion stated that the society took no responsibility for the condition of the property, that they were not saying the price was reasonable, and that the survey report was confidential.

Almost a year later after buying the house the foundations were found to be defective and repairs would be needed costing £18,000.

The surveyors admitted that they had been negligent in failing to notice the defects and saying that the house was adequate security for the loan. As a result of their admission, Mr Y won his case against them, as the court decided that it was common practice for borrowers to rely on the report from the building society's surveyor (see requirements for negligence, p. 167).

As a result of this case, most building societies are now offering a number of different types of surveys. As the minimum survey has been shown to be inadequate for the borrower the basic survey fee has been increased, but the survey report is now available to the borrower. For an additional payment most societies will arrange a full structural survey which will be the same as that done by an independent surveyor.

In practice most people will probably still opt for the minimum survey, so that if their loan application is turned down they will not proceed with the purchase. A full survey would then have been a waste of money.

(Note: chartered surveyors have to take out professional indemnity insurance against claims for negligence.)

4. If the loan is agreed, the purchaser can tell his solicitor to proceed with the purchase.

5. Preparation of the package of documents. The vendor's estate agents will have informed the vendor's solicitors of the prospective sale and the solicitors will have prepared a **draft contract**, either by looking at the title deeds if unregistered or by examining the land certificate if registered. They will also prepare a Property Information Form, a Fixtures, Fittings and Contents Form, and in order to save time they could also make the pre-contract searches and prepare answers to the enquiries before contract, so that they can send the results with the Package. See paragraphs 10 and 11 below.

If the land is already mortgaged then the vendor's solicitors will have first had to borrow the title deeds from the building society and will have asked for a redemption figure. If the land is registered they can prepare the contract by obtaining copies of the land registry entries called **office copy entries**, which contain details of the property, the vendor and mortgagee(s) or other interests in the property. The land charge search may also be made at this stage by the vendor's solicitors. See paragraph 12.

6. A package consisting of the draft contract, office copy entries, results of the pre-contract searches, completed Property Information Form and completed Fixtures, Fittings and Contents Form is sent to the purchaser's solicitors.

7. If the land is unregistered the vendor's solicitors will send, instead of office copy entries, either an **abstract of title** or an **epitome of title**. An abstract is a document written in a sort of legal shorthand, containing a summary of all the title deeds and events which link each deed, such as the death of an owner and the transfer after a grant of probate to the widow.

8. An epitome, which is a more modern version of the abstract, is a collection of photocopies of the relevant deeds, topped with a summary of dates and events and description of the deeds. In both cases, they need only go back to a good root of title of at least 15 years. A 'good root' is a deed which deals with the whole of the legal and equitable estate, such as a conveyance, and describes them in detail.

9. The purchaser's solicitors check the draft contract, to see that it contains all the particulars that they expect, e.g. that the price is the same as instructed, and that the fixtures are included or not (see p. 87).
  They also check the office copy entries or the abstract or epitome, to see that what they are buying is what they expect, e.g. a fee simple estate and not a lease, and that all mortgages have been disclosed. The vendor must undertake to redeem mortgages from the purchase moneys, so that the purchaser takes free of them.

10. The purchaser's solicitors also send a standard form called **inquiries before contract** to the vendor's solicitors. These contain questions on matters within the knowledge of the vendor and which the vendor is under no legal obligation to disclose, such as neighbour disputes, ownership of walls etc., availability of gas or electricity, shared facilities and outgoings. Additional questions not included on the printed form can be inserted, such as details of rewiring, or whether the central heating has been paid for. In the case of leases there are further questions concerning such matters as breach of covenant, details of the lessor, service charges and insurance cover. Failure to disclose some relevant fact or stating a downright lie could result in the vendor being sued for misrepresentation (see p. 72) or at worst the tort of deceit (see p. 72). There have been instances where there has been no actual dispute but there is a problem, such as noisy neighbours. Some potential purchasers are now going to the rather extreme length of employing private investigators to find out if there are any problems which have not been revealed by the vendor.

11. In addition to the above forms, which are sent to the vendor's solicitors, the purchaser's solicitors send two forms to the district council for the area in which the house is situated. The first is a **search** of the local land charges register, which is kept by the council. This search is made against the address of the house and not the owner. It deals with matters which are registered by local authorities under the Local Land Charges Act 1975, and various pieces of legislation. The charges may be financial, similar to a mortgage benefiting the council, when, for example, the council had to execute works on the property at their own expense, which has not been reimbursed by the owner. This could occur as a result of breaches of planning law or work contravening building regulations where the owner failed to comply with an order to demolish the work. Other charges are concerned with town and country planning, listed buildings, drainage schemes, civil aviation, land compensation etc. The registration of such charges acts as notice to the prospective purchaser, who then, if he continues with the purchase, takes the property, subject to the charge.
  The second form sent to the council is called **inquiries of district councils**. The matters dealt with in this are, *inter alia*, whether the roadway is maintained by the council, whether the property is drained into a sewer, whether there has been a breach of the building regulations or town and

country planning rules, if the area is gong to be redeveloped, or if the house is in a conservation, slum clearance or smoke control area.

12. If the land is unregistered, a **land charges search** is sent to the Land Charges Department, at Plymouth, to check that there are no encumbrances registered against the property, such as second mortgages, which have not been revealed by the vendor.

These encumbrances belong to other people who have some interest in the land being sold. In order to protect themselves, they should (but it is not obligatory) register their interests at the Land Charges Department. If they do not register, a purchaser who buys the land will usually take the land **free** from the encumbrance.

Examples of such interests include pending actions, e.g. where the land to be sold is subject to a dispute pending in court, and this includes a petition for the bankruptcy of the vendor, or a claim for transferring the property in a pending divorce. Where there has already been a court case concerning the land and judgment has been made involving the land, e.g. appointing a receiver, the writ or order enforcing judgment can be registered.

There are also six classes of land charges, which may be registered and are identified by the letters A, B, C, D, E and F. Class C includes legal and equitable mortgages not protected by the deposit of title deeds with the mortgagees, such as a second mortgage to a bank. This class also includes **estate contracts**, i.e. where the legal owner has contracted to convey the legal estate to another. This would include **options to purchase**, which can be very important in the building industry.

Class D land charges include restrictive covenants and equitable easements.

A class F land charge is a deserted spouse's right to occupy the matrimonial home, which is extremely important in recent years with the increase in divorces. Anyone buying from a separated couple must search in case a class F charge is revealed.

The search is made not against the land but against owners of the legal estates in the land.

Note: these interests would be noted in the office copy entries if the land were registered (and the interests had been registered). Thus one should also make an Index Map search at the Land Registry to check that the land is *not* registered.

13. If the purchaser's solicitors are satisfied as to the result of their searches and enquiries and have agreed the contents of the draft contract, they can **engross** it, i.e. make a proper final copy.

Conveyancing contracts are always engrossed in two identical parts. One is sent for signature to the purchaser, the other to the vendor's solicitors for their client's signature.

14. *Exchange of contracts.* Until contracts are 'exchanged', there has been no binding contract. It is a mere gentleman's agreement, which can be broken at any time by 'gazumping' (accepting another better offer) or varying the original offer, or withdrawing the property from the market. To emphasise that there is, as yet, no binding contract, most people are wise to make an offer 'subject to contract' (see p. 55), so that they will only enter into a binding contract if the searches, enquiries, survey and mortgage offer are satisfactory. Caution is important at the pre-contract stage.

Usually 10 per cent of the contract price is paid on exchange of contract by the purchaser as a sign of good faith and this will be forfeit if he fails to go through with the transaction.

The purchaser's signed contract is sent to the vendor's solicitors, and the vendor's to the purchaser's solicitors. This can be done by telephone according to the Law Society's formulae.

On exchange of contracts, there is a **binding contract** – and *not* before.

At this point, equity regards the purchaser as the landowner and thus insurance cover must be arranged on the property by the purchaser's solicitors.

Contract to completion

1. The date for completion is usually entered in the contract; otherwise most standard forms of contract automatically have a completion

date of 28 days following the date of exchange of contracts.

2. If the title was not proved before contract it must be done now. A form called Completion Information and Requisitions on Title is sent by the purchaser's solicitors to the vendor's solicitors. If the answers are not satisfactory, then the purchaser may terminate the contract. Nowadays, most solicitors prove title before contract, to save time.

3. The purchaser's solicitors send a **search** form to the Land Registry to search for encumbrances which may have been entered since the date of the office copy entries.

   If the land is unregistered, a search is made at the Land Charges Department against the vendor.

4. Each certificate is dated and provides protection for the purchaser against further entries being made, by which he would be bound, provided he completed within a certain number of days of the date of the certificate.

5. The vendor's solicitors prepare a completion statement, showing the amount due on completion. This will not necessarily be only the 90 per cent balance of the purchase price, but will also include sums for carpets etc., apportioned rates, ground rent etc.

6. The purchaser's solicitors will have drafted a document called a **conveyance** if the land is unregistered, or a **transfer** if the land is registered or an **assignment** if a lease (see Figure 4.9). This is sent to the vendor's solicitors for approval.

7. On approval, the purchaser's solicitors engross the deed. This deed, on completion, operates to pass all the legal estate from the vendor to the purchaser.

   The deed is sent to the vendor for his signature (execution). The deed may have already been signed by the purchaser, but this is only necessary if the purchaser has to be bound by restrictive covenants.

8. Provided the searches come back clear from the Land Registry or the Land Charges Department, the purchaser's solicitors can then proceed to complete the transaction.

9. (i) It is usual for completion to take place in person at the vendor's solicitors' office, but if the vendor is paying off a mortgage then completion is arranged at his mortgagee's solicitors' office. Nowadays, completion can be done by post, with the money being transferred by telephone through the banking system. This calls for split second timing, as the purchaser's money cannot be sent unless it is absolutely certain that the land will be conveyed to him that day.

   (ii) On completion the money is handed over in return for the transfer or conveyance, plus any other old title deeds, together with the keys to the property (the deeds represent legal ownership, the keys possession). Often the keys are held by the estate agent and he should therefore be telephoned to ask him to release them.

Post-completion

1. The purchaser's solicitors must get the transfer or conveyance **stamped** with evidence of payment of **stamp duty** or tax, payable on certain legal documents. In effect, it is a tax paid on certain legal transactions; e.g. at present, a purchaser buying a house costing £60,000 will have to pay £300 stamp duty as the first £30,000 is duty free.

2. Particulars of the transaction must also be given to the Inland Revenue.

3. Since 1990 all England and Wales is an area of compulsory registration, and in a normal sale and purchase of a fee simple or a lease with at least 21 years to run, the transaction must be registered. Thus, if the land is registered, certain documents have to be completed and submitted to the Land Registry, together with the transfer and any mortgage. The Land Registry then issues a **land certificate** if the land is unmortgaged and a **charge certificate** if the land is mortgaged.

4. Finally, if the property is mortgaged all the deeds must be sent to the mortgagee to protect his security.

Transfer of Whole (1)	**HM Land Registry** Land Registration Acts 1925 to 1986	**Form 19** (Rules 98, 109 or 115, Land Registration Rules, 1925)

Stamp pursuant to section 28 of the Finance Act 1931 to be impressed here.	When the transfer attracts Inland Revenue Duty the stamps should be impressed here before lodging the transfer for registration.

(1) For a transfer by a Company or Corporation form 19(Co) is printed. For transfer to joint proprietors form 19(JP) is printed.

(2) Please enter the administrative area (county and district, county, county or London borough etc.) in which the property is situated.

Administrative area (2)   Blistershire

Title number(s)   BL. 5678910

Property   Bleak House Lawyers Lane Leatherbottom

Date   1st August   1997   In consideration of Two Hundred Thousand

(3) Delete the words in italics if not required.

pounds (£ 200,000.00   ) *the receipt of which is acknowledged* (3)

(4) In BLOCK LETTERS enter the full name(s), postal address(es) (including postcode) and occupation(s) of the proprietor(s) of the land.

I/~~We~~ (4) URIAH HEAP OF BLEAK HOUSE
LAWYERS LANE LEATHERBOTTOM
BLISTERSHIRE BL1 2RX
ARMADILLO WRESTLER

transfer(s) to

(5) In BLOCK LETTERS enter the full name(s), postal address(es) in the United Kingdom (including postcode) and occupation(s) of the transferee(s) for entry in the register.

(6) On a transfer to a company registered under the Companies Acts, enter here the company's registration number for entry on the register.

(5)
PERCY PUMBLECHOOK OF DROWSY
COTTAGE SLEEPERS LANE SLUMBERWICK
SNORESHIRE SN15 5GQ BANANA RIPENER

(6) (~~Company registration number~~   )

the land comprised in the title(~~s~~) above referred to.

(7) Cross out, amend or add limitations or extensions as necessary (rules 76A and 77A).

This transfer is made with full title guarantee (7)

(continued overleaf)

Figure 4.9 Example of a transfer of land. High Court form LR19 reproduced by kind permission of The Solicitors' Law Stationery Society Limited.

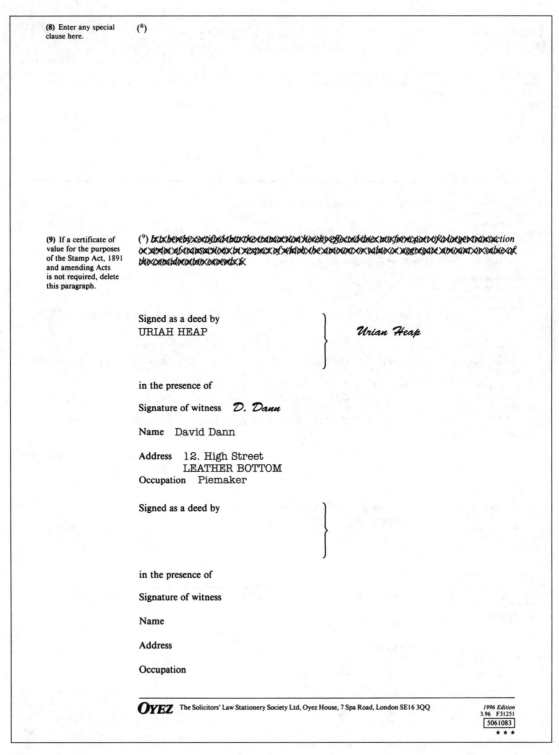

(8) Enter any special clause here.

$(^8)$

(9) If a certificate of value for the purposes of the Stamp Act, 1891 and amending Acts is not required, delete this paragraph.

$(^9)$ I̶t̶ ̶i̶s̶ ̶h̶e̶r̶e̶b̶y̶ ̶c̶e̶r̶t̶i̶f̶i̶e̶d̶ ̶t̶h̶a̶t̶ ̶t̶h̶e̶ ̶t̶r̶a̶n̶s̶a̶c̶t̶i̶o̶n̶ ̶h̶e̶r̶e̶b̶y̶ ̶e̶f̶f̶e̶c̶t̶e̶d̶ ̶d̶o̶e̶s̶ ̶n̶o̶t̶ ̶f̶o̶r̶m̶ ̶p̶a̶r̶t̶ ̶o̶f̶ ̶a̶ ̶l̶a̶r̶g̶e̶r̶ ̶t̶r̶a̶n̶s̶a̶c̶t̶i̶o̶n̶ ̶o̶r̶ ̶s̶e̶r̶i̶e̶s̶ ̶o̶f̶ ̶t̶r̶a̶n̶s̶a̶c̶t̶i̶o̶n̶s̶ ̶i̶n̶ ̶r̶e̶s̶p̶e̶c̶t̶ ̶o̶f̶ ̶w̶h̶i̶c̶h̶ ̶t̶h̶e̶ ̶a̶m̶o̶u̶n̶t̶ ̶o̶r̶ ̶v̶a̶l̶u̶e̶ ̶o̶r̶ ̶a̶g̶g̶r̶e̶g̶a̶t̶e̶ ̶a̶m̶o̶u̶n̶t̶ ̶o̶r̶ ̶v̶a̶l̶u̶e̶ ̶o̶f̶ ̶t̶h̶e̶ ̶c̶o̶n̶s̶i̶d̶e̶r̶a̶t̶i̶o̶n̶ ̶e̶x̶c̶e̶e̶d̶s̶ £

Signed as a deed by
URIAH HEAP   *Uriah Heap*

in the presence of

Signature of witness   *D. Dann*

Name   David Dann

Address   12. High Street
           LEATHER BOTTOM
Occupation   Piemaker

Signed as a deed by

in the presence of

Signature of witness

Name

Address

Occupation

OYEZ The Solicitors' Law Stationery Society Ltd, Oyez House, 7 Spa Road, London SE16 3QQ   *1996 Edition* 3.96 F31251 5061083 ★ ★ ★

Figure 4.9 cont'd

5 Pure personalty

Introduction – types of personal property

As we have already seen, pure personalty are **things** as opposed to land. They can be tangible or intangible, i.e. they can be things we can actually feel and touch, such as a brick or a lorry (chattels personal), or they can be things which we cannot touch, but which will, nevertheless, have value, such as a patent or a debt (see Figure 5.1).

Tangible things are also known as **choses** (things) **in possession**, as they are capable of being possessed, and if they are taken away from their owner, he can physically take repossession. Intangible things are called **choses in action** as the only way an owner can enforce his rights in relation to them is to take court action.

Chattels personal – choses in possession – tangible property – moveables – goods

Transfer of ownership

Ownership of chattels can be transferred very simply by merely delivering the goods into the hands of the transferee. There must, however, be an intention to transfer ownership, and not mere possession.

So, if I give you a present of a book, then that is sufficient to transfer ownership from me to you. If I lend you my book, I have no intention of transferring ownership. Therefore, despite delivery, I still own the book.

Contracts for the sale of goods

Because of the frequency of such transactions, the common law developed many rules relating to contracts for the sale of goods (chattels). In 1873, Parliament codified the law in the first Sale of Goods Act. This was updated and consolidated in the Sale of Goods Act 1979 (the Act) and further amended by the Sale and Supply of Goods Act 1994 and the Sale of Goods (Amendment) Acts 1994 and 1995. However, the Act specifically states that the common law rules of contract still apply, unless they are inconsistent with the Act (S.62(2)). However, clear express terms can in some circumstances override the statutory provisions.

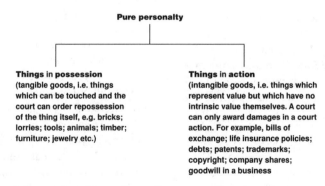

Figure 5.1   Categories and examples of pure personalty

The builder should be fully aware of the main principles of the Act, as all building materials, tools of trade, plant and equipment bought and sold may come under the ambit of this Act.

What are goods?

By S.61 goods include all personal chattels (see p. 111), but not things in action or money. Also, ships and things attached to the land, but which it was agreed would be detached before contract, are goods, e.g. timber.

The goods may be **specific**, i.e. identified and agreed upon at the time of the contract, e.g. my red Ford Sierra, registration number A123 XYZ. The goods may be **unascertained** such as a ten ton truck or 3,000 bricks, or even goods which have not yet been manufactured (*future goods*), e.g. a cement mixer, type No. 64C. **Ascertained** goods are goods identified and agreed upon after the contract, i.e. they were unascertained at the time of the contract, but became fixed or ascertained before delivery.

This Act does not apply to barter unless there is some element of part exchange and a price allocated to the traded-in goods. Nor does it apply to contracts for the supply of work and materials (the Supply of Goods and Services Act 1982 applies to these types of contract, see p. 57).

Forming the contract

All the usual elements apply and any form can be employed (see Ch. 3). But, by S.2(1) a contract for the sale of goods is one 'whereby the seller transfers or agrees to transfer the property in goods to a buyer for a money consideration, called the price'. Thus, the consideration is money in exchange for goods.

The price should preferably be fixed by the parties, although it can be fixed by reference to past dealings, or a stipulated method or even left up to a third party to decide. If no price or method of fixing is agreed, a reasonable sum should be paid (if it was obviously a contract but not if it was a gift).

The effect of breach of terms in a contract for the sale of goods

For general discussion of terms, see p. 56. The terms will be those expressed between the parties plus those implied by the Act (if not expressed by the parties). Unless the term is one which may not be excluded such as S.12(1) below, the parties may be able to override the Act. If no express mention is made, then the terms in the Act will be implied into the contract.

Breach of warranty

A warranty is a contractual term which is of minor significance. Under S.62, damages only can be recovered.

Breach of condition

A condition is a term of vital importance. If there is a breach, then the contract can be repudiated (set aside) under the Act, the goods can be rejected and the purchase price recovered.

By S.11 a plaintiff can choose whether to treat a breach of condition as a breach of warranty. This means he can keep the goods, but sue for damages. He has no choice in the matter, however, if he has accepted the goods or part of them and can only sue for damages, unless the contrary was agreed between the parties in the contract.

Acceptance

By S.35(1) acceptance by the buyer takes place

(a) when he intimates to the seller that he has accepted them or
(b) when the goods have been delivered to him and he does any act in relation to them which is inconsistent with the ownership of the seller.

Think about the many sale of goods contracts you enter into each week. Apply the two rules to each contract.

Furthermore, by S.35(4) 'the buyer is also deemed to have accepted the goods when after the lapse of a reasonable time he retains the goods without intimating to the seller that he has rejected them'. This is based on the equitable principle that it is unfair to other parties if you do not pursue your legal rights as soon as possible. 'Delay defeats equity.' What is reasonable depends on the circumstances and is up to the court to decide in a dispute. However, under subsection (5) whether a reasonable time has elapsed will depend also on whether the buyer had a reasonable opportunity to examine the goods to comply with subsection (2) below. But perishable goods such as fresh fish or ready-mixed concrete should obviously be dealt with quickly.

If goods are **delivered**, as for example when buying from a trade catalogue, by S.35(2) if the buyer 'has not previously examined them, he is not deemed to have accepted them under subsection (1) above until he has had reasonable opportunity of examining them for the purpose

(a) of ascertaining whether they are in conformity with the contract and
(b) in the case of a contract for sale by sample, of comparing the bulk with the sample' (see p. 115).

To avoid the common situation where the deliveryman asks the recipient of the goods to sign a note, which in some cases could state that receipt amounts to acceptance of the goods, by subsection (3) where the buyer deals as a consumer the buyer cannot lose his right to rely on subsection (2) above by agreement, waiver or otherwise.

The amended law now sets out more clearly the situations in which there has been an acceptance of the goods and thus the loss of the above rights.

The most important terms implied by the Act and to be performed by the seller

The right to sell (ownership): S.12(1): implied term

This section *cannot* be excluded by agreement in any contract.

In the case of a **sale**, the seller must have a right to sell, and in the case of an **agreement** to sell, the seller must have a right to sell at the time the property (the right of ownership) passes to the buyer. Thus, if John agrees to buy a lorry from Lorry Suppliers plc, then, without having to mention anything concerning ownership, he can assume that either the company already owns the lorry or they have a right to sell it in some other capacity (e.g. as an agent) or they will have a right to sell it when the time comes to transfer ownership. Obviously, if there is a breach of such a term, John would have no rights in the lorry, and when it is recovered by the true owner he can sue for the total purchase price of the lorry, even though he has been using it for some time and even though, normally, depreciation would be taken into account. By S.6 such a term is treated as a condition.

Sales by description: S.13: implied condition

A sale by description is one where the article is referred to by a description such as a Ford Sierra or a tin of Heinz Baked Beans selected by a buyer in a supermarket or half-inch thick staves or suit No. 123 in a mail order catalogue. A sale can be a sale by description even if the purchaser has seen the goods but is still expecting that they are of a certain type, e.g. your 1990 Porsche reg. No. ABC 123Y.

Section 13 implies a condition into such contracts that the goods shall correspond with the description. If they do not, the buyer can repudiate or sue for damages. This is so, even if the goods are otherwise of satisfactory quality. If for example the Porsche turns out to be a 1988 model, it is not what you had intended to buy. In *Arcos v. Ronaason* (1933), for example, a buyer ordered 'half-inch thick' staves. When they were delivered, the majority were between 'a half and nine-sixteenths of an inch thick'. There was nothing wrong with them, but they were not as specified. Therefore they could be rejected. This is very important in building work, where measurements are of crucial importance.

The word 'description' also covers the mode of packing, e.g. in Christmas wrapping; quantity, e.g. in boxes of a dozen; and the date of shipment, which can be very important for perishable goods.

If the sale is by sample and description, then it is not enough that the bulk of the goods correspond with the sample if the goods do not also correspond with the description.

Satisfactory quality: S.14

Section 14(1) in effect restates the common law rule 'caveat emptor', literally translated as 'let the buyer beware', i.e. the buyer should be careful when purchasing goods, examining them for obvious defects etc., as apart from below there is no **implied** term as to quality and fitness of goods in a contract for sale.

The following terms are only implied into sales made in the course of a business and do *not* apply to private sales.

Implied term S.14(2)

The Sale and Supply of Goods Act 1994 has amended this section. Goods sold must be of **satisfactory quality** (S.14(2)). This is defined under S.14(2A) as being goods that 'meet the standard that a reasonable person would regard as satisfactory, taking account of any **description** of the goods, the **price** (if relevant) and all the other relevant circumstances'. Furthermore, by section 2B 'the quality of goods includes their state and condition and the following (among others) are in appropriate cases aspects of the quality of goods:

(a) fitness for all the purposes for which goods of the kind in question are commonly supplied,
(b) appearance and finish,
(c) freedom from minor defects,
(d) safety and
(e) durability.'

The amended section now makes it quite clear that appearance and finish may make something of unsatisfactory quality, whereas under the old law it would not have necessarily made the goods of 'unmerchantable' (the old term) quality.

It should be noted that under subsection (a) the goods must be fit for *all* the purposes for which the goods are supplied. Many products have a number of purposes and if for some reason the product is not suitable for a well-known purpose it should be specifically brought to the attention of the purchaser. Failure to do this would make the seller in breach of this term.

Exceptions

S.14(2A) terms as to satisfactory quality are not implied into contracts in two situations (S.14(2C)): first, if the matter making the quality of goods unsatisfactory is specifically drawn to the buyer's attention **before** the contract is made; second, where the buyer examines the goods **before** the contract is made and the examination ought to have revealed those matters.

Note that the word 'supplied' is used so that the section does not just cover the goods sold but also ancillary items like free gifts. Also, second-hand goods are included in this.

Thus, goods sold 'as seen', 'with all faults' or even 'seconds' should be approached with care. 'You get what you pay for' really applies here.

Fitness for a particular purpose: S.14(3): implied condition

Once again, this condition is only implied into a contract made in the course of a business and does not apply to a private sale.

Thus, if the buyer makes known to the seller, either by telling him or by writing to him or even by implication, the purpose for which he wants to buy the goods, then S.14(3) implies a condition into that contract that the goods must be **fit for that purpose**.

This purpose would not have to be explained or implied if it was obvious what one was going to use the article for, e.g. a hot water bottle is designed to hold hot (but not boiling) water.

So for example in *McAlpine & Sons Ltd* v. *Minimax Ltd* (1970), McAlpines bought four carbon dioxide fire extinguishers from Minimax Ltd. A fire in a timber site hut in which there were two extinguishers was seriously worsened when the extinguishers were used. The extinguishers exploded causing far greater damage than if they had not been used. Obviously, the court held that they were not fit for the particular purpose for which they had been bought.

In building, contractors will often rely on the skill and judgement of manufactures selling goods to them. It must be made quite clear in the contract, therefore, that the buyer is relying on their skill in choosing a product for the particular job.

Sometimes there will be joint reliance, i.e. the builder will use his judgement in relation to choice of goods in one area but relies on the seller in another. There may nevertheless be a breach by the seller if this judgement appears to be wrong, even if the goods are otherwise of satisfactory quality, i.e. fit for the general purpose, but not for the particular purpose (see also p. 114).

Sale by sample: S.15: implied term

Where goods are ordered by express or implied reference to a sample, then S.15 implies a term that the bulk must correspond with the sample in quality. A reasonable opportunity must be given to the buyer to check this and the goods must not be unsatisfactory in a way which an examination would not reveal. In England and Wales such a term will be treated as a condition.

A sale by sample usually occurs where one is buying something which can vary in quality, such as facing bricks or wood or top soil. The buyer only wants to buy the goods if the goods delivered are all the same quality as that shown to him by the seller or the sales representative. Thus, when the goods are delivered, the buyer must (as soon as possible) examine the goods and check that they are of the quality he expects. If not he can reject the goods and recover the purchase price or keep them but ask for damages (in practice a discount or refund) unless the matter goes to court, in which case damages will be assessed.

Can the above terms be excluded?

Section 12, the condition as to title, can **never** be excluded by agreement, or in any other way.

In consumer sales, Ss. 13–15 cannot be excluded by the seller. A consumer sale is a sale made by a seller who is selling goods as a business to someone else for private use who is not buying in the course of business. Thus, to a certain extent, it pays to buy from people selling goods as their trade, even if they have only just started.

When buying from private individuals, e.g. where a student buys a motor bike from another, or where a building contractor (who is in business and thus not a consumer) buys from a supplier, Ss. 13–15 **can be excluded**. The exclusion may be by express agreement between the parties, or it may be inferred from previous dealings between them. To be valid, the exclusion must be reasonable. Thus, the buyer must have known about the exclusion and it must have been possible for him to have bought the goods elsewhere without such an exclusion, e.g. from the seller or from some other source. Also, an exclusion clause is more likely to be reasonable where the goods were manufactured to the buyer's specification. Nor must the buyer have been induced to agree to the exclusion, e.g. by giving him a discount.

Whether the exclusion of any other type of liability is valid will depend on whether it satisfies S.3 of the Unfair Contract Terms Act 1977 (see p. 60).

Other duties of the seller

When must the goods be delivered?

Time for performance is not of the essence, unless the parties make it so, either by express agreement or by implication. Thus the order of a suit for a wedding should be made conditional on receiving the suit by a particular date. If payment is to be required on a particular date, then this must be expressly stated in the contract (S.10).

If the goods have not been delivered within a reasonable time, then the buyer can introduce a new term making time of the essence, e.g. by giving the seller 14 days in which to produce the goods, failing which the contract will be set aside. The length of notice must be reasonable.

Must the seller deliver the right quantity of goods?

Yes. By S.30, if the seller delivers a larger or smaller amount than was ordered, then the buyer can reject all the goods, accept all the goods or accept the right amount and reject the rest. But if the buyer accepts the larger amount, he must pay for the extra amount at the contract rate.

Section 30 does not apply if the differences are so tiny as to be inconsequential.

Duties of the buyer

Paying for the goods

He should normally pay on delivery unless the contrary has been agreed.

Taking delivery

If he fails to take delivery of the goods, he is in breach of contract.

Remedies

If the seller has suffered loss through non-acceptance or late payment, he can sue for damages. He can also sell the goods (even though strictly speaking they are already sold) if the goods are perishable, such as ready-mixed concrete. If the goods have been accepted but not paid for, there is a breach of condition and the buyer can recover the goods and claim damages if he has suffered loss.

Transferring ownership of goods from seller to buyer

On Monday Sam agrees to sell his car to Bill for £2,000, which he pays for by cheque. Bill says he will pick it up on Saturday as he is away all week on business. On Wednesday the cheque is cleared. On Thursday the car develops some electrical problem and catches fire. Whose responsibility is it? Who suffers the loss? (see Figure 5.2). There is no easy answer.

The answer in any event depends on the question: when does the property (title or ownership) in the goods pass to the buyer? This is because, by S.20, the **risk** (responsibility for the goods) passes to the buyer at the same time as property passes **unless otherwise agreed**.

We are also helped by the fact that the car is classified as 'specific' goods (see p. 112), for the Act lays down rules which are applied depending on whether the goods are specific or unascertained.

Thus to answer the question we would have to know what the parties agreed about the passing of property, because by S.17, property passes in specific goods **when the parties intend** it to pass. So Sam and Bill could agree that property would pass on Wednesday or when the cheque was cleared or when he was going to pick it up on Saturday. The risk would pass at the same time unless the contrary was agreed. If there is a delay in delivery (or collection), responsibility falls on the one who was at fault.

The rules

Unascertained goods: S.16 amended by the Sale of Goods (Amendment) Act 1995 (this section is now subject to S.20A (see below))
For example 10 tonnes of slag, sand, potatoes etc. Property cannot pass until the goods become earmarked or set aside for the particular contract, when they then become ascertained.

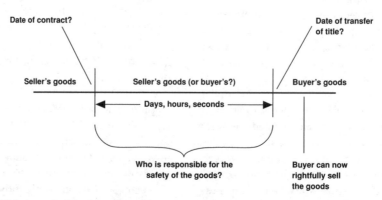

Figure 5.2 Why it is important to determine when property passes

Specific or ascertained goods: S.17
For example my brown Metro reg. A456 BRO. The property passes when the parties intend it to pass, i.e. there must be an agreement, either express or one implied from previous dealings. So S.17 could be applied to the example involving Sam and Bill.

If there is no agreement or contrary intention shown, then the following rules must be applied: S.18

Rule 1

Unconditional contracts for the sale of specific goods in a deliverable state. Here, the property passes when the contract is made, even though time for delivery or payment is postponed.

Thus, this could have occurred in our example, in which case the car would have become the property of Bill at the time of the agreement to buy. He should therefore have taken out insurance immediately, even though he was not picking up the car until Saturday (see Figure 5.3).

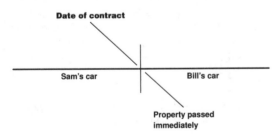

Figure 5.3 Rule 1

Rule 2

Sales of specific goods where the seller must do something to put them into a deliverable state. In this case, the property in the goods cannot pass until the goods have been put into a deliverable state and he must then give the buyer notice of this.

So Sam may have agreed to put new tyres on the car and property would not pass until he had done this and told Bill (see Figure 5.4).

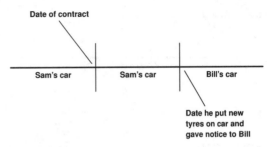

Figure 5.4 Rule 2

Rule 3

Sales of specific goods where the seller has to do something to ascertain the price, such as weighing, testing, measuring them etc. The seller must do this and give notice, and property will then pass to the buyer (see Figure 5.5).

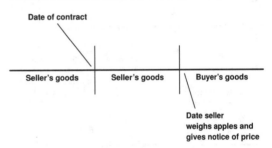

Figure 5.5 Rule 3. Note that this does not apply if the buyer has to do this

Rule 4

Goods sold on approval, or on a sale or return basis. Property will only pass to the buyer when either of the following occurs.

First, if he signifies that he has accepted the goods, such as by telling the seller or if he adopts the goods, such as by using them (or selling them to a third party) (see Figure 5.6).

Second, if he keeps the goods beyond the time laid down or, if no time was specified, after a reasonable period without signifying that he is not going to accept them (see Figure 5.6).

Rule 5(1)

Unascertained or future goods sold by description. For example 10 tonnes of next harvest's corn.

Contract for sale on approval

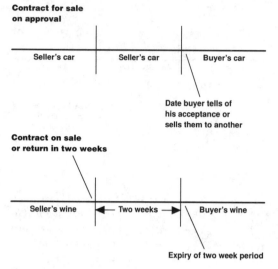

(a)

Contract to buy 10 tonnes of sand from bulk

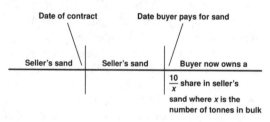

(b)

Figure 5.7 Rule 5

Property passes when the goods in a deliverable state are unconditionally appropriated to that contract with the consent of the other and notice is given of that fact (see Figure 5.7).

Figure 5.6 Rule 4

Contract to buy 10 tonnes next harvest's corn

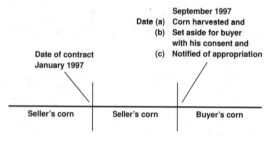

Rule 5(2)

Delivery of goods by the seller to the buyer or carrier without reserving a right of disposal. In such cases the seller is deemed to have unconditionally appropriated the goods to the contract (see above and Figure 5.7(b)).

Rule 5(3)

*Undivided share in a specified quantity of **unascertained** goods in a **deliverable state forming part of a bulk**.* For example 10 tonnes of sand from Owen and Co. Builders Merchant. If this bulk is 'identified either in the contract or by subsequent agreement between the parties and the bulk is reduced to (or less than) that quantity, then if the buyer is the only buyer ... to whom goods are then due out of the bulk:

(a) the remaining goods are to be taken as appropriated to that contract at the time when the bulk is so reduced and

(b) the property in those goods then passes to that buyer (see Figure 5.7(b)).

Thus if, for whatever reason, the sand in the builder's yard is reduced to below the amount required to satisfy the contract, then the minute it is so reduced, the remainder becomes the property of the buyer. This is known as 'ascertainment by exhaustion'.

S.20(A): New rule regarding undivided shares in goods forming part of a bulk

This new section was introduced as a result of consultation with buyers of commodities who, after paying for goods comprising a cargo, were then left without remedy when the owner of the goods became insolvent. Property had not passed to the buyer, as the goods had not been ascertained, as covered by Rule 5(1) above. So he had lost his money **and** his goods and would be an ordinary creditor in the insolvency proceedings. However, as far as builders are concerned, the rule could have great benefits, as it would also cover the situation where he and others buy sand or bricks etc. which are part of a bulk and have not yet been set aside for the different purchasers.

The word 'bulk' is defined to mean a mass or collection of goods of the same kind contained in a defined space or area and such that the goods are interchangeable with any other goods of the same number or quantity (S.61(1)).

S.20(A) thus applies to a contract for the sale of a specified quantity of unascertained goods provided the following conditions are met:

(a) the goods or some of them form part of a bulk which is identified either in the contract or by subsequent agreement between the parties (e.g. 4 tonnes of sand from the heap in that builder's yard);

(b) the buyer has paid the price for some or all of the goods which are the subject of the contract and which form part of the bulk (so the builder has paid in advance for the sand) (S.20(A)(1)) (see Figure 5.7(a)).

Thus, in such situations, unless the parties agree otherwise, as soon as those conditions are met, i.e. goods are identified and paid for, the property in an undivided share in the bulk is transferred to the buyer and the buyer becomes an owner in common of the bulk (S.20(A)(2)). In other words, our builder, with many other buyers who have pre-paid, will become owners in common of the total bulk, despite the fact that his physical share of the sand cannot be identified and the sand remains at the builder's yard.

His share of the bulk will be in direct proportion to the amount he has paid (S.20(A)(3)).

This is so unless the aggregate of the undivided shares of buyers of the bulk would be more than the whole of the bulk. In such case the buyer's share would be reduced proportionately, so that the aggregate of the undivided shares is equal to the whole bulk (S.20(A)(4)). So, for example, if Owen Builders Merchants have 60 tonnes of sand in their yard and A, B, C and D have bought and paid for 10, 20, 30 and 40 tonnes respectively, then they will actually own one-tenth, one-fifth, three-tenths and two-fifths of the sand or 6, 12, 18, 24 tonnes, totalling 60 tonnes in all.

If the buyer has only paid for some of the goods due to him, any delivery to the buyer out of the bulk shall be attributed to the goods in respect of which payment has been made (S.20(A)(5)). So, a builder cannot say that he owns sand in the yard if he has only partly paid for his order and an appropriate portion has already been delivered to him. The sand delivered to him is his; the rest remains the property of the seller for the time being.

Delivery includes an appropriation of goods to the contract in such a way that the property has passed to the buyer (S.6(1)). So the builders merchant may have telephoned to say that they have loaded a lorry with the sand ordered. This would satisfy the requirement regarding appropriation.

Part payment will be treated as payment for a corresponding part of the goods (S.20(A)(6)).

Deemed consent by co-owner to dealings in bulk goods: S.20(B)

Obviously, in order to allow trading of the bulk goods to continue without interference from buyers who now own shares in the bulk, S.20(B)(1) provides that any such buyer shall be deemed to have consented to deliveries of goods from the bulk to any other owner in common (S.20(B)(1)(a). Also a buyer is deemed to have consented to any dealing with or removal, delivery or disposal of goods from the bulk by any other owner in common (S.20(B)(1)(b)).

In such cases no-one can sue anyone who has acted in accordance with subsections (a) or (b) above.

Nothing in S.20(A) or S.20(B) shall impose an obligation on a buyer of goods out of bulk to compensate any other buyer of goods from that bulk for any shortfall in the goods received by that other buyer. Nor shall these sections affect any contractual arrangement between buyers of goods out of bulk for adjustments between themselves or affect the rights of any buyer under his contract (S.20(B)(3)).

(Note: the above rules are only applied in the absence of agreement on such matters.)

Reservation of title under S.19

Because of the problems caused by the insolvency of sellers who had gone into liquidation after payment but before property had passed, many contracts have reservation of title clauses. Under S.19 of the Sale of Goods Act 1979, in a contract for the sale of specific goods or where the goods have been subsequently appropriated to the contract, the seller may, either by an express term of the contract or at the time of appropriation, reserve title to the goods until certain conditions are fulfilled. Thus, despite delivery to the buyer or to a carrier, the ownership of the goods **does not pass** until the conditions are met by the seller. A common condition imposed is that the goods must be paid for in order for the property to pass.

A term of this nature was included in the leading case *Aluminium Industrie Vaassen BV* v. *Romalpa Ltd* (1976) under which property was not to pass until the purchaser, R Ltd, had paid **all** the debts owed to the seller ABV. Until the sums had been paid, R Ltd was to hold the goods for ABV in a fiduciary position. The contract permitted such goods to be sold but unfortunately did not expressly deal with what was to happen to the proceeds. R Ltd sold some of the goods and then went into liquidation. A sum of £35,000 was held in a separate bank account by R Ltd's liquidator as representing the proceeds of sale of the goods belonging to ABV. ABV asked for a tracing order to recover the money from this account. The Court of Appeal allowed this as it was said that the duty to account for moneys acquired by such a sub-sale was part and parcel of the fiduciary relationship. (Tracing is an equitable right of persons in such situations as above to recover from the holder assets to which they are entitled or other assets which have been bought using the proceeds of sale of the original assets.)

Thus, a builder may have to buy building materials subject to such terms. How does this fit into a building contract situation?

It should be noted that once goods become incorporated into the building works they become fixtures and thus property in the materials passes to the owner of the land (see p. 87). Also, by Cl. 16.1 of the JCT contract, property passes to the employer if the builder has unfixed materials on site and has been paid for them under an interim certificate. Furthermore, by Cl. 16.2, if the value of materials stored off-site has been included in

an interim certificate by virtue of Cl. 30.3 (see post), such materials become the property of the employer.

By Cl. 30.3 of the JCT contract, an amount in an interim certificate may include the value of goods not yet delivered to site, provided the goods have been ordered in writing by the contractor or subcontractor and the contract expressly provides that the property (ownership) in the goods shall pass unconditionally to the contractor or the subcontractor, as the case may be, not later than the event set out in Cl. 30.3.2 and 30.3.3. These clauses state that nothing must remain to be done to the materials prior to incorporation into the works and they should have been set aside as ready for the works, marked with the name of the employer and the works destination. Furthermore, building contracts are contracts for work and materials and so will be covered by the Supply of Goods and Services Act 1982 (see p. 57).

Such reservation of title terms under S.19 may thus give the supplier of the materials a form of security, so that on bankruptcy or liquidation of a building firm, the suppliers may claim priority over other creditors.

Destruction of specific goods

If the goods are destroyed without the seller's knowledge **before** the contract is entered into, the contract is **void**.

If the goods are destroyed **after** the contract is made but **before** the **risk** has passed, then the contract becomes **void**.

If the goods are destroyed **after** the risk has passed, the buyer bears the loss. Obviously, insuring against loss is vital (for when the risk has passed, see p. 117).

Conclusions

Thus, building contractors' main concerns with contracts for the sale of goods are as follows.

Time: S.10
Date of delivery must be specified or agreed between the parties (see p. 115). Date for payment must also be agreed.

Quality or correspondence with description or sample: Ss. 13–15

As a building contractor's contract is not a consumer contract, these implied conditions can be excluded. This should be checked by reading the small print before agreement; otherwise it would be difficult to prove that there had not been an agreement on the exclusion.

On delivery of the goods, they should be checked immediately, preferably while the delivery men are still there, so that if they are not as required, the goods can be returned.

Sometimes the supplier inserts a clause giving a time in which goods can be returned, e.g. within 24 hours of delivery, so there must be no delay for otherwise there will be no chance of rejection and only damages can be claimed. In a building contract, it is more important to have the correct supplies at the right time than to have the chance of money compensation at a later date. Immediate rejection gives the building contractor a chance to get supplies elsewhere.

Transfer of ownership under hire purchase and consumer credit agreements

Not everyone is in the fortunate position of being able to buy goods outright for cash. Many people choose to buy their goods by paying in instalments. There are three different ways of buying goods in this way.

By hire purchase

Under this arrangement, the seller of goods **hires** them to the purchaser and the contract of hire gives the purchaser the option to buy after a certain number of instalments have been paid (usually the last). If the purchaser exercises the option there is a transfer of ownership of the goods to him. Thus, in a hire purchase agreement, the ownership of the goods remains with the seller unless and until the hirer exercises his option to purchase (see Figure 5.8).

Under the Supply of Goods (Implied Terms) Act 1973, as amended by the Consumer Credit Act 1974 and the Sale and Supply of Goods Act 1994, there are terms implied into a hire purchase

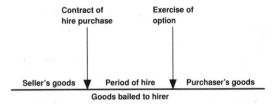

Figure 5.8 Diagram to illustrate a hire purchase agreement

agreement similar to those contained in Ss. 12–15 of the Sale of Goods Act 1979. Thus, by S.8, there is an implied term that the seller will have a right to sell the goods at the time property is to pass to the hire-purchaser. Also by S.8, there are implied terms concerning freedom from encumbrances and enjoyment of quiet possession. By S.9, goods bailed or hired by description must correspond to their description. By S.10 there are implied conditions as to satisfactory quality and fitness for a particular purpose. By S.11, there is an implied condition concerning sales by sample that the goods must correspond with the sample in quality (see pp. 113–15).

Credit sale agreements

Here the purchaser contracts to buy the goods but he is to pay for them in instalments. The Sale of Goods Act 1979 covers such transactions, and the ownership of the property passes to the purchaser in accordance with S.18.

The Sale of Goods Act 1979 sections concerning implied terms as to title, sales by description and satisfactory quality and sales by sample therefore also apply (see pp. 113–15).

Conditional sale agreements

Here there is a contract to buy the goods but it contains a term by which the ownership of the goods does not pass until **all** the instalments are paid. Whilst the Sale of Goods Act 1979 applies as to the implied terms of quality etc., because of this conditional aspect as to the passing of ownership, S.18 does not apply.

Protection of consumers who buy goods on credit

Traditionally, contracts for the sale of goods on credit have been weighted against the consumer who is generally in a weaker position for a number of reasons. This is largely due to the fact that the suppliers have drawn up standard contracts protecting their own interests and the consumer has little choice in using that standard contract (see p. 60).

The Consumer Credit Act 1974 has introduced a number of statutory protections for the consumer. For example, agreements must be in a particular form clearly setting out important terms; where an agreement is entered into at some other place than the supplier's place of business, then the purchaser has an opportunity to change his mind, etc.

Unfortunately this Act only applies to agreements with consumers and a consumer cannot be a corporation nor must they be buying in connection with business. Thus, builders who are buying goods for their business will generally be unaffected by these rules.

Contracts for the exchange or barter of goods

It is not uncommon nowadays to find that someone is offered goods in exchange for other goods or services, especially in times of high unemployment. The Sale of Goods Act 1979 as amended only applies to goods in exchange for money. However, the Supply of Goods and Services Act 1982 as amended applies to contracts of exchange or barter.

Thus, terms as to title (S.2), compliance with description (S.3), quality and fitness (S.4) and sales by sample (S.5) are implied by the Act (see below).

As far as part exchange is concerned, it seems from the few decisions made that part exchange contracts are covered by the 1979 Act, especially in relation to the new or more expensive item. As for the traded-in old goods, it also appears that the 1979 Act applies.

Contracts for the supply of work and materials

We saw earlier that the 1979 Act did not apply to contracts for work **and** materials. Thus, building contracts are not covered by that Act despite the fact that a large part of the contract's performance depends on the incorporation of goods belonging to the builder. The rest of the contract is to supply work and labour.

The Supply of Goods and Services Act 1982, however, does apply to such contracts.

By S.2 there is an implied condition that the transferor of goods (repairer, builder etc.) has the right to transfer **ownership** of goods. If there is a breach then the contract may be repudiated and a full refund may be claimed. This is so even if the goods have been incorporated into the building already. Extra damages may be claimed if, for example, someone else has to carry out the job at a higher cost. S.2 applies to all such transfers in business or in private.

Section 2 also implies warranties regarding freedom from encumbrances and quiet possession (see p. 93).

There is no equivalent to S.18 of the Sale of Goods Act 1979 and thus normally property will pass on completion of the contract. However, in the case of building work, once materials have been incorporated as fixtures, property then usually passes to the landowner.

Section 3 implies a condition that, where the goods are transferred by description, then the goods will correspond with the description. Perhaps this will end the days of paying the book price for a motor part which has been 'modified' to fit.

If goods are transferred in the course of a business, then under S.4 there is an implied condition that the goods must be of satisfactory quality and they must also be fit for any particular purpose for which these goods have been supplied. Thus, a builder in supplying materials for building work must make sure that they are of satisfactory quality. If particular materials have been specified in the contract, then the builder must still make sure that the materials are of satisfactory quality but does not have to ascertain that they are fit for a particular purpose as the contract specifications have overridden this requirement. It is still possible, however, that he could be found to be negligent if he failed to point out to the client that the materials were not suitable for the particular purpose. To cover himself he should disclaim liability for the materials, keeping evidence of the client's knowledge of the disclaimer, trusting that the disclaimer would be regarded as reasonable under the Unfair Contract Terms Act 1977. Otherwise he should probably refuse to take on the contract in the first place. In *Young & Martin Ltd* v. *McManus Child Ltd* (1969) M Ltd, a building contractor, subcontracted roofing work to Y Ltd, specifying a particular type of tile. Y Ltd then subcontracted the work to another firm B. B bought the tiles direct from the manufacturers S. The tiles had not been made properly and they were defective. As a result the roofs of 17 houses needed replacing. M Ltd sued Y Ltd, who were held to be in breach of an implied term (at common law) that these tiles should have been of satisfactory quality. Y Ltd would, of course, have then been able to recover the damages paid from B under an indemnity in the sub-subcontract and B could have claimed for breach of S.14 of the Sale of Goods Act against the manufacturers (see Figure 5.9).

See Cl. 8 of the JCT contract as to work and materials and Cl. 36 where the supplier of goods is a nominated supplier. In the latter case the building contractor is not liable for special purposes of specified materials supplied by a nominated supplier. He is, of course, still liable if they are of unsatisfactory quality for general purposes as in *Young & Martin Ltd* v. *McManus Childs Ltd* (1969).

If goods are referred to by sample then under S.5 there is an implied condition that the bulk must correspond with the sample. The transferee must have a reasonable opportunity of comparing the bulk with the sample and the goods must be free from any defect rendering them unsatisfactory which would not be apparent on reasonable examination of the sample. If a supply is by both sample and description then the transferor must comply with S.5 and S.3.

S.5(A) (incorporated by the Sale and Supply of Goods Act 1994) allows for a modification of the remedies available to someone affected by a breach of one of the statutory conditions. Should a transferee have the right to repudiate the contract as a result of a breach of S.3 or S.5, but in fact the breach is so slight that it would be unreasonable for him to do so, then unless there is a contrary intention, if the transferee is *not* a consumer, the breach may be treated as a breach of warranty and not a condition.

We have only dealt here with the goods element of a building contract under the Supply of Goods and Services Act 1982. The part dealing with implied terms as to the work and labour element is dealt with in Chapter 3 on contract.

Passing of property and risk

Contracts for the supply of building materials must be read in conjunction with the main contract, so that insurance is maintained to cover risk of loss, whether the materials are on site or off and whether property has passed or not (see p. 120).

Payment

By S.109 of the Housing Grants, Construction and Regeneration Act 1996, a party to a construction contract is entitled to payment by instalments, stage payments or other periodic payments unless it is specified in the contract that the work will, or is estimated to, last for less than 45 days. The parties are free to agree the amounts and timings of the payments. If there is no such agreement

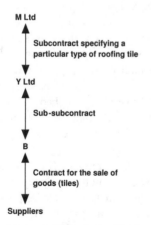

Figure 5.9 *Young & Martin Ltd v. McManus Childs Ltd* 1969

then the provisions of the Scheme for Construction Contracts apply (see p. 45). There are other sections which deal with dates for payment, notice of intention to withhold payment, the right to suspend performance for non-payment and the prohibition of conditional payment provisions (see Ss. 110–113).

Transfer of possession of chattels

Like land, physical things can be in someone else's possession. This can be for a variety of reasons. The possessor may have hired the thing, or borrowed it, or be doing something to it, such as mending it, or may be holding it as security for a loan. This temporary transfer of possession is called a **bailment**. The transferor is called the **bailor** and the transferee the **bailee**. After the particular purpose has ended, the goods are transferred back. It should be noted that the bailor need **not** be the owner of the goods. He may be a mere possessor such as an employee with a company car who bails the car to his garage for repair. This sort of bailment is probably the most common. The bailee is under a duty to look after the goods in his care, and must hand them back to the owner or his agent unless payment has not been made, in which case he has a **lien**, i.e. a right to hold the goods pending payment.

If someone finds goods, and keeps them in his possession, he will become a bailee of those goods, pending their return to the owner.

Types of bailment

Hire

For example, television, ships, telephones, video recorders, motor cars, lorries, JCBs etc. Hire agreements are also called leases or rental agreements or even finance leasing. Legally, however, they are treated as contracts of hire. The important element is that there is a complete transfer of possession to the hirer.

Obviously, the hirer, as bailee, must look after the goods during the hire period, but he is not responsible for damage or loss to them, unless caused by his own negligence. If the bailee tries to sell the goods, then the hire agreement automatically comes to an end, and the bailor can seek repossession.

If the hire agreement could last for more than 3 months for a rental of less than £15,000, then this is a consumer hire agreement if made with a consumer under the Consumer Credit Act 1974 and statutory formalities have to be complied with. The person hiring the goods should nevertheless make sure that they are aware of the contents of the agreement.

Hirers have also been given greater protection in the Supply of Goods and Services Act 1982.

Under this Act, special terms are now implied into the contract of hire. This means that, without having to mention them, important duties are imposed on the owner of equipment, who lends goods for a consideration. The relevant sections are Ss. 6–10.

These sections closely follow the implied terms contained in the Sale of Goods Act 1979 (see pp. 113–15).

By S.7 there is an implied condition that the bailor has the right to transfer possession to the bailee. Thus, the bailor must either own the goods or he must have lawful possession. There is also an implied warranty that the bailee will enjoy quiet possession except in so far as possession may be disturbed by the owner of any charge or encumbrance disclosed to the bailee prior to the contract. By S.7(3) the previous sections do not affect the right of the bailor to repossess the goods under an express or implied term of the contract, e.g. if the hirer does not pay the rental.

If the property has been hired by description, then by S.8 there is an implied condition that the goods comply with that description. This can be very important in car fleet hiring agreements.

Section 9 follows closely S.14 of the Sale of Goods Act 1979. Thus, S.9(1) states that the hirer must beware. Section 9 only applies to goods bailed in the course of a business whereas Ss. 7, 8 and 10 apply to all contracts of hire. Section 9(2) imposes an implied condition that the goods supplied must be of satisfactory quality.

Thus, the goods must not only be fit from the beginning of the contract, but they must remain so for a reasonable time. However, the Act does not impose an implied condition that the goods must be of satisfactory quality for the duration of the contract of hire. This is usually because the owner expressly contracts to service and maintain the goods. The inclusion of such a clause should therefore be checked. There is no implied right to have unlimited maintenance throughout the contractual period.

Note that S.9(3) does not apply to goods examined by the bailee **before** contract or to defects which were specifically drawn to the bailee's attention prior to contract.

By S.9(4) and S.9(5) there is an implied condition that where goods are bailed for a particular purpose which was made known to the bailor either expressly or impliedly, then those goods must be reasonably fit for that particular purpose. This will be so unless circumstances show that the bailee did not rely on the skill or judgement of the bailor.

Note that S.9 also applies if the goods are hired from a finance house although supplied by the retailer who sold the goods to the finance house before hire. Sections 9(4) and 9(5) will also apply even if a particular purpose was made known only to the 'credit broker', i.e. the retailer who arranged for the goods to be sold to the finance house before hire.

Loan

A loan may be gratuitous, i.e. with no payment, or it may arise from a contract, in which case there must be some sort of consideration.

Here the borrower must use the goods loaned, provided he complies with the conditions of the loan, e.g. he must not drive the motor car in the dark at night.

During the bailment, the borrower is responsible for the goods and must take reasonable care of them, but he is not responsible for 'fair wear and tear', i.e. the normal wear associated with normal use. He is liable for his own negligence and dishonesty as well as that of his employees, particularly if he did not use reasonable care when

employing them. He is never liable for Acts of God or war or robbery with violence.

At the end of the loan, the goods must be returned, to comply with the terms of the contract, e.g. 'give it back to me next week'. If there is no contract, because the borrower is giving no consideration to the lender, his duties will be determined by torts which deal with the unlawful keeping of or interference with chattels. Such torts are beyond the scope of this book.

Pledge or pawn

Chattels may be pledged or pawned (the words are almost interchangeable) as a method of raising money. The goods are transferred into the possession of the lender (pawnee) by the owner (pawnor), where they act as security for a loan made to the pawnor.

The owner of the goods is then entitled to have the goods returned to him, provided he repays the loan with interest within a specified period. The lender of the money becomes a bailee of the goods and must make sure that the goods are not damaged or lost during the bailment through his negligence. Until the loan is repaid, the lender retains a **lien** over the property, and after a specified time, can then sell the goods to recover his loss. If there is any money over he must then pay the pawnor the balance.

Professional pawnees are called **pawnbrokers** and must be licensed under the Consumer Credit Act 1974. They usually have a symbol of three balls outside their windows.

Goods being prepared or otherwise worked on

As stated before, bailments are very commonly made when getting property repaired. Often this involves incorporating new parts or equipment into the property. For example, if my garage repairs my car, I will expect to pay for all the goods and materials used in the repair such as spark plugs, fan belts etc.

Under the Supply of Goods and Services Act 1982 statutory terms are now implied into contracts for the supply of work and materials (see p. 122).

Carriage of goods

Goods carried by owner
Such goods are carried at his own risk and he should therefore have sufficient insurance cover. If goods are being carried by a builder's own lorries, then the driver's duties and the builder's liability depend on the instructions given, the driver's contract of employment, the state of the lorry and the goods and statutory limitations, e.g. under the Road Traffic Acts. Thus, if equipment is damaged in transit and there was no fault, then the builder must claim for his loss from his insurance company. If the driver was at fault, then this may be grounds for dismissing or even suing him. If someone else was at fault, then they could be sued in tort.

Goods carried by seller or agent
In this case the Sale of Goods Act 1979 applies (see p. 113).

Goods carried by an independent carrier
The law depends on whether or not the carrier is a **private carrier** (which is more usual) or a **common carrier**. Private carriers are simple bailees, whereas common carriers have greater duties. The law is now supplemented by the Supply of Goods and Services Act 1982 (see above).

1. Common carriers. A common carrier is in business to transport anybody or anything by land, sea or air, subject to a few exceptions. He must therefore carry anything for a reasonable sum. If he refuses to do so, he can be sued or even prosecuted.

Furthermore, a common carrier must ensure the safety of the goods entrusted to him, unless the damage was caused by an Act of God, enemy action, because of an inherent defect in the goods, or the fault of the consignor. For this reason he must insure the goods, and he will be liable to the consignor for any loss. This is strict liability and he will not be able to rely on these exceptions if he was negligent. However, the common carrier may exclude this liability by inserting an exemption clause, restricting or excluding his strict liability, subject of course to the Unfair Contract Terms Act 1977.

Obviously as a result of the harsh liability imposed on a common carrier, there are virtually no common carriers left, and most carriers have in their conditions that they are not holding themselves out to be common carriers.

2. Private carriers. They are bailees and are therefore liable only for negligence. If they wish to exclude their liability they can, within the confines of the Unfair Contract Terms Act 1977.

The following is a summary of the main rights and duties of a private carrier (in addition to those implied by the Supply of Goods and Services Act 1982).

- He must deliver the goods to the consignee or some other specified person at the correct address in accordance with his instructions.
- He must obey any change in instructions given.
- He may act as an **agent of necessity** and sell the goods he is carrying, if the goods are in danger of perishing. In such cases, it must be impossible for him to contact the owner of the goods, and this is unusual nowadays with modern forms of communication.
- He does not hold out that his vehicles are safe to carry the goods.

Choses in action

For example debts, shares, goodwill, negotiable instruments, patents, designs, trademarks, copyrights, life insurance policies and legacies.

Transfer of ownership

Because they are not physical things, it is impossible to transfer possession of choses in action.

Transfer of ownership must be done by **assignment**, either in accordance with S.136 Law of Property Act 1925 or by the rules in equity, unless there are special rules appropriate for that type, e.g. negotiable instruments and shares.

Why would anyone wish to transfer a chose in action? As you can see from the above examples, all have the potential to provide possibly large sums of money and therefore buying power. So whilst a debt, for example, seems to be a strange sort of property, banks and finance houses make their money by owning debts. Should a creditor be short of money then he can sell his debt to someone else to get immediate cash to solve his financial problems.

Example

Creditor is owed £10,000 by debtor payable one year from now with interest from today's date. Creditor needs his money, but cannot call on debtor to pay it yet. He therefore wishes to sell his debt to X to realise some, but not all, of the money. Obviously, X is not going to pay the full amount as there would be no financial benefit to be gained when debtor pays the debt on the due date.

How does he transfer the debt?

1. Transfer under S.136 Law of Property Act 1925. The transfer must be in writing and must be of the whole amount, not part only. The transfer document must be signed by the transferor and express notice must be given of the transfer to the debtor. Thus, creditor must transfer, in writing, the whole debt to X and notify debtor that on the due date he must pay the sum owed to X and not creditor.

Because the transfer has been done properly, if debtor defaults, then X can sue debtor, without needing to get creditor to join in the action as co-plaintiff (see Figure 5.10).

2. Transfer in equity. If the transfer did not meet the requirements of S.136, e.g. because it was not in writing, then there *must* still be an intention to transfer the debt for it to be valid in **equity**. Notice will still be necessary, but this need not be express, for otherwise, debtor will not be aware that his debt has been taken over by X and may pay creditor instead (creditor may keep the money and X would then have to sue him).

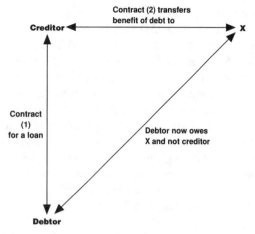

Figure 5.10 Transfer of a debt

If debtor defaults, X must ask creditor to join him as co-plaintiff in the action to recover the money.

Intellectual property

Patents, registered designs, copyright and trade marks come within a category of personal property now called **intellectual property**, as they are all concerned with ideas or creative work. They are all choses in action. Because of their potential value and because of the difficulties of protecting such intangible forms of property, the law has evolved complex rules to enable the creators or owners of the property to safeguard their rights. In particular, they need to be able to stop other people from copying their work or using or manufacturing their inventions without permission. Some protection requires the registration of the invention or design. Nearly all the law is now found in statute law. Because of the international implications, if protection is required in other countries, then international conventions may have to be complied with and legal action would also have to be started within those countries to enforce property rights. This is one area of the law which will eventually benefit from the United Kingdom's membership of the EU and protection may then be granted in all member states.

The builder will mostly be concerned with copyright, in particular the copyright of plans and possibly patents. Registered designs protected under the Registered Design Act 1949 as amended are more concerned with articles which have an artistic design applied to them, such as a dinner service pattern or a drinking glass design. In certain cases, a design may be covered by both copyright protection and protection under the Registered Design Act 1949. Trade mark protection is more the concern of manufacturers of goods, and this is an area which is giving many problems at present with foreign manufacturers making cheap copies of manufactured goods and then applying a forged registered trademark to the copy.

Copyright

Relevant law
The law is found primarily in the Copyright, Designs and Patents Act 1988 (the Act), as amended.

What property rights are owned?
By S.2 and S.16 the owner of a copyright has the exclusive right to copy, give copies to the public, perform, show or play the work in public, broadcast or adapt the copyright work. He may give permission, called a **licence**, to others to reproduce the material and if a fee is paid this is called a royalty. By S.90(3), the licence must be signed by the owner or his agent.

In addition, **moral** rights may be created in favour of the author or commissioner of the work, even though he is not necessarily the owner of the copyright.

What material may be the subject of copyright?
In Part I of the Act, S.1(1) describes copyright as a 'property right which subsists ... in the following descriptions of work:

(a) **original** literary, dramatic, musical or artistic works;

(b) sound recordings, films, broadcasts or cable programmes; and

(c) the typographical arrangement of published editions.

Lengthy definitions of the above words are contained in Ss. 2–8 of the Act and, in general, their meanings are far wider than a layman would expect. Copyright does not come into being until it is recorded in some way, e.g. in writing. Thus, mere ideas cannot be protected. There must be permanence in their manifestation. One should not talk about good ideas in public as they would be impossible to protect.

Thus, the builder will deal with written or spoken literary work which may include a table or compilation or a computer program (S.3). The Copyright (Computer Programs) Regulations 1992 Reg. 3 has extended this to cover 'preparatory design material for a computer program'. In *Anacom Corporation Ltd* v. *Environmental Research Technology* 1994, an electronic circuit diagram was held to be a literary work because it could be read by someone capable of understanding such material.

The builder will also deal with artistic work. This is defined by S.4 as covering graphic works, photographs, sculpture or collage, irrespective of artistic quality; a work of architecture, being a building or a model for a building; or a work of artistic craftsmanship. It may be argued from the use of the generic words 'artistic work' and the lack of the words 'irrespective of artistic quality' that works of architecture, buildings or models must have some artistic quality. So, a builder could be prevented from copying an architect-designed house but not Aunt Matilda's potting shed (unless it had artistic quality!). 'Graphic work' includes any painting, drawing, diagram, map, chart or plan. The builder obviously uses or produces plans and drawings, models, buildings and also prepares brochures which may contain photographs, tables etc. In *British Northrop Ltd* v. *Texteam Blackburn Ltd* 1974 the drawings used to design rivets were held to be protected by copyright law.

The builder may also be concerned with the creation or use of copyright of films, video tapes, sound recordings, cable or other broadcasts when for example, presenting safety lectures to employees or the public or in preparation of in-house information. Ss. 5–7 cover these areas.

Finally the builder will often use 'typographical arrangements of published editions' when using books. The use of the word original in relation to literary, dramatic, musical and artistic work does not mean that the work must be totally novel like an invention. Rather, it means that the work must be manifested in an original way from the new author. Nor does it necessarily require a large amount of skill or effort.

What acts are prohibited by the Copyright Act?

Copyright is infringed if a copyright work is copied, shown or broadcast or adapted in any way unless licensed to do so by the copyright owner (S.16). By S.17 copying means reproducing the work in *any* material form, even in handwriting, and includes storing the work in any medium by electronic means, e.g. photocopying, videoing or facsimile (fax). Still photographs from films or a temporary display on a visual display unit (VDU) might fall into this category.

Also there can be **secondary** infringement of copyright work by importing an infringing copy (without licence) unless it is for private or domestic use (S.22). Similarly, there is secondary infringement if someone, without licence, possesses, sells, lets for hire, exhibits in public or distributes, provides the means for making infringing copies or permits the use of premises or provides apparatus for infringing performances of a copyright work (Ss. 23–24).

There are a number of exceptions to the restricted acts. For example 'fair dealing', which is not defined but would include photocopying of literary or artistic work or of a published edition for the purposes of research or private study, is specifically permitted (S.29).

For how long does the copyright last?

Generally, under the Act the term of protection for copyright of new literary, dramatic, musical and artistic work now lasts for the life of the author/designer/maker plus 70 years calculated from the end of the year of death. Similarly for new sound recordings or films, protection lasts for 70 years from the end of the calendar year in which it was made. Protection of typographical arrangements of published editions lasts for 25 years from the end of the years of publishing.

How do you create a copyright?

It is created automatically on creation of the copyright work, e.g. on finishing a book or the designs for a house. Nothing need be done or registered with anyone. Readers will be familiar, however, with the symbol © which appears in many publications. This is not necessary for legal protection in the United Kingdom, but it is necessary to gain protection in certain countries party to the Universal Copyright Convention. That convention requires that all publications should have the copyright symbol ©, together with the name of the copyright owner and year of first publication. Of course, the protection afforded by other countries, will only be as good as that given to one of their own nationals, no more and no less.

A work qualifies for copyright protection if the author was at the 'material' time a 'qualifying person', i.e. he must fall within a wide category of people which includes *inter alia* British citizens, British nationals (overseas) or someone who is domiciled or resident in the United Kingdom (including a corporate body) (S.154).

The qualifying person must be the author at the **material time**. So, in relation to a literary or artistic work, if it is an unpublished work then it will be at the time the work was made or, if published, when the work was first published (S.154(4)). In the case of a film such as a company video, it will be when the film was made (S154(5)(a)).

The relevant work will qualify for protection if it is first published in the United Kingdom. Qualification may be extended by Order in Council to other countries such as colonies.

Who owns the copyright?

Unless varied by agreement, as, say, in a contract to write a book for a publisher, the copyright will normally belong to the author, who is the first owner. However, under S.11(2), if a work is made by an employee **in the course of his employment**, then the employer is the first owner of any copyright, unless there is a contrary agreement. (See

p. 227 as to who is deemed to be an employee.) It should be noted that there is no longer a section in the Act which automatically vests ownership of copyright in a commissioner of work. So if you are asked to draw up designs and plans for someone, you will be the first owner unless you agree to the contrary (and sensibly receive higher payment for losing this valuable property right).

Transfer

S.90(1) states that copyright is transmissable by assignment, by testamentary disposition (by will) or by operation of law (mnemonic OAT), as personal or moveable property. Operation of the law would include bankruptcy and intestacy (death with no will). S.90(3) requires that any transfer is not effective unless it is in writing signed by or on behalf of the assignor (the copyright owner or his agent). The assignee will then have full rights to the copyright. All or part of the copyright may be assigned, for the whole or part of the term of protection. Thus, many 'best selling' authors sell the film rights separately but retain the copyright of the published and the original literary work.

Moral rights

We have seen that very often the author of a work will not necessarily be the owner of the copyright. Nevertheless, it is obviously to someone's benefit to be identifed as the writer or the designer of a work, e.g. when applying for a job, and it is important that this right should be protected in law as otherwise someone else may falsely claim to be the author and reap the reward of another person's creativity and reputation. This right is called a paternity right.

By S.77 the author of a copyright literary, dramatic, musical or artistic work and the director of a copyright film has the right to be identified as the author or director of the work in certain circumstances. S.77(5) also gives this right to the 'author of a work of architecture in the form of a building' whose right may be identified 'on the building'. I am sure you have seen many plaques recording the details of the architect. This right is not infringed unless it has been asserted in the terms of S.78. Section 78(2) states that the assertion

may be made firstly, generally, or secondly, on an assignment of the copyright, by including in the instrument (document) a statement that the author or director asserts his right to be identified, or thirdly by an instrument in writing signed by the author or director. If you look at the front of this book you will see an example of my assertion.

The court will take account of any delay in asserting the right when considering what remedies to award in a subsequent court action (S.78(5)). So there could be a late assertion but this would affect one's remedies.

It should be noted that by S.79 the right to be identified as author or director does not apply to computer programs, the design of a typeface, any computer-generated work, to anything done by or with the authority of the copyright owner, where the work was produced in the course of employment and original ownership vested in the person's employer, where copyright would not be infringed by the various exceptions under the Act, e.g. fair dealing, or where it was included in a reference work with the author's consent.

Apart from the Act, various other common law remedies have existed to protect one's moral rights to works. Thus, if someone pretended to have designed a building, the real designer could sue for the tort of passing off, provided he could prove he suffered financial damage. Also if a contract existed between parties and there was a clause that no change could be made without the author's consent, any such changes could be the foundation of a breach of contract action.

Remedies for infringement

Anyone infringing a copyright may be sued by the copyright owner (S.96(1)). This will be in the Chancery Division of the High Court. By S.96(2) all the normal rights of relief in relation to property such as damages, injunctions and account are available to the plaintiff. An account can amount to more than damages as the actual profits made are awarded to the plaintiff. Also an Anton Piller order could be made. Not strictly a remedy, this is a High Court order which permits the plaintiff to enter the premises of the defendant and seize the disputed material. To get this the plaintiff must have persuaded the court that to wait for a writ or injunction to be served would probably cause the defendants to destroy or remove the evidence. This order was named after the case *Anton Piller KG* v. *Manufacturing Processes Ltd* 1976. Also an order for delivery of the infringing copies may be made. The copies may be ordered to be destroyed. It should be noted that the old rule regarding the restriction on injunctions in relation to buildings which have already been started is no longer law. So, it is now possible to obtain an injunction to order the demolition of a building infringing copyright work.

In addition to the above rights, S.107 imposes criminal liability on persons who **in the course of a business** and without licence offer for sale or hire articles that infringe copyright, if the persons doing so knew or had reason to believe the copies were infringing copies. Exhibiting such work commercially is also prohibited.

Patents

Merely inventing something is not enough to protect it from being used or copied. So it is important to obtain a patent in order to prevent others from reaping the benefit of one's work. A patent is a valuable piece of property, as those who want to use or manufacture the invention must first obtain a licence and pay a fee for the privilege.

Where are patents obtained from?
In the United Kingdom patents are applied for at the Patent Office and, if granted, will give protection in the United Kingdom only. Protection in other countries is only afforded to those who have obtained patents in those countries. There are a number of international conventions and treaties which give certain advantages to patentees who have already obtained a patent in one signatory country, e.g. the Patent Co-operation Treaty 1970, the European Patent Convention and the Community Patent Convention applicable to the European Union. One application only need be made to the European Patent Office in Munich to give protection in all EU countries.

Relevant law
As a result of the European Patent Convention 1973 and the Community Patent Convention 1975, UK law on patents has been considerably altered. The Patents Act 1977, now amended by the Copyright, Designs and Patents Act 1988, which codifies the law on patents contains new principles of law. The procedures are governed by the Patent Rules 1990. Other applications are governed by the appropriate domestic national law or as influenced by the conventions.

What inventions may be patented?
Section 1(1) Patents Act 1977.

In order to be able to get a patent, an invention must:

(i)   be new;
(ii)   involve an inventive step;
(iii) be capable of industrial application;
(iv)   *not* be concerned with a discovery, scientific theory, literary, dramatic, musical or artistic work, scheme for performing a mental act, playing a game, or doing business or a program for a computer or presentation of information (S.1(2)).

A patent will not be granted if it would encourage offensive, immoral or anti-social behaviour or if it concerns certain biological processes (S.1(3)).

For how long does the patent last?
A patent lasts for 20 years and is no longer renewable. The period runs from the date published in the *Patent Office Official Journal*.

To keep the patent alive, fees must be paid annually from the fifth year on.

How do you get a patent?
Applications may be made personally. Alternatively, patent agents or lawyers specialising in patents can register on the owner's behalf.

The application should be made in the prescribed form and filed at the Patent Office. It must contain the following:

(i) a **request** for the grant of a patent;

(ii) a complete **specification** containing a description of the invention;

(iii) a **claim** setting out the scope of the patent – it must not be too narrow, thereby allowing others to copy the general principle involved, nor must it be too wide, so that a patent may be refused because it covers matters already known;

(iv) any **drawing** referred to in the description or claim;

(v) an **abstract** giving technical information.

There follows a lengthy procedure starting with a **preliminary examination and search**. These check that the formalities have been complied with and examine **prior art** documents to discover whether the invention is truly 'new'. The results are given to the applicant. If he chooses to carry on with his application, the application, claim, preliminary search report etc. are **published**. This gives the applicant protection against anyone using this material during this stage (providing he eventually is granted a patent). A **substantive** examination is then made, and if all the requirements of the patent rules are met, a patent is granted to whoever is entitled.

Who owns the patent?

The inventor can apply for a patent by himself or jointly with another person. But the patent will only be granted to the inventor(s) or the person entitled by law or by contract (e.g. a research scientist's employer) or to their successors-in-title, i.e. the persons entitled after the death of a person or on winding up of a company.

By S.39 if an invention was discovered during the course of an employee's normal duties then the invention will belong to the employer.

Transfer

A transfer is by assignment and this must be in writing, signed by the parties or their agents. All transfers or other dealings such as mortgages must be registered at the Patent Office. Failure to register will make the transaction ineffective against subsequent transferees who took a transfer without knowledge of the earlier dealing.

Infringement

This is defined by S.60 of the Patents Act 1977 and relates to persons claiming the right to manufacture or deal in a product or claiming a right to use or deal in a process. There are a number of exceptions such as the private, non-commercial manufacture of a product. The London Patents County Court has introduced an alternative dispute resolution scheme (see p. 43).

Proceedings may be brought in the Patents Court (part of the High Court) or Patents County Courts or may be referred to the Comptroller of the Patent Office, who acts as an arbitrator. He has fewer powers, and can only award damages or make a declaration. Any other remedy must be pursued in the Patents Court.

Negotiable instruments

These are another type of chose in action and include **cheques** and promissory notes, both of which are special forms of bills of exchange. Bank notes (but not coins) are also negotiable instruments (instrument – a document).

A negotiable instrument is a document which represents someone's claim to a particular amount of money and which has a special characteristic called **negotiability** (see below). The document itself has virtually no value, but it is the owner's evidence of ownership of whatever sum is written on it. Thus, a £5 note is in itself worth about 2p. But, it is known that it can be used to buy £5's worth of goods.

Transfer of ownership of negotiable instruments

As we have already seen, most choses in action have to be transferred by written documents signed by the owner or his agent. Imagine doing this for a £5 note!

A negotiable instrument, on the other hand, is transferred very simply **either** by merely handing it over to the new owner (delivery) just like a parcel **or** by endorsing **and** delivering it. Endorsing means writing one's signature on the reverse (or elsewhere) of the document. Thus, our £5 note can be transferred by mere delivery. So could a cheque. But, if the owner wishes to transfer the cheque *again* he must endorse it and then deliver it to its new owner. The new owner can then collect the money from the drawer's bank account despite the fact that the cheque was originally payable to some other person. The fact that the new owner is entitled to the money is due to the concept of **negotiability**, as the cheque was not only physically transferred to him, but he also got all the legal rights associated with ownership of the cheque.

Negotiability

Negotiability means, therefore, that on the transfer of a negotiable instrument the transferee gets:

- all legal rights to it
- with no defect in the title (even if the transferor's title was defective)

provided

- he acted in good faith,
- he was not aware of the defects such as if it was stolen by the transferor
- **and** he gave value for it.

Example (see Figure 5.11)

Sue pays her £20 newspaper bill by cheque. The transfer of the cheque is by delivery only. It has not yet been 'negotiated'.

Mr Smith the newsagent is now the owner of the cheque. Mr Smith is short of cash and has no time to go to the bank. He goes into the butchers, endorses the reverse of Sue's cheque by writing his signature on it, and gives it to the butcher in exchange for cash. Mr Smith has thus negotiated the cheque and the butcher now has full rights to it, because he acted in good faith, had no notice of any defects in title (of which there were none) and because he gave value (the cash) for it. The butcher can now pay the cheque into his own bank account.

Advantages of negotiable instruments

What is so useful about negotiable instruments is that they provide a relatively safe method of transferring sums of money around. Provided certain safeguards are maintained, only the persons properly entitled to the money may actually collect the money involved.

Cheques

Probably the most common type of negotiable instrument used in business and in private is the cheque, which is a type of bill of exchange. The law governing cheques is chiefly found in the Bills of Exchange Act 1882 and the Cheques Acts 1957 and 1992.

A bill of exchange is defined by the 1882 Act as 'an unconditional order in writing addressed by

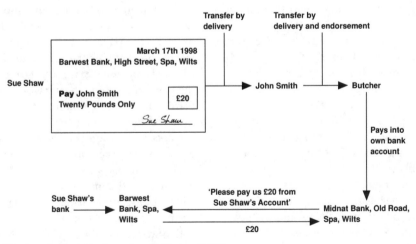

Figure 5.11 Diagram to illustrate the concept of negotiability

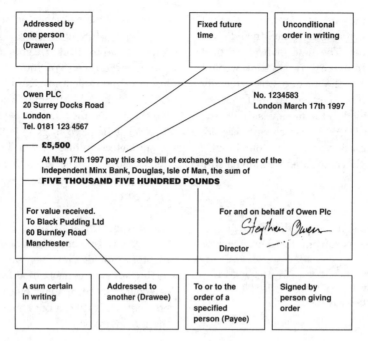

Figure 5.12 Bill of exchange

one person to another, signed by the person giving it, requiring the person to whom it is addressed to pay on demand, or at a fixed or determinable future time, a sum certain in money, to or to the order of a specified person or to bearer'.

They are commonly used in international trade and depend on one party being owed money by another, who is then asked to pay another party a sum of money, thus avoiding two transfers of money.

In the example (see Figure 5.12) Owen plc owes the Independent Minx Bank £5,500. Black Pudding Ltd owes more than this to Owen plc.

Owen plc therefore draws the bill, as drawer, on Black Pudding Ltd as drawee to pay on 17 May 1997 the payee bank. If the drawee accepts the bill, then it will undertake to pay the debt on the due date and Black Pudding Ltd becomes primarily liable on the bill.

During the intervening two months, the payee bank will hold the bill and if it requires cash may negotiate it to others by delivery and endorsement, and whoever is the holder on 17 May may present the bill for payment.

A cheque (see Figure 5.13) is defined by S.73 1882 Act as a 'bill of exchange drawn on a banker and payable on demand'.

Most cheques are never negotiated; they are merely cashed by the payee, and little can go wrong.

But, if there are insufficient funds in the drawer's bank account, the bank is under no duty to pay out to the collecting bank and the cheque will probably be returned marked 'return to drawer'. The payee now has the problem of collecting the money on the dishonoured cheque. The drawer is under a duty to pay the money owed to the payee. But, if the cheque was a mere gift and there was no consideration on the other's part, then the donee cannot sue on its dishonour.

Nobody need accept a cheque in payment of a debt. But, if it is accepted and paid by the drawee banker, then this is goods evidence that the payee has received the money. For this reason it is vitally important that bank statements should be scrutinized cross-checked with completed stubs, both of which should then be filed for a sufficient period. Further checks can be made by sending

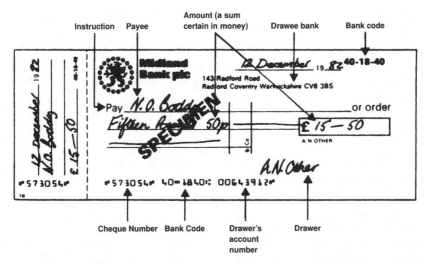

Figure 5.13 Cheque (this cheque is reproduced with the permission of Midland Bank plc from their booklet 'How to use a Bank Account')

covering letters containing the cheque number, and writing the cheque number on the invoice or bill and the date on which the cheque is sent.

Drawing the cheque

Date

The correct date is primarily important for record purposes. If the date is missing, the bank does not have to pay out on the cheque, and often, the payee will return the cheque to the drawer. Any holder of the cheque, however, may fill in the correct date, if they wish.

Once a cheque is in the hands of the payee or a holder to whom it has been negotiated, it should be paid as soon as possible into their bank account or cashed. Not only are such cheques liable to get lost, but banks will not usually pay out on a cheque if it is more than 6 months old. If left longer, the payee would have to go cap in hand to the drawer for a new cheque, as the drawer is under no duty to furnish a new one, although, of course, this does not absolve his indebtedness.

Some people post-date cheques for many reasons, e.g. because their account is not yet in funds or they want to pay for goods to be delivered on a particular date, which, if not done, they then have an opportunity to stop the cheque. The payee

may refuse such a cheque, as strictly speaking it does not fulfil the statutory definition, as it is not payable on demand. If he does accept the cheque, but pays it into his bank before the due date, then the paying bank should hold the cheque and pay on the correct date. If it pays before, however, the bank should not debit the payer's bank account and indeed would be liable for any loss if the payer subsequently stopped the cheque.

Payee's name

The cheque can be made payable to the payee or to his order or to 'bearer'.

The correct name should therefore be used for the payee, particularly when dealing with a company or firm, although often the banks will still pay on a cheque where the payee's name has been misspelt.

If the words 'or order' are crossed out by the drawer, the cheque cannot be negotiated to anyone else by endorsement, and the cheque ceases to be a negotiable instrument but a mere simple chose in action. Thus, only the payee should claim payment on the cheque and, if it was negotiated to anyone else by endorsement, then they would get no better rights under it than those of the endorser. Furthermore, the transferee would have to join with the payee to sue if it was dishonoured.

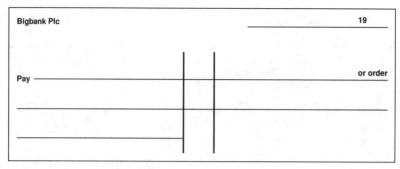

Figure 5.14 General crossing of a cheque

If the cheque says 'pay bearer' then any holder of the cheque may claim its value, and there is no need to transfer it by endorsement; delivery would be enough, like a bank note. It is for this reason that few bearer cheques are seen, because of the dangers of theft. Few people realise, however, that endorsing a cheque 'in blank', i.e. by writing their signature without specifying a new payee, has the same effect. Thus, if one needs to negotiate a cheque, an endorsement in blank should only be done actually at the time of transfer, or more safely, the signature should follow a direction to pay a specified transferee, e.g. pay John Smith signed B T Smale. A thief would then have to pretend to be John Smith before he could be paid.

The amount of money

Part of the statutory definition of a bill of exchange is that it must be for a sum 'certain in money'. As far as cheques are concerned, the sum is written in words first, then in figures in the right-hand box. The two should always be checked to be the same. If there is a difference, then under the Act the amount in words will be the amount payable. Most people, however, would send such a cheque back for alteration.

If the drawer realises he has made a mistake in any part of the cheque, he can alter the mistake and initial the alteration.

To prevent unauthorised alterations, the drawer ought to make it as difficult as possible, by drawing a line after the words or figures, as shown in Figure 5.13. No gaps should be left, preventing extra figures being inserted.

Crossings

Most people use crossed cheques without realising their significance.

A general crossing is merely two parallel lines drawn across the cheque (with or without the words '& Co' – see Figure 5.14).

A crossed cheque may only be paid through a bank account and cannot be cashed over the counter. Thus, the holder of a crossed cheque has first to pay it into his bank account. If he has no bank account, he must negotiate it to someone who has. Their bank then collects the money from the paying bank.

Sometimes the crossing is 'special' when the name of a bank is specified. In such cases, the money must be paid through the specified bank (to be credited to the payee bank account). This is very common when collecting for charity. To save administration expenses, the charity requests that any donation cheques should be crossed and a particular bank named in the crossing (see Figure 5.15).

A crossing with the words 'not negotiable' written across it means that the cheque cannot be negotiated at all. If it is attempted, the transferee gets no better right to it than the transferor. So it is good sense to refuse to accept such cheques except as the original payee.

A cheque crossed with the words 'A/c payee only' or 'account payee' means that the collecting bank must only credit the named payee and no other. It is no longer negotiable and since the Cheques Act 1992 it is not transferable at all.

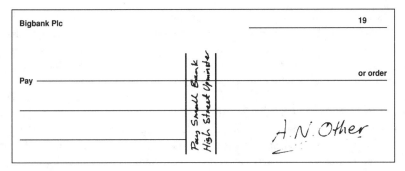

Figure 5.15 Special crossing

Ending the bank's authority to pay cheques

Whilst the customer's account is in credit, the bank is under a duty in contract to pay out on any cheque properly drawn on it. However, unless previously agreed, the bank does not have to pay if the account is even the slightest bit overdrawn. In business, this can be not only embarrassing, but financially damaging if a company or individual appears to be unable to pay their debts. If the bank had no cause whatsoever to stop payment, the customer can claim damages in contract if he is in business. A private individual may, however, be able to sue for libel, if the payee is told that the customer's cheque has been returned to him through lack of funds, provided that this could be said to be a defamatory statement.

Cheques may be 'stopped', i.e. payment countermanded. This is more expensive than most people imagine, as administration charges have to be paid. Written instructions usually have to be given. If the cheque was supported by a bank guarantee card, the cheque cannot be stopped.

Other circumstances which also terminate the bank's authority to pay out include notification of the customer's death or mental incapacity. In practice this can be very important and banks should be immediately informed of the death of a customer; otherwise, if a cheque book falls into the wrong hands, the bank may pay out on a forged cheque, leaving behind a tangle of legal liability which no-one wants to have to sort out at a time of bereavement.

If the bank is served with a garnishee order (see p. 139), the bank account is frozen and no payment can be made without a court order.

If an insolvency petition is presented against a customer, the bank must not pay out, if notified of the petition.

Similarly, if a notice of a company's liquidation has been given, the bank's duty to honour the company's cheques stops.

Debt collection

As a debt is such a valuable type of property, it is therefore surprising how many businessmen, builders included, fail to understand the importance of efficient collection of debts. By this I mean the prompt rendering of accounts, the issuing of reminders and the following up of unpaid bills. It is essential to have an efficient book keeping system and a trained and experienced book keeper. Computer accounting will of course maximise efficiency. Many businesses fail, particularly those of 'small' builders, because the builder tries to do all the work, including that of the book keeper. His main concern should be to maximise his time doing the job he is trained for, building. By delegating the paperwork to other appropriately qualified and experienced workers, he can relieve himself of anxiety, and can concentrate on making his business prosperous. The wages of such workers may reduce his own salary to start with, but in the long run, the firm will be more financially stable, with a steady stream of work and a steady flow of income.

In which courts does one sue?

Debts are civil liabilities which must be recovered in the civil courts, i.e. in the Queen's Bench Division of the High Court, or in the county court. (For the criteria to determine in which court to sue, see p. 29.)

The procedure in the High Court is quite complex despite recent reform, and when suing for large sums of money, it is advisable to seek the assistance of your solicitor. Similarly, when suing for sums over £3,000 in the county court, it is better to use your solicitor. But, if suing for sums less than £3,000, there is a SMALL CLAIMS PROCEDURE which is an arbitration procedure designed for the layman to 'sue-it-himself'. Advisory booklets and the necessary forms are available from the local county court offices and the procedure is relatively simple and informal. The address of your local courts are found under COURTS in the telephone directory.

It should be noted that before actually commencing any action in court, a final letter setting out your claim, whether written by yourself or from your solicitor, will often achieve the effect of encouraging payment. The letter should state the details of the debt, e.g. the work involved, the amount required, dates and other relevant details and that, if the debtor fails to pay within a specified period such as 14 days, then court action will be commenced. It may also be advantageous to point out that if judgment is made against the debtor he will then find it extremely difficult to get credit in the future.

Course of action

Debts are **liquidated** amounts, i.e. they either are fixed or can be determined by simple arithmetic calculation. This means that in both the county court and in the Queen's Bench Division a method of action can be used where, if the defendant fails to put in a defence, then judgment can automatically be entered against him. These are called **default actions**, as judgment can be given in **default of defence**.

1. County courts. Certain documents must be filed with the county court, i.e. a **request for summons** and **particulars of claim** with copies for each defendant and a fee must be paid. On receipt the court office enters a plaint onto the court records.

The office then arranges service of a **summons** on the defendant, together with a copy of the **particulars of claim**, so that he knows for what he is being sued. Attached is a reply form. Service is usually done through the post, but it can be arranged for a **bailiff**, a court officer, to deliver the summons in person (this sometimes also encourages payment). A plaint note is sent to the plaintiff.

If, as is usual in debt cases, there is no defence, the debtor will not bother to file one, and at the expiration of 14 days from the date of service, the plaintiff, i.e. the creditor, will be able to file another document, pay an additional fee to enter judgment automatically. This also applies to the **small claims procedure**.

2. Queen's Bench Division. A slightly different method is used to start the action. The plaintiff's solicitor will prepare a **writ** in duplicate and will send the writs to court for **issuing**, i.e. to get the action on the court record. The writs are then sent back, duly stamped, and the solicitor then has four months in which to serve the writ on the defendant. Until a few years ago, it was common to serve writs in person, but the rules have now been changed, so that most are now served through the post (see Figure 1.3).

The writ will either have a **statement of claim** actually written on it or there may be only a brief outline of what the claim is all about, in which case a separate **statement of claim** has to be served on the defendant. With the writ is served a form called an **acknowledgement of service** which also contains a guide on how to complete the form and the penalties which attach, if it is ignored.

If the acknowledgement is sent back to the court indicating that the defendant does not wish to defend, or if he ignored the acknowledgement altogether, then judgment can be applied for and entered automatically.

If there is a defence

If in either court the defendant feels that he has a good reason for not paying the bill, then he is well advised to immediately seek legal advice, as a defence or possibly even a **counterclaim** will need to be filed and a court hearing will be necessary to settle the dispute. This is very common in building disputes, where one party has not paid another for some good reason.

Enforcing judgment

Where judgment has been awarded against the defendant, either automatically or by a judge (or District Judge in a small claims case) the problem remains: how is one going to recover the money from the judgment debtor?

Before bothering to pay any more court fees to enforce judgment, it is worthwhile seeking an **oral examination** of the judgment debtor first.

The debtor is summoned to attend before a District Judge or Master, to answer **on oath** questions concerning his assets and liabilities. If he fails to attend the court, he will be in contempt, and can be appropriately dealt with as the court thinks fit. At worst he could be committed to prison.

The advantage of this interrogation is to find out what property belongs to the debtor, and whether he has a job, as some methods of enforcing judgment will not be appropriate. Very often, the debtor will make an offer to pay at this stage. Even if he offers only instalments, this may nevertheless be quite acceptable. It is better to make sure the money comes in, even if it is in 'drips and drops'. Occasionally, the enormity of having to attend court will make the debtor pay the whole sum at the oral examination. Enquiry agents (private investigators) can also be used to find out such information.

Methods of enforcing judgment

High Court – Queen's Bench Division

1. Writ of fieri facias (Fi.fa), execution against goods. This writ is obtained by the plaintiff on payment of a fee, ordering the sheriff's officers to enter the debtor's premises and to seize goods to the value of the debt plus costs. The goods are usually then sold and the creditors reimbursed. The sheriff's officers can take anything belonging

to the debtor (one of the reasons for the oral examination) but not his tools of trade, clothing or bedding, fixtures or third-party goods, nor land or equitable interests in chattels.

2. Garnishee order (attachment of debts). At the oral examination, the debtor will have revealed whether he is owed money. His main debtor will usually be his bank, providing his account is in credit. The garnishee order is obtained from the court and sent to the bank ordering it to pay the debt directly to the court. On receiving the order, the bank account is in effect 'frozen' and the plaintiff can then ask for payment from the court.

3. Charging order. The plaintiff can request this order from the court which 'charges' certain property belonging to the debtor. Once the charge has been placed on the property, it acts as security for the payment of the debt; the charge is only removed if the debt is paid. If it is not paid within a specified period, the property can then be sold to satisfy the debt. It is rather like the court ordering a house to be mortgaged.

If the charge is made against land, e.g. the debtor's house, then this should be registered as a land charge at the Land Registry (see p. 104).

A charging order may be made against the debtor's stocks and shares, and even his money.

4. Attachment of earnings. At the oral examination, it will have been ascertained whether or not the debtor is in employment. If he is, then the creditor can apply to the **county** court for an order to the employer to deduct money from the debtor's salary or wages in instalments.

5. Writ of sequestration. This can only be used in relation to the non-performance of an injunction or an order for specific performance. In this case, the writ directs up to four nominees, one of whom is usually the sheriff's officer, to enter the debtor's land. They then receive any rents, pensions or other income due from this property, and they may even take charge of his goods.

Their presence, and the right to continue to receive the income, continues until the debtor is no longer in contempt.

6. Appointment of a receiver.   A receiver may be appointed to take charge of the debtor's property until such time as he pays the debt. Usually, this means that the debtor's affairs are in chaos and need to be properly looked after.

County court

1. *Warrant of execution*
   This is similar to writ of fieri facias (see above)
2. *Attachment of earnings*
   (See above)
3. *Charging order*
   (See above)
4. *Garnishee order*
   (See above)

Bankruptcy

This is not a true method of enforcing judgment, but it can be used to recover part (or occasionally all) of the debt owed. Also, the threat of making a person bankrupt may sometimes galvanise the debtor into action, and he may actually pay his debts.

Relevant law
The law is found in the Insolvency Act 1986 Second Group of Parts and supplemented by the Company Directors Disqualification Act 1986 together with the Insolvency Rules 1986 and precedents made in relation thereto. Only humans may become bankrupt. However, the Insolvency Act (the Act) brought together the law for both individual and corporate insolvency. Thus, in business a sole trader whose business becomes insolvent may have insufficient personal funds to pay business debts and be forced into bankruptcy. Similarly, partners could be forced into bankruptcy if there are insufficient partnership funds to pay their debts. In the case of bankrupts the aim is to free them from their debts and enable them to start afresh. Registered companies, on the other hand, are wound up on insolvency and thus cease to exist (insolvency means that a person is unable to pay his debts when required to do so).

What is bankruptcy?
A bankruptcy is a court procedure whereby the property of an insolvent person is administered under the authority of a court and distributed between his creditors.

During bankruptcy, the bankrupt is subject to a number of disqualifications depending on his position and status. For example, no bankrupt may be appointed to be a justice of the peace or a mayor or local authority councillor. Nor may someone be elected to Parliament. If a director of a company becomes bankrupt he becomes disqualified from office and cannot take part in the management of the company except with leave of the court. His bankruptcy, however, protects the debtor from further legal proceedings.

The bankrupt will be automatically discharged from bankruptcy three years later, which will release him from his obligations incurred prior to bankruptcy and will, in most cases, restore his old position, freed from any disqualification he may have suffered.

Voluntary arrangements
The Insolvency Act 1986 encouraged the use of the voluntary arrangement. Under Part VIII Ss. 252–263 and the Insolvency Rules 5.1–5.30 the person in debt has an opportunity to avoid bankruptcy by proposing to his creditors that they come to a voluntary arrangement between them whereby he agrees to pay off his debts at mutually convenient intervals. Such agreements have always been and are still available quite separately from the procedure under the Act. But this procedure is now more usual and is outlined briefly here.

The debtor may himself make such a proposal, whether or not he is an undischarged bankrupt. Also, on the hearing of a bankruptcy petition, the court may itself suggest that a voluntary arrangement is more suitable and refer the debtor to an insolvency practitioner (a lawyer who deals in this type of law) for a proposal to be made. Even before the proposal is put to the creditors, a debtor can request an interim order under S.252 of the Act which will stay any bankruptcy or other legal proceedings against his property. This takes the pressure off him so that he can

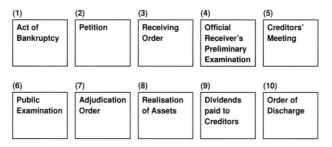

Figure 5.16 Ten stages of bankruptcy

prepare his case properly before his meeting with his creditors.

In his proposal the debtor must state why a voluntary arrangement is a good idea and why the creditors are likely to agree to the arrangement. The proposal must contain details of the debtor's assets and an estimate of their values, the extent, if any, to which they are charged (mortgaged) in favour of the creditors and the extent (if any) to which certain particular assets will be excluded from the voluntary arrangement; property to be included; amount of the debtor's liabilities; what guarantees he has given; proposed duration of the arrangement; dates of and amounts of distribution; amount to be paid to the nominee (the Insolvency Practitioner); banking and investment proposals; credit facilities; name, address and qualifications of the supervisor of the arrangement (that he is a qualified and proper nominee). Following the agreement between the debtor and Insolvency Practitioner nominee, the interim order (see above) should be sought. The nominee then makes a report to the court recommending and giving details of the creditors' meeting. The nominee notifies all the creditors sending them, *inter alia*, copies of the proposal and a claim form for them to state how much they are owed. The proposal may be approved, approved with modifications or rejected. Generally a resolution must be passed by at least a three-quarters majority. The voluntary arrangement is registered in a Register of Voluntary Arrangements which is open to public inspection.

The petition

Bankruptcy petitions may be presented to the High Court and to certain county courts provided the debtor is domiciled or present in England and Wales at the date of the petition or has been ordinarily resident or carried on business here for the previous three years.

Petitions may be presented by a creditor, the debtor himself, the supervisor of a voluntary arrangement or any person bound by the arrangement or, if there have been criminal proceedings, by the Official Petitioner or anyone specified in a criminal bankruptcy order. Before presenting a petition, a check should be made at the proposed court, to ascertain whether it has the jurisdiction to deal with the matter.

Creditor's petition

In a creditor's petition, the creditor must state:

1. that the debt is more that £750 (the bankruptcy level);
2. that the debt is a liquidated sum, i.e. fixed either by the contract or on judgment, and is unsecured, e.g. by a mortgage (or if secured the creditor must surrender the security);
3. that the debtor appears either to be unable to pay his debts or to have no reasonable prospect of paying them and there is no outstanding application to set aside a statutory demand in respect of the debt (S.267(2)(c)).

A statutory demand is a demand in a prescribed form under the Act requesting the debtor to pay the debt or secure or compound for it to the creditor's satisfaction.

Certain statutory proofs are necessary to show that the debtor appears to be unable to pay his debts.

If it appears to the court that the debtor will be able to pay his debts or he has offered to and can secure or compound for a debt and the offer has been unreasonably refused, then the court can dismiss the petition.

A creditor's petition must be supported by an affidavit sworn by the creditor or someone who knows the facts.

Debtor's petition

The debtor's petition can only be founded on the ground that he is unable to pay his debts (S.272(1)). He must provide a statement of affairs with details of his assets and liabilities. For 'small bankruptcies', i.e. where the unsecured debts are less than £20,000 and the debtor's estate appears to be £2,000 or more, the court may prefer to order an inquiry into the debtor's affairs by an insolvency practitioner whose report may then recommend a voluntary arrangement (see above). An interim order may then be made (see above). If these are not appropriate then a bankruptcy order would be made.

Supervisor's petition

This would occur where the debtor has failed to carry out his responsibilities under the voluntary arrangement he made with his creditors or gave false information to the creditors (S.276).

Criminal bankruptcy order

This would only occur where someone has committed an offence and has been found guilty in the Crown Court. Such an offence must have caused damage or loss of more than £15,000 and the court then makes a criminal bankruptcy order. The Official Petitioner or anyone named in the order may in such case petition the Bankruptcy Court which must then make a bankruptcy order.

The bankruptcy order

At the hearing of the petition, if the court is satisfied as to the truth of the facts contained in the petition, then it may make a bankruptcy order for the protection of the debtor's estate or make a bankruptcy order but issue a certificate for the summary administration of the estate if it is a 'small bankruptcy' (see above) or make a

bankruptcy order but appoint the former supervisor of the voluntary arrangement as trustee (provided he had given notice of his intention to do this in the petition) or appoint an insolvency practitioner to make a report.

The bankruptcy order makes the Official Receiver (a Department of Trade official) the receiver of the debtor's property, without whose permission (or that of the court) no action can be taken against the debtor. The debtor, from this point, has **no** control over his own property.

The court sends a copy of the order to the Official Receiver. The Official Receiver may give notice to a local newspaper and must register the order under the Land Charges Act 1972 with the Chief Land Registrar (see p. 107). He also sends a copy of the order to the debtor. He may also give notice to the *London Gazette*.

The bankruptcy order may be rescinded in a number of situations, the most obvious being if the debts are paid in full or if a voluntary arrangement has been made.

During the period between presenting the petition and making a bankruptcy order, the court can appoint the Official Receiver as interim receiver, having control of the debtor's property and being able to collect rents or other income of the property. He is not a trustee unless and until the debtor is adjudicated bankrupt. Alternatively, if an insolvency practitioner has been ordered to make a report then he could be appointed interim receiver. Also the debtor, creditor or appointed insolvency practitioner can apply for an interim practitioner to be appointed. The applicant must be supported by an affidavit (a sworn statement) giving the grounds for the application, whether the Official Receiver has been informed where a voluntary arrangement has been made and an estimate of the debtor's assets. A copy is sent to the Official Receiver so that he can attend the application. If the court is satisfied, it can make any order it wishes. In particular it can specify what proposals must be dealt with by the interim receiver. The interim receiver's appointment continues until either the petition is dismissed or one or other of the orders is made at the hearing of the petition or on application.

If it was not his petition, within days of the bankruptcy order the debtor must submit a **statement of affairs** to the Official Receiver.

The statement must contain details of the debtor's assets and liabilities, the creditors' names and addresses and whether they hold any security for their debts. The statement must be supported by an affidavit swearing that the statement is true.

The Official Receiver gives notice to all the creditors of the first **creditors' meeting**. Details of the meeting must be advertised in a local paper and at least 21 days' notice given (see Figure 5.17).

First creditors' meetings: S.293

The creditors must prove their debts, i.e. by sending a form to the Official Receiver or to the trustee in bankruptcy, with an affidavit verifying the debts mentioned in the form if the Official Receiver requires an affidavit or in other cases an unsworn claim. If necessary, evidence of the debt may be required.

A debtor who has not done so is unable to vote at any creditors' meeting. Creditors may vote in person or by proxy which must be in a prescribed form.

Resolutions are normally passed by simple majority, i.e. they are ordinary resolutions. Some resolutions require a simple majority of creditors, others a 75 per cent majority in value of creditors as well, i.e. a special resolution. These are required for such matters as removing a special manager who was appointed by the Official Receiver to deal with the debtor's estate or business.

The main purpose of the first creditors' meeting is to appoint a trustee. Additionally, a creditors' committee can be set up.

Public examination

A public examination of the debtor may be held at court after the bankruptcy order has been made if at least half the creditors require this.

The examination requires the attendance of the debtor unless, for example, he is too ill to attend. Failure to attend is an arrestable offence. The public examination may be dispensed with in certain circumstances.

The creditors or their agents, proxy holders or holders of a creditor's power of attorney may attend the examination and question the debtor about his affairs.

The debtor answers on oath and must answer the questions, even if they may incriminate him.

Realisation of assets

The trustee must then set about realising the debtor's assets with the intention of repaying the creditors (hopefully in full). The law regards the trustee as the **owner** of the debtor's property and he must therefore act in good faith and

- may sell the property privately or by auction
- take receipt of debts owed to the debtor and
- take action to recover debts owed to the debtor: S.287.

The trustee may also take other action with the permission of the committee of creditors, e.g. refer a dispute to arbitration or carry on the debtor's business with a view to winding it up.

The property which may be used to pay the creditors obviously includes money, goods, freeholds and leaseholds in land and easements and profits belonging to the bankrupt at the commencement of the bankruptcy: S.436 and S.283.

To protect the parties to the JCT Standard Form of Contract, Cl. 16.2 deals with goods stored off site, such as at the builder's premises, in the following manner: 'Where the value of any materials or goods intended for the works are stored off-site, has in accordance with Cl. 30.3 been included in any interim certificate under which the amount properly due to the contractor has been paid by the employer, such materials and goods shall become the property of the employer and thereafter the contractor shall not, except for use upon the works, remove or cause or permit the same to be moved or removed from the premises. . . .'

Thus, on payment under an interim certificate relating to unincorporated material, the ownership of such goods becomes that of the employer, even though possibly stored at the builder's own premises. Clause 16.2 continues thus: 'but the contractor shall nevertheless be responsible for any loss thereof or damage thereto and for the cost of

---

THE INSOLVENCY ACT

1986

In Bankruptcy

In the Uptown County Court   No.123 of 1997

RE: COSMO DE'ATH carrying on business as a funeral director under the name of Cosmo Coffins from and residing at 222, Death Row, Uptown, Somercester. The First Meeting of Creditors of the bankrupt, against whom a Bankruptcy Order was made on 1st April 1997, will be held at 11 a.m. on 15th August at my address.

Date 15th July 1997

L.M. Arkenshaw

Official Receiver

Hanley House

High Street

Uptown UP1 3SC

Figure 5.17   Notice of first creditors' meeting

storage, handling and insurance of the same until such time as they are delivered to and placed on or adjacent to the works. . . .'

Thus, if such goods are lost to the employer through the bankruptcy of the builder, he may prove for these goods as a debt.

It should be noted that under Cl. 30.3: 'the amount stated as due in an interim certificate may in the discretion of the architect/supervising officer include the value of any materials or goods before delivery thereof to . . . the works . . . provided

30.3.1) the goods are intended for incorporation in the works
30.3.2) nothing remains to be done to the materials . . .
30.3.3) the materials have been and are set apart at the premises where they have been manufactured or assembled or stored and have been clearly and visible marked . . . so as to identify the employer, where they are stored on the premises of the contractor, and in any other case the person to whose order they are held and their destination as the works.'

In addition to the property mentioned above, the trustee can exercise certain statutory powers and **reclaim** certain property which the bankrupt has already transferred. These include transactions defrauding creditors (Ss. 423–425) or transactions of property at an undervalue (S.339). These are where the bankrupt has transferred property intending to avoid paying his creditors or with the intention of preferring one creditor over others (S.340).

The trustee in bankruptcy may also **disclaim onerous property**: Ss. 315 and 321. Such property must be a burden to the bankrupt and of no benefit at all to him. Thus, a lease subject to onerous covenants could be disclaimed, as could unprofitable contracts. A builder may, for example, have attempted to take on uneconomic contracts purely in an attempt to improve his cash flow situation. Disclaimer by the trustee would operate to terminate the bankrupt's rights and duties under

the contract and no further work would need to be done.

The trustee in bankruptcy does not acquire rights over the following:

- the bankrupt's tools of trade including books, vehicles and anything needed by him for work;
- his or his family's clothes, bedding, furniture and household equipment up to a value of £250;
- property given to the bankrupt **until** he became bankrupt – this will usually occur under a will or a settlement, e.g. if Blackacre Farm is given to Joe for life until he becomes bankrupt and thereafter to Mark on Joe's bankruptcy, Mark gets the farm as the bankruptcy accelerates Mark's interest;
- old age pensions and other state benefits;
- property held by the bankrupt as a trustee.

The court must make an order before a bankrupt's income can be used to pay creditors and in so doing they must take into account the requirements of the bankrupt and his family.

Thus, contracts do not terminate automatically on bankruptcy but, if they are onerous, may be disclaimed by the trustee in bankruptcy.

Because the employer is particularly vulnerable if a builder goes into bankruptcy (or if a company is wound up through insolvency) most contracts should contain some clause to cover such eventualities.

Clause 27.2 in the JCT Standard Form of Contract states: 'In the event of the Contractor becoming bankrupt or making a composition or arrangement with his creditors (or if a company, being wound up) . . . the employment of the Contractor under this Contract shall be forthwith **automatically** determined but the said employment may be reinstated and continued if the employer and the Contractor, his trustee in bankruptcy, . . . receiver, . . . shall so agree.'

Clause 27 continues, to allow the employer to employ other persons to complete the works where the contract has been determined under Cl. 27.

Payment to creditors

Any creditor wishing to claim in a bankruptcy must have first proved his debt in the required form (see p. 143).

Interim payments called dividends are paid at frequent intervals pending a final dividend which is paid when all the bankrupt's property has been sold. Ideally the creditors will be paid in full. However, the very nature of insolvency indicates that there will not be enough money to be shared among the creditors. For this reason there is a statutory order of priority for the payment of debts. This does not take into account the debts of a secured creditor, who on non-payment of his debts may take his security to satisfy the debt, e.g. sell the mortgaged property. Any money surplus to the debt is paid into the estate to help pay the other creditors.

The statutory order of priority is (mnemonic A-SPOD):

(i) administration expenses
(ii) specially preferred debts
(iii) preferred debts
(iv) ordinary debts
(v) deferred debts.

Administration expenses

This includes the expenses of the Official Receiver and trustee in bankruptcy, and must be paid in full (S.324(1)).

Specially preferred debts

These must be paid in full:

(i) funeral and testamentary expenses of a deceased insolvent debtor;
(ii) property belonging to a friendly society or trustee savings bank in the possession of the bankrupt who was an officer of the society or bank.

Preferred debts: Schedule 6

Provided the above have been paid in full, the following debts must be paid equally without preference. If there are insufficient funds to pay all the preferential debts then they must be paid a proportion (i.e. they must be abated) (S.386). These include:

1. Taxes deducted out of wages or subcontractor's remuneration under S.549 Income and Corporation Taxes Act 1988 by the bankrupt as employer under PAYE schemes for 12 months prior to the bankruptcy order.
2. Wages and salaries owed to employees for four months prior to the bankruptcy order not exceeding a sum set by the Secretary of State.

    By virtue of the Employment Rights Act 1996 additional sums may be claimed from a bankrupt employer for such matters as paid time off, guarantee payments, pay while suspended on medical grounds (see p. 231) from the Insurance Fund, if necessary (see p. 237).
3. All holiday pay due to an employee prior to the bankruptcy order.
4. Social security contributions due to the DSS from the bankrupt as employer.

Ordinary debts

These are all the remaining unsecured debts other than deferred debts, including unpaid corporation tax, income tax or capital gains tax.

Mention should be made of the unusual position of landlords who, although mere unsecured ordinary creditors, are allowed to **distrain** upon (take possession of) goods of the bankrupt to cover the amount of rent owed. On discovery that the tenant is insolvent, he may seize the tenant's goods at any time whether or not a bankruptcy order has been made. However, if he distrains after the commencement of bankruptcy, he is only entitled to claim six months' rent due prior to the date of the adjudication of bankruptcy by this method. If any further sums become due after adjudication, the landlord may then exercise his right of distress (i.e. he can distrain) for all new sums accruing (S.347). He cannot distrain after the bankrupt has been discharged. The Collector of Taxes also has a right of distress regarding unpaid taxes under the Taxes Management Act 1970.

Deferred debts

These may only be paid after **all** the other debts have been paid in full.

1. A loan to a partnership whereby the interest is paid by reference to varying profits of the business: S.2 Partnership Act 1890.
2. Money due from the sale of goodwill of a business in the form of a share in the profits: S.3 Partnership Act 1890.
3. A loan by a spouse: S.329.
4. Claims arising under S.6(3) or S.1(3)(a) Financial Services Act and S.49 Banking Act.

Criminal offences: Ss. 350–362

Until the bankrupt is discharged there are many actions which, if committed by him, will amount to a criminal offence. The following are examples.

Under Ss. 350–362 Insolvency Act 1986

- Failing to disclose all of his property to the trustee or official receiver: S.353.
- Concealing or destroying or falsifying any document or making a false entry: S.361.
- Misleading the creditors by false representations so as to obtain their agreement with reference to his affairs: S.356.
- Obtaining credit for more than £250 without informing the lender of his bankruptcy: S.360.
- Carrying on a business or trade in a name different from the one under which he is made bankrupt without telling clients/customers of his bankruptcy: S.360(1)(b).

- In the period of two years prior to the petition materially contributing to or increasing the extent of his insolvency by gambling or by rash or hazardous speculation: S.362(1).

In addition to the above there are many other offences that may be committed regarding the bankruptcy. For example, a bankrupt may also have committed criminal offences before or after a bankruptcy order by obtaining property by deception, obtaining a pecuniary advantage by deception under the Theft Act 1968 as amended.

Discharge

A bankrupt may apply to court for his own discharge at any time after being adjudicated bankrupt. The court will not grant the discharge in certain situations, e.g. if the bankrupt has committed an offence (see above).

Under S.279 of the Insolvency Act 1986, the bankrupt is entitled to be automatically discharged three years from the bankruptcy order or, if the estate was being administered summarily, two years.

Once discharged, the bankrupt is free of nearly all his debts, i.e. those which could be proved in his bankruptcy. Certain debts will still be enforceable, e.g. ones incurred by fraud: S.281(3).

6 Legal personality

The concept of legal personality

In English law, only 'legal' persons have rights and duties in law. Thus, whilst animals can benefit from our law, such as protecting them from being mistreated, no duties can be imposed on them, only their owners, for they are not legal persons.

Types of legal persons

The are only two types of legal persons (see Figure 6.1): **natural** and **juristic** (or **artificial**) persons. Natural legal persons are of course human beings. Juristic or artificial are **corporations** (from corpus – a body).

Humans

Potentially, humans have full rights and duties, unless they become insane or bankrupt. Whilst under age, i.e. under 18, they have restricted rights and duties, e.g. they cannot vote, cannot sit on juries, can only enter into certain contracts and cannot marry without consent (under 16 they cannot marry at all). Bankrupts also have restricted rights (see p. 140).

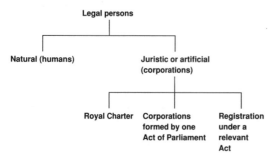

Figure 6.1 Types of legal persons

A person operating a **business** on his own is said to be a **sole trader**. He is self-employed and he and the business are one and the same thing. If the business does well, so will he in his private capacity. If the business should fail, however, then he may become bankrupt, as his private money may have to be taken to pay his business debts.

Being self-employed, he will not enjoy as many benefits of the welfare state as would an employed person, such as sickness benefit. He should therefore take out adequate insurance cover to provide income for himself and his family should he become ill, have an accident or die (see p. 274).

On his death, the business will usually come to an abrupt end. This can cause great hardship for his next-of-kin, his employees, whose jobs disappear overnight, as well as persons who had entered into contracts with the deceased, and who now have little hope of the contract being performed. The deceased's will must also be very clear as to what should be done with the business. His executors have no implied rights to do more than carry on trade for a short time with a view to selling the business as a going concern. If the deceased left no will, then as the administrator (usually the surviving spouse) cannot take out a grant of letters of administration for at least a fortnight, a fortnight will pass by with no-one having the authority to do anything in relation to the business.

Corporations

These can be defined as **artificial legal persons**. Examples of corporations include the monarchy, nationalised industries if any, limited companies, building societies, local authorities and the Chartered Institute of Building.

Corporations have all the general rights and duties which a sane adult would have, except those which are impossible because they are inanimate. Thus, when entering into a contract with a limited company, even though the negotiations are made with a shop assistant, for example, the actual contract is made with the company, as *it* is the legal person.

Special characteristics of corporations

Separate entity
Corporations are thus said to have a **separate legal personality** to that of the persons who set it up or its shareholders. As it is a separate legal person, one should not strictly talk of 'owning' a registered company as you cannot own people, although this is an expression commonly used. A person who has a major shareholding in a company should be described as such. But of course it was always common in family businesses for one shareholder to hold say 99 per cent of the shares and one can understand why it was said that he 'owned' the company. The law has now been changed to acknowledge the factual position by allowing a **private** company to be formed by only one person (S.1(3A)). The concept that the company and the shareholders are two entirely separate entities was first put forward in the leading case *Salomon* v. *Salomon & Co. Ltd* (1897).

When does it die?
Unlike humans, a corporation does not die unless some step is taken to bring it to an end. In the case of chartered corporations this would be on revocation of the charter. In the case of registered companies, this would be on winding up, or on being struck off the company register. Statutory corporations would require the repeal of the Act that created it. (Compare this situation with sole traders on p. 148 and partnerships on p. 157.)

Owns property in its corporate name
A corporation may acquire and own all types of property in its own name, e.g. Bloggs Ltd will own land and not Mr and Mrs Bloggs, even though they may be the only shareholders in the company.

Contracts are in its own corporate name
A corporation enters into contracts in its own name. Examine your contract of employment and you will see that your employer, if a company or local authority, is referred to by its corporate name.

Necessity for agents
Corporations have to act through human agents because they have no physical existence. These are usually called directors.

Liability in law
Corporations can be prosecuted and can sue or be sued in their correct corporate name. Thus, for example, prosecutions under the Health and Safety at Work etc. Act 1974, will usually be against the company and not against its personnel, unless there is an exceptional reason. It is of vital importance for legal proceedings to be undertaken using the correct corporate name and this should be ascertained before commencing legal action. In the case of registered companies, a search can be made at either Companies House, London, or at the Registrar of Companies in Cardiff (see p. 153). But note the comments concerning business names on p. 153.

Powers of the corporation
Chartered corporations are similar to humans and have the right to enter into any sort of contract or do anything, provided it is not illegal or contrary to public policy.

On the other hand, a statutory corporation or one created under the Companies Act can only do what it is authorised to, either by its governing Act of Parliament or the objects clause of its memorandum of association (see p. 153).

It used to be the rule that anything done outside its authority was deemed **ultra vires**, i.e. beyond its powers and **void**. A void act gave the parties **no** rights and duties in law. So, at one time if your company contracted to do something which they had no power to do, and the other party failed to pay for this work, your company could not sue because the basic contract would be void through being ultra vires. Conversely, the other party could not force your company to perform its part of the agreement.

Two things have changed this rule. First, most of the authority given to corporations is drawn in the widest terms, to ensure that all their acts are valid. Second, in relation to registered companies, under S.35(1) as substituted by the 1989 Companies Act, 'The validity of an act done by a company shall not be called into question on the ground of lack of capacity by reason of anything in the company's memorandum.'

Should one require knowledge of a company's memorandum of association this can be gained by a search at Companies House. The register is recorded on microfiche and a small sum has to be paid. Otherwise, a company search agent can be employed to do the search for you.

Methods of creating modern corporations

This process is called **incorporation**. There are three methods: by Royal Charter, by an individual Act of Parliament or by registering under an appropriate Act.

Royal charter

These may be granted by the Crown, after a petition has been made requesting incorporation by charter. This method is now used for educational and professional bodies such as the Chartered Institute of Building, the Royal Institute of British Architects, the British Broadcasting Corporation, the Bank of England, universities, local authorities who have borough status etc. (see Figure 6.2).

Special Act of Parliament

Obviously, for a proposed corporation to have its own Act of Parliament, it must be of great public importance to warrant the time spent in Parliament debating the Bill. The normal rules relating to direct legislation apply. Examples of corporations created in this way included all the old nationalised industries as well as most local authorities. Thus, the Coal Authority was created by the Coal Industry Act 1994, and local authorities were created by virtue of the Local Government Act 1972. These types of corporations are sometimes referred to as public corporations.

Note: the last government had a **privatisation** policy and corporations such as British Telecom

Figure 6.2 Royal Charter of the Chartered Institute of Building (photographer Kingsley Jones, courtesy of the CIOB)

were taken out of the public sector and made a registered company in the private sector. An Act of Parliament was necessary. Because of its size and importance and because the shares were offered to the public, it became a **public limited company**. Thus, a public **limited company** *is* a corporation, but not a **public** corporation.

Registration under Acts

The most common sort of corporation created by registration is the registered **company**. The relevant Act is the Companies Act of 1985 as amended. Building societies are created by registration under the Building Societies Act 1986 as amended.

As 'ordinary' people can set up registered companies, and these are the most important sort of business organisation, we shall examine these in more detail.

Types of registered company

It should be noted that the word company does not in itself indicate more than a collection of people gathered together for a common purpose. Indeed, the word used in a name of a business is misleading, e.g. Jones & Company Builders, or Jones & Co. Builders. Both of these examples are *not* limited companies (see below). They will probably be either sole traders or partnerships (although they could also be unlimited companies, but these are unusual).

Companies registered under the Companies Act may be either unlimited liability or limited liability companies. The liability is that of the members (shareholders) and refers to whether or not they have to pay the company's debts if the company is unable to do so.

Unlimited liability

Here, the members have to pay the debts of the company if its assets are insufficient. There is **no** limit on the amount they have to contribute until all the debts are satisfied. The company has the ordinary characteristics of a corporation, however. Also, there are a number of other special advantages. For example, there is no necessity to publish accounts annually.

Limited liability by shares

Here, the members' or shareholders' liability is limited to the amount they have left unpaid on their shares. This is best illustrated by an example.

Tiny Construction Company Ltd has been set up with a share capital of £1,000 divided into £1 ordinary shares. There are four members holding a quarter of the shares each. Three of them have paid for their shares immediately; the fourth, just out of college, has been allowed by the others to pay only half.

Nominal Share Capital	£1,000 in £1 shares			
Issued Share Capital	£1,000	Tom	250	shares
		Dick	250	shares
		Harry	250	shares
		Ringo	250	shares
			1000	
Paid-up Share Capital	£875	Tom	250	
		Dick	250	
		Harry	250	
		Ringo	125	
			875	
Uncalled Share Capital	£125			

The company never becomes profitable and ceases trading owing substantial amounts to many creditors.

Tom, Dick and Harry lose all the value of their shareholding, but **no** more, as their liability was limited by shares. Ringo, on the other hand, is called upon to pay the extra £125, on top of losing the value of the £125 he has already paid. But, he still benefits from the limited liability in that he need contribute no more than £125.

It should be remembered that the company **must** pay its debts – if it can. But, if it cannot, the members will *not* have to put their hands in their pockets, unless they have not paid for their shares.

For this reason, all 'limited' companies must warn potential clients and creditors of the members' limited liability, by incorporating the word limited into its name. Public limited companies must, under the Act, go further. They must incorporate the words public limited company or plc or the Welsh equivalent into their names. Certain companies can seek permission from the Department of Trade to be allowed to drop the word limited, provided the objects of the company are for scientific, commercial, artistic, religious or charitable purposes.

Limited liability is of great advantage to the shareholder. He will not necessarily go bankrupt if the company in which he has invested has gone into liquidation.

If he has invested a small amount in a large public limited company, then he will have to face the loss of that money. However, if he has set up his own company with, say, his wife as the other shareholder, when it comes to borrowing money, in practice banks or other financial institutions will usually insist on personal guarantees or second mortgages on his personal property to overcome the limited liability. Thus, because of this private security, he could still go bankrupt if the company failed.

Liability limited by guarantee

Here, the members guarantee that if the company cannot pay its debts at a particular time, such as winding up, then they will pay a guaranteed amount each. Their liability is thus limited to the amount guaranteed and they need pay no more. They are often used to form clubs, charities etc. as no business capital is required.

Registered companies classified as public limited companies or private companies

Under the Act, registered companies must now be classified in a way which is more akin to company law in the EU.

Public limited companies

A public company is a company:

- limited by shares or guarantee but with a share capital;
- that must state in its memorandum of association that it is a public company;

- whose nominal value of the company's issued shares must be at least £50,000 (at present), a quarter of which must be paid up;
- having the words **public limited company** or plc, or in Welsh **cwmni cyfenedig cyhoeddus** or CCC, in the company name;
- with a minimum of two members;
- quoted on the Stock Exchange so that the public can buy its shares;
- which starts business on issue of a registrar's trading certificate.

Private companies

A private company is **not** a public company. Therefore it follows that it may be:

- an unlimited company, *or*
- a company with a share capital of less than £50,000 *or*
- a company in whose memorandum there is no statement that it is a public company.
- It may now have only one member.

It is an offence to directly offer a private company's shares to the public (S.143(3) Financial Services Act 1986). Thus, such companies are never quoted on the Stock Exchange. Nevertheless, an **over the counter** market has developed in London, to which people wishing to purchase shares in private companies may apply. Also there is now an alternative investment market. Obviously, only those companies wishing to sell will place their shares in this way. The Public Offers and Securities Regulations 1995 now govern this area.

Family-run companies, which are probably the majority of private companies, usually restrict the sale of shares by inserting a clause in the articles of association (see p. 153). Such a restriction may require the seller to offer the shares first to members of the family or to the directors. This method prevents strangers being able to take control of a family business.

Unlike public companies, there is no longer the need for two shareholders. One will suffice.

If it is a limited company, it must have **Limited** or **Ltd** or its Welsh equivalent in its name.

Unlike a public company, it can start trading immediately on issue of its certificate of incorporation.

Formalities of registering a company

A registered company can be set up quite easily using a company agent who will supply the necessary documents and seal for a relatively small fee. He can even supply an 'off the shelf' company which is already in existence. If the name is not suitable, it can be changed to one that is. More complex requirements should be dealt with by a solicitor specialising in company matters.

The documents are registered at the Companies Registration Office, Crown Way, Maindy, Cardiff CF4 3UZ. If the requirements of the Act are complied with, the Registrar of Companies issues a certificate of incorporation. This is the company's 'birth certificate' as it becomes a legal person from that day. If it is a public company, it also requires a trading certificate in order to start business (most businesses will have already been trading as private companies or partnerships).

The most important of the documents required on registration is the memorandum of association which deals with the external affairs of the company, and the articles of association which deals with internal affairs, such as meetings, shares etc. (see Figure 6.3). The memorandum has certain compulsory clauses concerning the company name; domicile of office (England, Wales or Scotland, as it affects the relevant law); objects, i.e. the acts a company may perform; liability (limited or not); whether public or private; and the number, type and value of shares.

One important aspect of forming a company is the choice of name. The rules relating to company names are found in the Business Names Act 1985. A company cannot register a name which is identical to that of an existing company. The name chosen must not be offensive, nor must it mislead people into thinking that it is connected with the government, a local authority, royalty etc. Other legislation prohibits the use of the work 'bank' or certain charities' names. If the name is similar to that of another, whilst it may still be registered, the first company may start a **passing-off** action and sue the new company in tort if he is carrying on a similar business. The use of one's own name is usually defended by the courts even if confusion would result. The company registrar can require a company to change its name if it is too like that of another or if the name is so misleading as to be harmful to the interests of the public.

Often, companies (or any sort of business) trade under a different name for some reason. In such cases, all business stationery **must** carry the true names and addresses, e.g. if Paul Smith and Andrew Long are in partnership with the title Brickbuilders and Co. then their letters, invoices etc. must have their own names on them and the address of the partnership office.

Outsiders dealing with the business can, in writing, insist on the names and addresses of the businessmen or company name and address.

If a company wishes to change its name for any reason (often where it is associated with some financial disaster) then it may do so, by the directors calling a general meeting and proposing a special resolution. If two-thirds of the shareholders entitled to vote agree, then the name can be changed.

Other aspects of forming a registered company

Taxation
Registered companies are subject to corporation tax. The directors of the company, however, will pay normal income tax. It may be better therefore to set up business as a sole trader or as a partnership, as there are certain tax advantages to be gained for new businesses formed in this way, in which trading losses can be offset against income for, at present, the first four years of business. This income includes salaries paid for the previous three years.

Financial advice is thus absolutely necessary at the time of business formation.

Accounts
Copies of annual accounts must be filed with the Registrar of Companies. These must be properly audited in accordance with the Companies Act. They are available to the public for inspection. Abbreviated accounts may be filed in relation to small private companies.

Example of a memorandum of association for a private limited company

The name of the company is 'Thom and Jerry Limited'.

The registered office of the company will be situate in England.

The objects for which the company is established are 'to act as buyers, sellers, agents, managers and converters of real and personal property and businesses of all kinds, including the building and/or conversion of houses, flats, offices, warehouses, factories and the like and the letting thereof at such places as the company may from time to time determine, and the doing of all such other things as are incidental or conducive to the attainment of the above objects'.

The liability of the members is limited.

The share capital of the company is two thousand pounds divided into two thousand shares of one pound each.

WE, the several persons whose names and addresses are subscribed are desirous of being formed into a company, in pursuance of this memorandum of association, and we respectively agree to take the number of shares in the capital of the company set opposite our respective names

NAMES ADDRESSES AND DESCRIPTIONS OF SUBSCRIBERS

	Number of shares taken by each subscriber
Obadiah Thom 123, The Street, Splott, Surrey Builder	One
Cosmo Jerry, The Nook, Knotty Ash, Surrey Builder	One

Dated this 22nd day of May 1984.

Witness to the above signatures –
       Shirley Owen,
       63, Chemical Road,
       Trapp,
       Surrey.

Figure 6.3 Example of memorandum of association

While this may seem to be onerous, it is one way of concentrating the mind on the necessity for proper account keeping, the absence of which is so often the reason for a business failure.

Winding up of a registered company

A registered company may be wound up in three ways:

- by the court or
- by a voluntary winding up which may be either a members' or a creditors' winding up.

By the court

A petition may be presented to the Companies Court (part of the High Court) or to a county court having jurisdiction in bankruptcy for companies with a paid up share capital of less than £120,000.

There are six grounds under S.122 on which such a petition may be based:

(i)   the company has passed a special resolution for it to be wound up by the court;

(ii) the company, being a public company, has not received a certificate of ability to commence business and has been registered for more than a year;

(iii) the company has not commenced business within the year after incorporation or has suspended business for a year;

(iv) the membership has fallen below two;

(v) the company cannot pay its debts;

(vi) the court thinks that it is just and equitable to wind it up.

Inability to pay its debts

A company is deemed to be unable to pay its debts in similar circumstances to those of an individual or partnership in bankruptcy (see p. 140).

The procedure is very similar to that of a bankruptcy and the Insolvency Rules apply. Thus there is a petition; hearing in court; winding-up order; statement of affairs to Official Receiver; Official Receiver's report; creditors' meeting; appointment of a liquidator; liquidation committee; vesting of control or possession (but not ownership) in liquidator and payment of debts.

Voluntary winding up

These types of winding up are desirable as there are fewer formalities.

By S.84 of the Insolvency Act 1986, a company may be voluntarily wound up if:

- the articles specified a date or event in which the company should be dissolved and the company in general meeting passes an ordinary resolution to be wound up voluntarily;
- the company resolves by special resolution to be wound up voluntarily (no reason need be given for such a resolution);
- the company passes an extraordinary resolution that because of the company's liabilities it cannot continue in business and that it should be wound up.

Thus, if a company is insolvent, it does not have to be wound up through the courts.

The course of a voluntary winding up, however, depends on whether the company is solvent or not. If it is solvent, the company is wound up by a **members' voluntary winding up**. No-one, other than the members and a liquidator appointed by them, is involved. By S.89 a **declaration of solvency** must be filed with the Registrar of Companies within a certain number of days of passing the resolution to wind up the company. Any director making such a declaration without good reason may be imprisoned for six months and/or fined. The declaration is not necessarily to the effect that the debts will be paid immediately but that the company will be able to do so, in full, within a period, being not less than 12 months from the beginning of the winding up.

From the time of the resolution to wind up, the company must cease business unless it is necessary to be able to sell the business as a going concern; the company remains in existence, however, until actual dissolution (see below). Share dealings must of course cease as must the powers of the directors.

The liquidator appointed by the company in general meeting and who must be an insolvency practitioner holds regular general meetings, annually, giving details of the course of the liquidation. On finally settling the company's affairs, a final general meeting must be called, gazetted and notice must be given of the accounts and the decision of the meeting to the Registrar of Companies. Gazetting means putting a notice in the *London Gazette*, a publication used for many formal, legal and official announcements.

The company is then automatically dissolved three months after the registration of the returns of the final meeting.

If no declaration of solvency has been filed, then the winding up is a **creditors' voluntary winding up**. A creditors' meeting is called and gazetted and advertised in local newspapers (see Figure 6.4).

At the meeting a liquidator and a liquidation committee are appointed. Statements of the company's affairs, a list of creditors and the amount of their debts are presented. If the company had already appointed a liquidator, the creditors' appointment takes precedence although an application may be made to court for a decision if there is a dispute.

SPLODGE LIMITED

NOTICE IS HEREBY GIVEN pursuant to Section 98 of the Insolvency Act 1986 that a meeting of the Creditors of the above-named Company will be held at Splodge Lodge, 42, High Road, Uptown, Somercester, UP3 5HB, on the 9th August 1997 at 11.00 a.m. for the purposes mentioned in Section 99 to 101 of the said Act.

Creditors wishing to vote at the meeting must lodge their proxy together with a full statement of account at the registered office – Splodge Lodge, 42, High Road, Uptown, Somercester, UP3 5HB, not later than 12 noon on 8th August 1997.

For the purposes of voting a secured creditor is required (unless he surrenders his security) to lodge at Splodge Lodge, 42, High Road, Uptown, Somercester, UP3 5HB, before the meeting a statement giving particulars of his security, the date when it was given and the value at which it was assessed.

Notice is further given that a list of the names and addresses of the Company's creditors may be inspected, free of charge, at Splodge Lodge, 42, High Road, Uptown, Somercester, UP3 5HB, between 10.00 a.m. and 4.00 p.m. on the two business days preceding the date of meeting stated above.

Rodger Splodge
Director

1st July 1997

Figure 6.4 Notice of creditors' meeting

The course of liquidation is much the same as under a members' voluntary winding up (above).

The Crown

Before concluding corporations, a mention should be made of the special position of the Crown or monarchy, which is a type of corporation represented by one person only, the reigning monarch.

Traditionally, the monarch is outside or above the common law. Indeed it used to be said that the Sovereign could do no wrong. Even today in her private capacity, the Queen cannot be sued except using special legal methods.

However, in the twentieth century, the monarchy in its public capacity, responsible not only for royal households but also government ministers, their departments, the civil service, the armed forces and the police, is no longer immune from action.

The Crown Proceedings Act 1947 brought in a limited amount of liability in law for the Crown. The Crown can now be sued on its contracts and for torts committed by its employees or agents; e.g. nurses, police, government department employees etc. The correct department has to be sued, or failing this, the Attorney-General (the chief law officer of the Crown and head of the Bar).

The Health and Safety at Work etc. Act 1974, for example, specifically mentions the Crown, but limits the enforcement powers. Crown notices are served instead of normal notices and the Health and Safety Executive will take the matter further if there is no compliance.

Hospitals, however, now have to follow the requirements of the Act as a result of the National Health Service and Community Care Act 1990.

The armed forces are in a special position because of the nature of their work. Whether or not they are liable for something will depend on relevant legislation. Under the Crown Proceedings (Armed Forces) Act 1987 forces personnel may now sue the Crown for injuries caused in their work.

Unincorporated associations

These are associations of individuals from two upwards who get together for a common legal purpose other than for profit and who intend to create legal relations. They may be simple unincorporated associations such as clubs and societies who have not formed companies, or they may be special types, such as partnerships and trade unions, governed by special Acts of Parliament. An association of the above type is **not** a corporation and is **not** a separate legal entity. Thus, the club *is* the membership, the partners are the partnership etc.

Nevertheless, because one is dealing with a collection of people, the law has modified its approach towards unincorporated associations in a number of ways.

Simple unincorporated associations, e.g. The Trubshaw Hamster Appreciation Society

Builders are most likely to meet these as clients, or they may belong to one socially. These are not partnerships because they are not in business, nor are they companies because they have not incorporated themselves under the Companies Act.

The members of the society **are** the club. It has **no** separate legal personality from its members. They should have a set of rules, however, even if they are only oral. Thus, should a civil wrong be committed by the club or a member on its behalf, all the club members are liable. Similarly, if the club enters into a contract, then strictly speaking all are liable on the contract.

However, the chances of tracing all the members at a particular time, in order to sue them, is extremely difficult and undesirable. Thus, the courts allow the club to be sued (or to sue) in a **representative action**. This means that the plaintiff may trace a few members, who he then sues as **representatives** of the club. Judgment against them is judgment against the club, and if there is no club property to satisfy judgment, then the property of the representatives may be taken. The representatives, of course, can then claim back from the other club members a contribution to their loss. But this is easier said than done in practice.

Most clubs will have a **committee**. If they should do or authorise something on behalf of a club, then they will be primarily liable as if they were representatives as above.

Property of the club belongs to all the members. But in most cases, if there is more than a few hundred pounds' worth, then trustees should be appointed to deal with it properly on behalf of the club.

Anyone signing or otherwise making a contract on behalf of a club make themselves primarily liable for that contract, with a right of contribution and indemnity from the other members.

Thus, one can see that membership of a club has its drawbacks. For the builder, when undertaking work for a club, it is probably sensible to make sure that he has a sufficient advance payment before starting the job. This can also be said of small registered companies as clients, and the builder ought to check on their creditworthiness. There are firms (business organisations) which undertake enquiries into financial status, and these can be very useful in business, when dealing with an unknown client.

Partnership

Definition

A partnership is defined by S.1 Partnership Act 1890 as 'the relation which subsists between persons carrying on a business with a *view of profit*'.

So any business relationship between two or more people, including companies, which falls within the definition will be a partnership. There must be a business, at least two people and an agreement to share profits. Thus, if you and a friend agree to buy a dilapidated house, do it up and split the profits, then that is a partnership. Partnerships may be set up for a particular project (a consortium), or for a particular length of time or they may carry on for an unspecified period.

Not legal entities

A partnership is **not** a corporation and thus the partners are the business. If the business does not do well, then the partner's own private assets can be taken to pay the business debts. If the partnership is sued, then the individual partners must be sued (see the Civil Liability (Contribution) Act 1978 on p. 163). However, the courts allow actions to be brought or defended by the partnership in the partnership name: RSC Ord. 81 r.1, CCR 5 r.9.

The partners are self-employed for tax and social security purposes, and once again, adequate insurance should be taken out by each of the partners to cover not only their own life and health but also those of the other partners.

THIS PARTNERSHIP AGREEMENT is made on the thirtieth day of January 1998 BETWEEN BIGGLES SMITH of 9 Splott Lane Ewell Surrey and COSMO BLOGGS of 37 Coronation Street Ewell Surrey hereinafter called the partners

IT IS AGREED as follows:

1. The partners shall carry on business in partnership as builders under the firm name of Bloggs, Smith & Co. or such other names as the partners wish at 86 Brain Drive Ewell aforesaid and at such other place or places as the partners may from time to time approve.

2. The partnership will commence on the date of this agreement and shall continue until terminated in accordance with this agreement.

3. The partners shall be entitled to the capital and profits arising from the partnership in equal shares and they shall bear the losses equally.

4. The bankers of the firm shall be Eastbank PLC of 2, Coronation Street Ewell aforesaid. Cheques drawn in the name of the firm must be signed by both partners.

5. The usual books of account shall be properly kept.

6. The partners shall devote their whole time and attention to the business of the partnership.

7. Each partner is entitled to 4 weeks' holiday each year.

8. Neither partner shall without the consent of the other engage in any business other than that of the partnership or engage or dismiss any employee of the partnership.

9. Each partner shall be entitled to draw £400 as salary from the partnership bank account each month.

10. On April 1st each year an account shall be taken of all assets and liabilities of the partnership and a balance sheet and profit and loss account showing what is due to each partner in respect of capital and share of profits shall be prepared and signed by both partners who shall be bound by the contents unless an error is found and rectified within three months.

11. The partnership may be terminated by either partner giving to the other at least three months' notice in writing.

12. If any disputes arise as to the meaning of this partnership agreement or any rights or duties of the partners under it, such disputes shall be referred to an arbitrator to be nominated by the partners or in default of such nomination to an arbitrator to be appointed by the President of the Chartered Institute of Arbitrators. The decision of the arbitrator shall be binding on both partners.

**SIGNED AS A DEED by Biggles Smith
in the presence of:–**

**SIGNED AS A DEED by Cosmo Bloggs
in the presence of:–**

Figure 6.5 Partnership agreement

Formalities

The essence of a partnership is contractual. Thus, a partnership can be created very informally as in the example above. A verbal agreement may be sufficient or the partnership may be inferred from the parties' conduct. A person may even be estopped from denying that he is not a partner of someone else, if that is how he has been behaving.

Most sensible and businesslike people, however, enter into a written agreement which will form the basis of their working relationship. The best type of written agreement would be a proper partnership agreement which would contain terms covering all the sorts of matters which could create problems in the future (see Figure 6.5).

A partnership agreement is particularly important because, in the absence of any agreement between the parties, the terms of the Partnership Act 1890 apply and this can cause individual hardship.

Partnership Act 1890: rights and duties in the absence of contrary agreement either oral or written

Under the Act, partners are entitled to share profits and losses equally, and must make equal contributions to capital: S.24(1). In practice, most people make unequal capital contributions and therefore **must** make a contrary agreement to make sure their share of the profits is in proportion to their contribution or work-load.

Each partner is entitled to share in the management of the firm: S.24(5). Once again, a sensible partnership will have partners of different ages and experience to make sure of the continuity of the firm. By agreement, new young partners being taken on will usually not be allowed to take part in running the business.

The Act makes no allowances for a salary to be paid pending profit sharing at the end of the year. The partnership agreement should therefore alter this position: S.24(6).

If the partnership has no time limit, then any partner may give notice to dissolve the partnership **at any time**: S.32. It is important therefore to enter into a fixed term partnership which will then end only at the expiration of the period: S.26.

The death or bankruptcy of a partner will bring a partnership automatically to an end. This position should be altered by agreement, to allow the other partners to carry on the business. No allowance is given in the Act for retirement.

Liability of partnership

Each partner acts as an agent for the firm and binds the firm by any acts made on its behalf: S.5.

The old rule that only one action can be brought for a claim in contract is now altered by the Civil Liability (Contribution) Act 1978 (see p. 163). Thus, at common law, liability in contract is joint and only one action could be brought necessitating knowing all the partners names to make sure of getting judgment against all of the firm (and thus being able to enforce the judgment against all of them). Now, judgment against one partner, if not satisfied, will allow the continuance of the action or the starting of another action against the other partners. In practice, however, a representative action can be used and judgment can then be given against all the partners.

In tort the parties' liability is joint and several, i.e. they are liable together and separately.

Other aspects of forming a partnership

Certain types of businesses used **only** to be allowed to operate as partnership (e.g. solicitors, accountants, architects) as incorporation was not allowed by the Companies Act 1985. This is now permitted in certain circumstances.

Partnership names

See Business Names Act. Where a partnership has more than 20 partners, the names of *all* need not be included on their documentation. But a list of names must be kept at the principal place of business.

Ownership of property

See p. 96.

Taxation

New rules have been introduced by the Finance Act 1994 for partnerships formed and carrying on business on or after 6 April 1994 and from 1996–7 for all partnerships. Each partner must file his own tax return showing his profits from the partnership, together with a statement of the partnership income for that financial year. Good advice should be sought in order to comply with these new rules.

7 The law of torts

Types of wrongs

Crimes

There are many things which people should not do, because they are crimes, i.e. wrongs which are so bad that the commission of them may lead to **prosecution** by the State, and possibly some sort of punishment (see Ch. 1). Punishments may take the form of fines, imprisonment, suspended sentences, community service orders, driving bans etc. (see p. 37).

The most serious crimes, such as murder, were developed by the common law, and for this reason the Crown prosecutes the offender on behalf of the State. Such criminal cases are referred to in the law reports as '*R* v. ?', the title *R* being short for Rex (King) or Regina (Queen) and v. short for versus (against). Less serious cases are usually prosecuted by the police, and the name of the chief constable for the area will be used in the title of the case. Prosecutions by private citizens are possible but these are less common.

Before such prosecutions can be dealt with in the criminal courts (magistrates' courts or Crown Courts), the offender will usually have to be apprehended by the police. The case for the prosecution will then be prepared, either by the Crown Prosecution Service which was introduced by the Prosecution of Offences Act 1985, which employs barristers and solicitors, or by the police themselves, in less serious cases.

There are many other types of offences which are dealt with in the criminal courts which perhaps do not strictly fall within our own concept of what is criminal, e.g. traffic offences, hygiene regulation breaches, breaches of the Health and Safety at Work etc. Act 1974, breaches of the Construction Regulations, Town and Country Planning Act etc. These offences have been created by statute, and are designed to improve society in general. As they are statutory creations, many are not the responsibility of the police, and other agents of the state have to apprehend offenders and **prosecute**. Who does what depends on the enabling Act, and reference should always be made to the appropriate Act to see that any prosecutions are **intra vires**, i.e. within the powers given to them by the Act. Thus local authorities will prosecute people making a statutory noise nuisance and who have breached a noise abatement notice, under the Environmental Protection Act 1990 as amended by the Noise and Statutory Nuisance Act 1993, or for breaching building regulations. The Inspectorate of the Health and Safety Executive will prosecute under the Health and Safety at Work etc. Act 1974.

Whilst the main purpose of a successful prosecution is some sort of punishment, the criminal courts are now empowered to award compensation to the victim of the crime. In the magistrates' court this right is limited. For minor offences, this will be helpful to the victim as it obviates the need for an expensive civil court action.

Torts

There is another category of wrongs, however, which does not fall within the public legal sphere. These are torts which are **private** or **civil wrongs**. Whether or not someone has the right to **sue** another for such a wrong depends on the individual tort and the circumstances. Generally, torts are a product of the common law and not of statute (although there are some exceptions).

Torts are not treated in the same way as crimes, because the tort is usually offensive only to one person, or a small section of society, and not to society in general. If someone walks across my lawn, or calls me a thief, or spits on me, it will only be upsetting to me, unless their behaviour crosses the thin dividing line that separates a 'pure tort' from a **crime**. Thus, torts are in the province of private law, and are dealt with in the civil courts (county or Queen's Bench Division). But some behaviour can be **both** a **tort** and a **crime**. For example, if a workman is injured on site, and there was a breach of the Construction Regulations by his employer, his employer may be **prosecuted** for that breach, as it was a crime. **Also**, he may be sued in the civil courts, either for the tort of breach of statutory duty or for breaking his employer's duty to take care of his employees, i.e. a sort of negligence.

In tort, there is no 'punishment' for the tortfeasor as for a crime. If he is found to have committed the tort, then he will be made to pay compensation or damages to the other; there are alternative remedies, such as injunctions. The following are torts: negligence; defamation (libel and slander); trespass to land; trespass to the person (assault and battery); trespass to chattels; malicious prosecution; false imprisonment; breach of statutory duty; *Rylands* v. *Fletcher* liability; nuisance (private).

Breaches of contract

If someone does something which is against the terms of their contract with another, then liability is once again civil (private). Generally, the state is unconcerned with contract disputes and rights and duties stem from the agreement made between the parties (the contract). Thus, only the contractual parties have rights and duties, and no-one else. These presuppose that there was a valid, binding contract to start with, and the rules of contract law must be applied. On the whole, contract law has been formulated by the common law. But in the nineteenth and twentieth centuries, consumerism has led to Acts of Parliament being passed, to give greater protection to certain

types of contracting parties, such as a buyer of goods.

If there is a dispute, then the parties generally **sue** in the civil courts (county or Queen's Bench Division).

Some breaches of contract would amount to torts or even crimes. For example, if a builder builds so badly that the house is unsafe and kills people, he could be in breach of his contract to build in a workmanlike manner. Therefore, he could be prosecuted for breaching building or construction regulations, and could be sued for negligence or breach of statutory duty.

Whether a plaintiff sues for breach of contract or a tort, the causes of action are dealt with using the separate principles of law.

Breach of contract, if proved, generally leads to a claim for damages, although there are other remedies, e.g. a decree of specific performance, or an injunction.

The torts which we will examine are those which are more likely to be committed in construction work. Each one will be dealt with separately. However, there are some common aspects to each tort. But before we look at each tort, let us first examine the different types of liability in tort (see Figure 7.1).

Intention

In criminal law, many crimes require the presence of two factors in order to achieve a successful prosecution – the *mens rea* and the *actus reus*. These can be roughly translated as the evil intention and the act itself. So if someone is to be found guilty of murder then the prosecution must show sufficient evidence that the accused intended to kill the victim and that he did indeed carry out his intention. The requirement of intention to do something in relation to a tort is not quite so common as in criminal law. Most of the torts dealt with in this book do not need *any* intention to commit the tort or cause the harm; merely carrying out the behaviour which results in the harm is enough. So, generally, it will not avail the builder to say 'but I didn't intend such and such to happen'.

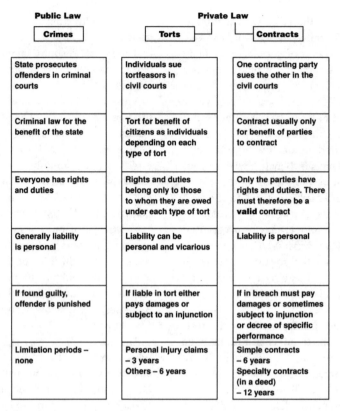

Figure 7.1 Characteristics of crimes, torts and contracts

Motive

Similarly, on the whole, the motive of someone who commits a tort is usually irrelevant. Often, someone genuinely wishes to help in a situation and it would be wrong to deny the victim damages merely because the tortfeasor had a good motive. In nuisance, however, malice on the part of the defendant may just tip the case in the plaintiff's favour (see p. 195).

Negligence

In tort, the word negligence has two meanings: first, in relation to the specific tort referred to on pp. 167–84; second, it relates to a state of mind which may be relevant with respect to torts in general. In this context, negligence indicates not an intention to commit the tort, but a sort of

carelessness which may be a factor in deciding whether a tort has been committed.

Types of liability for torts

Personal liability

Here, the person who actually committed the tort (the **tortfeasor**) is sued and will have to pay **all** the damages.

If other people are involved in committing the **same** tort, then they will be **jointly** liable, i.e. they are both personally responsible for the **same** tort. In such cases, they may both be sued, or either may be sued. They are both liable for all the damage.

Sometimes, the resulting damage may have been caused by different torts. For example, a drunken driver may have caused an accident which is made worse by the fact that his car was negligently repaired by his garage. The sufferer can sue both parties for two separate tortious acts (both negligence). They are **concurrently** liable, and both are personally liable for the victim's injuries.

By S.3 of the Civil Liability (Contribution) Act 1978, where judgment for a debt or damages has been awarded against a person who is jointly liable with another, then this is no **bar** to another action or the continuation of the same action against that other person. Thus, if Arthur, Bill and Charlie (a building partnership) are sued for a tort or a debt for which they are jointly liable, and Arthur admits that he is liable and judgment is entered against him, the action against Bill and Charlie can continue. Before this Act, judgment against one tortfeasor could have prevented further actions in certain cases.

On being held liable in an action or on settling a claim made against him a defendant may recover a contribution from any other person who is responsible for the same damage. The responsibility of such other persons may be by virtue of liability in tort, in contract, breach of trust or in any other way (S.6(1)).

In claiming such a contribution, the defendant must do this within **two** years of the date on which the right to recover contributions arises: S.1 Civil Liability (Contribution) Act 1978. However, most contribution proceedings are part of the main action. Thus, if the plaintiff fails to join all possible co-defendants, the defendant himself may serve a third-party notice on other persons who are jointly or concurrently responsible for the damage.

The amount that must be contributed by others will be assessed by the court in much the same way as under contributory negligence (see p. 185). This works well if all the parties remain solvent and have sufficient insurance cover for such eventualities. Unfortunately, in the building industry, often contractors and employers may go into liquidation and there is actually no-one to share the payment of damages.

Thus, in a building contract, if a contractor is sued for a tort and he regards the employer as partly to blame for the resultant damage, he may serve a third-party notice on the employer and join him as a co-defendant in the action. Alternatively, if he settles the claim out of court, he may then seek a contribution from the employer within two years of the date of the cause of action.

The Law Commission may review the law on joint and several liability if it is considered to be necessary.

Vicarious liability

Where someone authorises another, either expressly or impliedly, to do something, and that person in so doing commits a tort, then the person authorising the behaviour will be **vicariously** (*Vicarius* – Latin, a delegate), i.e. indirectly, liable for that tort.

Thus, employers may be vicariously liable for their employees. Vicarious liability is **additional** to personal liability. So, for example, an employer may be personally negligent for failing to warn his operatives of a dangerous substance used on site. He may be vicariously liable if one of his foremen inadequately supervises workmen dealing with that substance, so that an accident occurs, injuring those workmen.

Vicarious liability is joint and the provisions of the Civil Liability (Contribution) Act 1978 apply. The injured person can sue one or the other, or both the employer and the foreman. If only the employer is sued, he may then try to recover **all** the damages paid from the employee, as an employee owes a right of **indemnity** to his employer (but see p. 267). If the employer partly contributed to the commission of the tort, then he can only claim a contribution (i.e. **part**) from the employee. If the employee alone is sued, the employer can claim nothing from the employee, as he has lost nothing, nor can he claim anything from his employee (unless the employer was partly to blame, in which case the employee can bring in the employer as a co-defendant in the course of the action).

Obviously, the employer is not always vicariously liable for his employee's torts. The following must therefore be noted.

The tort must have been committed 'in the course of his employment'

An employer is never responsible in tort for the private behaviour of his employees. Thus, a quantity surveyor who runs someone over on a Sunday social outing will be solely and personally liable for that act, even if he was driving a company car. The law will find the employer vicariously liable when his employee is doing his job – but badly.

How wide is 'in the course of the employment'?

Obviously, vicarious liability is an additional problem for the employer. Not only is he responsible for his **own** torts, but he may also be responsible for the torts of his many employees. To minimise his liability, he will therefore try to restrain the behaviour of his employees. For example, he will lay down terms and conditions of employment in the employee's contract of employment. He may specify forbidden acts, and these may be backed up by written warnings or notices to the employees.

Unfortunately, while these restrictions are designed to avoid vicarious liability, the courts are reluctant to protect the employers in such situations. Thus, often, where an outsider would have presumed that the employee had the authority to do the forbidden act, which amounted to the tort, then in all likelihood the employer will still be liable.

If the employee is doing something totally outside the scope of his employment, or as it is known 'on a frolic' of his own, such as leaving work without permission, then any tort committed is his sole responsibility. For example, in *Hilton* v. *Thomas Burton (Rhodes) Ltd* 1961, some demolition workers were allowed the use of their employer's van. After working some time, they decided to travel seven miles to a cafe for tea. When they were almost there, they changed their minds and started back to the demolition site. One was killed in an accident caused by the other's negligent driving. The employer was held **not** to be vicariously liable, as they were 'on a frolic'.

Employers, by law, must have insurance cover, to cover claims by injured workmen. Building contractors usually have greater insurance cover than most. Thus, if they are sued for vicarious liability, the claim will (if proved) be met by the insurance company. But if the workman was 'on a frolic', then he alone will have to shoulder the claim. If he is insured, e.g. under a third-party motor insurance policy, which is compulsory when driving, then he is also protected from severe financial loss. But if he was driving without insurance, or the tort was committed in an uninsured situation, then he will have to bear the cost, possibly, for the rest of his life.

It is clear, therefore, that employees are wise to obey their employer's instructions, to make sure of the existence of vicarious liability, and thus share the financial burden.

Furthermore, failure to insure for third-party claims whilst driving, is foolhardiness of the utmost degree. Being prosecuted for driving whilst uninsured is nothing compared with a lifetime of weekly payments to an injured victim.

What is meant by 'employment'?

There can be no vicarious liability on the part of an employer for someone who is not an employee (unless he was authorised in some other way, e.g. as an agent).

Thus, one must distinguish between employees who have a contract **of** employment or a contract **of** service with their employer from independent contractors who are employed under a contract **for** services.

This seems simple enough, but modern practices, especially in the construction industry, very often combine elements of both types of contract, and the result, legally, can be confusing. The following example indicates the probable legal situation.

A builder is vicariously liable for his own workmen's torts (see Figure 7.2). If the builder subcontracts the roofing work to another company, the builder will not necessarily be responsible for the torts of the subcontractor's employees. The roofing company will, of course, be responsible for the torts of its workers.

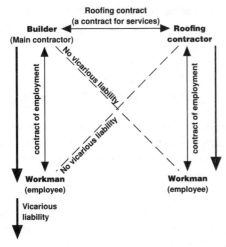

Figure 7.2 Diagram to illustrate the concept of vicarious liability

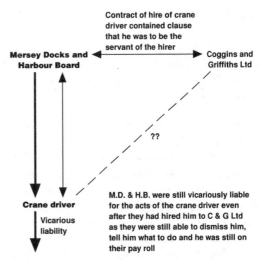

Figure 7.3 *Mersey Docks and Harbour Board* v. *Coggins and Griffiths Ltd*

The trouble is that it is often difficult to decide whether someone is an employee or an independent contractor. Remember, we are concerned with the tort situation, and not with the contractual.

There are many factors to be taken into account. Someone is more likely to be an 'employee' in tort:

- if he is described as such in his contract (but see below);
- if he is paid by an 'employer' who pays his national insurance contributions and taxes;
- if his 'employer' has the right to dismiss him, and to tell him what to do (but not necessarily how to do it) (the control test).

Even if the contract describes the worker as an **employee**, the court will decide on the true nature of the relationship. A similar problem was faced in *Ferguson* v. *John Dawson and Partners (Contractors) Ltd* 1976. A labourer was taken on the 'lump' system, to be paid at an hourly rate. As he was on the 'lump', **no** tax was deducted from his wages, nor did he receive national insurance cards, or a written statement of his contract terms. He was injured when he fell from a roof, and claimed damages for breach of statutory duty, laid down by the Construction (Working Places) Regulations 1966. Thus, he had to prove that he was an employee. On examining **all** the relevant circumstances, the court found him to be an employee.

Sometimes, an employee of another may become the responsibility of the person for whom he is actually working. This can happen where a workman is lent or hired to another company. The leading case on this is *Mersey Docks and Harbour Board* v. *Coggins and Griffiths Ltd* 1947 (see Figure 7.3).

A crane driver employed by Mersey Docks and Harbour Board was hired to Coggins and Griffiths Ltd. During the course of operating his crane for the company, he negligently injured someone. Who should be sued? The company for whom he was **actually** working or his real employers, Mersey Docks and Harbour Board? The contract of hire between them had said that the crane driver was to become the servant of the hirer (C & G Ltd). Nevertheless, the court found that he was still the employee/servant of the board for the purposes of vicarious liability in tort. This was because the board alone could dismiss him and tell him what to do, and they were still paying his wages.

Thus, applying these criteria described previously, labour-only subcontract workmen would probably, in tort, be the employees of the main contractor, for the purposes of vicarious liability. But, of course, each case is decided on its merits.

Liability for independent contractors – exceptions

As stated above, generally an employer is **not** vicariously liable for the acts of an independent contractor, but this all depends on the circumstances of each case.

Employer's personal negligence in choosing or instructing independent contractors

However, even if he may not be vicariously liable, he may nevertheless be held responsible for the acts of an independent contractor. This is because there may be a breach of his own **personal** duty of care. Thus, he will be personally liable if he chose an incompetent contractor.

In building work, therefore, a main contractor should exercise reasonable care in choosing subcontractors. There is less room for such error in the case of nominated subcontractors, where the employer will have played his part in the choice. But remember this is tort, and we are concerned with the liability of the main contractor.

Thus, the standard of care will be approached in the usual way (see p. 175). If the work is of a specialist nature, then the contractor should see that the job is given to a competent subcontractor. He would be negligent if he failed to check that the subcontractor was competent, or if he failed to instruct him properly on what was required.

In *Kealey* v. *Heard* 1983, an owner of adjoining houses contracted with a number of different specialists to convert the houses into flats. One of these specialists was a plasterer, who was injured on scaffolding erected by another unknown workman. The plasterer sued the owner for, among other things, negligence. The Queen's Bench Division held that he was liable. He owed a duty to take care and control of the building work on his property. He had not discharged this duty by appointing an independent contractor who he thought to be competent. He should have supervised the work properly.

(Note: the occupier's duty to his lawful visitors under the Occupiers' Liability Act 1957, see p. 184, is discharged by S.2(4)(b) if the damage caused is due to faulty workmanship of construction, maintenance or repair done by an independent contractor and the occupier had, in all the circumstances, acted reasonably in choosing or supervising that contractor.)

The employer's absolute or strict personal duty to see that the independent contractor takes care

If a tort or statute imposes a strict duty on someone, such as an employer, then there can be **no** defence to say that the independent contractor actually committed the dangerous act.

1. Statutory duties are absolute. Each piece of legislation must be construed in order to see whether there is strict liability or not.

2. Rule in Rylands *v.* Fletcher (see p. 190). The employer, as the occupier, is responsible for independent contractors.

3. Public nuisance. Employers are responsible for independent contractors' torts committed on a highway which amount to a public nuisance.

4. Private nuisance. An occupier may be responsible for independent contractors (being within their control), for nuisance they commit on a highway, or interference with rights of support, like demolishing a semi-detached house, and probably for other types of nuisances.

5. Employer's Liability (Defective Equipment) Act 1969. If damage is caused by a faulty piece of equipment, then, under the Act, the **employer** is liable for injury to his employee, *provided* the equipment was faulty through the fault of another (e.g. the manufacturer).

6. Liability for a negligent act, which is an essential element of the delegated task. If a person contracts with another to do something which is, in itself, completely negligent, or partly negligent, or which will necessitate some negligent act, then that person will be liable for the independent contractor's act.

7. Damage by fire.

8. In common law the employer is always under a duty that care will be taken in relation to his employees. Whether he does something or whether it is done by his independent contractor may not be relevant.

Strict liability

Usually, in tort, there has to be some element of fault on the party of the tortfeasor. In other words, he should have taken greater care than he did when he committed the tort.

Examples of strict liability

There are some exceptions to this rule, however, when a person will be strictly liable even though there was **no** fault or intention on his part. Merely by doing certain things is enough to give rise to an action in tort.

Defamation

Thus, writing something libellous about another which is not true is grounds for an action for defamation. This is so even if the maker of the statement meant no harm, or thought the statement to be true. Such a tort is said to be actionable *per se* (in itself).

Trespass

This is also a tort actionable *per se*. Merely entering on another's land is sufficient to give grounds for an action in trespass. No damage need be shown, nor intention, and mistake is no excuse.

Rule in Rylands *v.* Fletcher

There is strict liability for people allowing dangerous things to escape from their land (see p. 190).

Vicarious liability

Also, vicarious liability is strict liability. This means that, provided the elements of the tort are proved against the tortfeasor, then the employer/authoriser will be liable, even though there was no fault on his part.

Statutory breaches

In the case of statutory duties, there may be strict liability for failure to abide by the statute, e.g. certain breaches of the Nuclear Installations Act 1965, Water Industry Act 1991, Merchant Shipping (Oil Pollution) Act 1971 and Part II Environmental Protection Act 1990.

Liability for independent contractors

See also the examples given under the preceding heading, where persons are strictly liable for the torts of independent contractors.

Premises close to highways

There is also strict liability for dangerous premises situated close to a highway which, through disrepair, injure passers-by (see p. 193).

Specific torts: negligence

This is probably the most well known and certainly the most important of all the torts. Most people know of someone who has sued for damages following a car accident, an accident at work or a hospital operation that went wrong. In each case, the most likely cause of action was negligence, although they might have been able to sue for breach of statutory duty as an alternative (see p. 209).

In order to be able to sue successfully for negligence resulting in physical damage to person or property, three essential elements must be proved. The lack of any one of these elements will make the action fail.

It should be pointed out that there are additional special rules, however, in relation to negligent statements and negligence resulting only in financial or economic loss, unconnected with physical damage, and these are dealt with on pp. 173 and 174.

The vital elements

1. The defendant must have owed the plaintiff a 'duty of care' at the time of the alleged negligent behaviour, *and*

2. There must have been a breach of the duty of care, *and*

3. Actual damage must have been caused as a result of the breach.

However, the presence of the vital elements can only be determined after the event. Thus, whilst one can recommend an ideal pattern of behaviour in a given situation, this may still not prevent there being negligence at a later date.

This is because the tort of negligence is not a set of fixed rigid rules like the construction regulations. It is very fluid and can change and adapt to meet society's needs. Of course, if there is a case of negligence proved in circumstances similar to those in which you are operating, then obviously you should examine your procedures before someone gets injured.

The duty of care

For many years before 1932, many successful cases of negligence were taken to court – but these were usually based entirely on there having been a precedent covering the special circumstances of the case. For example, perhaps in 1850 someone was injured by a stagecoach being driven negligently. If that man was able to sue, then a victim injured at a later date in similar circumstances would also have been able to recover damages in negligence.

Thus, despite the existence of liability for negligence, what did not exist was a standardised way of deciding whether, in **any** given factual situation, a legal duty to take care existed, giving someone the right to sue for the tort of negligence. Remember, no law can exist without there being a duty and a corresponding right.

Donoghue v. Stevenson – formulation of a test 'the neighbour principle'

So until 1932, there was no such test to find whether a duty relating to negligence existed. Then the famous Scottish case of *Donoghue* v. *Stevenson* was taken to the House of Lords, purely on a preliminary point of law.

What Mrs Donoghue's lawyers wanted to know was, could a ginger beer manufacturer owe a consumer of his ginger beer a duty *vis-à-vis* the ginger beer to take care of that consumer so that he would not suffer injury by drinking it?

What happened was this. Mrs Donoghue had gone to a park cafe with a friend. The friend bought her a bottle of ginger beer, which unbeknown to all contained a decomposed snail, the bottle being of the old opaque type. As a result of drinking the ginger beer and being shocked at the sight of the snail, she was absent from work with gastroenteritis and shock.

Before her lawyers would risk pursuing the case on the facts, they asked the court to rule on whether, if they proved those facts, the case would be successful. This of course would depend on whether the manufacturer was under **any** legal duty to the consumer.

If we look at Figure 7.4 we can see what the judges were trying to do. They knew of the other legal relationships (of contract) between manufacturer/cafe proprietor and cafe proprietor/friend. But these were of little help to Mrs Donoghue who had no contract with anyone. Also, even if the friend had been the one to suffer and had sued in contract, the damages that he would have received would have been based on normal contractual rules (see p. 79) and not on the more true reflection of what was suffered as in negligence.

The House of Lords finally decided by a bare majority that, if the case proceeded and the facts were proved, Mrs Donoghue would succeed (subsequently the case was never tried on the facts and Mrs Donoghue never recovered any compensation).

In the Lords' judgment, a very important principle emerged. This was described in Lord Atkin's speech and is usually referred to as the 'neighbour principle'.

He said that 'You must take reasonable care to avoid acts or omissions which you can reasonably foresee would be likely to injure your neighbour.

'Who, then, in law is my neighbour? . . . Persons . . . so closely and directly affected by my act that I ought reasonably to have them in contemplation as being so affected when I am directing my mind to the acts or omissions which are called in question.'

Donoghue v. *Stevenson* 1932

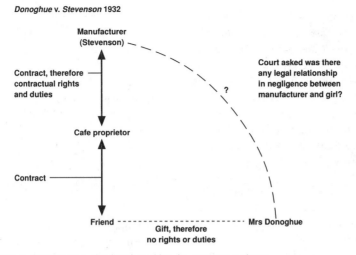

Figure 7.4 *Donoghue* v. *Stevenson* – the legal position in contract and tort

The first part of the quotation is Lord Atkin's requirements for the tort of negligence. The second part shows that neighbours may only be people who are in a close legal relationship with the defendant. This was Lord Atkin's attempt to define 'proximity', a word used in cases prior to *Donoghue* v. *Stevenson* which had proved a stumbling block through lack of a suitable definitive test.

In 1970 in *Home Office* v. *Dorset Yacht Club Ltd*, Lord Reid gave even greater importance to the 'neighbour principle' by saying that 'it ought to apply unless there is some justification . . . for its exclusion'.

If one applies Lord Atkin's neighbour principle strictly one can see that it would then be applied extremely widely, allowing cases to be brought in any foreseeable harm situation. However, the courts did indeed limit the extent of the principle in certain situations (see p. 172).

Then came Lord Wilberforce's dicta in *Anns* v. *Merton London Borough Council* 1978 which restated the principle in a more modern way. He said that whether a duty of care exists can be 'approached in two stages. *First*, one has to ask whether as between the alleged wrongdoer and the person who has suffered damage, there is sufficient relationship of *proximity* or *neighbourhood* such that in the reasonable contemplation of the former,

carelessness on his part may be likely to cause damage to the latter, in which case a prima facie duty of care arises.

Secondly, if the first question is answered affirmatively, it is necessary to consider whether there are any considerations which ought to negative or to reduce or limit the scope of the duty or the class of person to whom it is owed or the damages to which a breach of it may give rise.'

This decision, which struck horror in local authorities throughout the country, instigated a period in which the duty of care 'net' was flung far and wide so that virtually any foreseeable problem could be covered by a duty situation unless there were policy reasons to forbid its existence.

Slowly it appeared that the courts were more reluctant to use this wide approach and there was a narrowing of its application. There was an even greater polarisation of physical injury cases in which a duty of care was almost always owed (except perhaps the nervous shock situations as at Hillsborough) as opposed to monetary or economic loss. A series of cases spelt a disenchantment in the Anns formulation, particularly in some cases involving similar facts to Anns. See *Governors of the Peabody Donation Fund* v. *Sir Lindsay Parkinson* 1984, *Leigh & Sillavan Ltd* v. *Aliakmon Shipping Co. Ltd* 1986 and *Yuen Ken Yew* v. *A.G. of*

Hong Kong 1987. Finally the House of Lords in *Murphy* v *Brentwood District Council* 1990 specifically overruled the Anns decision on similar facts. Another House of Lords decision of that year which has been followed in later cases was *Caparo Industries* v. *Dickman*. This case involved financial loss. However, the court's restricted re-formulation of the neighbour principle has found favour since. The Lords decided that three requirements were necessary:

- there must be reasonable foreseeability of the relevant loss;
- it must be just and reasonable that a duty should exist and
- there must exist a sufficient relationship of proximity between defendant and plaintiff.

The Lords intended that these requirements should be applied in new fact situations only. It therefore seems that simple personal injury duty situations will be decided on the basis of cases following the *Donoghue* v. *Stevenson* approach. Anything new at the moment seems to apply the *Caparo Industries* v. *Dickman* approach. Other countries have not been so disenchanted with the Anns approach and it will be interesting to see how the law progresses in the next few years.

The proximity requirement

'Persons so **closely** and **directly** affected by my act' per Lord Atkin in *Donoghue* v. *Stevenson*. 'A sufficient relationship of proximity between defendant and plaintiff', *Caparo Industries* v. *Dickman* 1990.

The courts therefore require that there should be a sufficient closeness of relationship (proximity) between the wrongdoer and the plaintiff. In some cases, the proximity aspect has been sufficient to override any other objections (such as those of policy) to there being a duty of care in existence. It may even override the objection that, in normal circumstances, damages for pure monetary loss cannot be claimed in negligence (see p. 174). For example in *Junior Books Ltd* v. *Veitchi Ltd* 1982, specialist flooring subcontractors negligently laid a floor in the plaintiff's factory. The

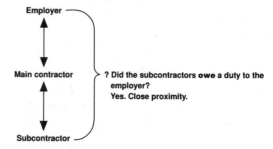

Figure 7.5 *Junior Books* v. *Veitchi*

floor required replacing and the plaintiff asked for damages to cover that cost as well as the cost of removing the machinery and loss of company profits during the period of relaying. It should be remembered that, as they were subcontractors, there was no privity of contract between them and the employer, and thus no chance of suing them in contract. (For some reason not mentioned, the employers chose not to sue the main contractors in contract or delict (Scottish tort).)

However, the House of Lords decided that the parties were in a sufficiently proximate position to allow a duty of care to exist, as the subcontractors should have known the consequences of negligently laying the floor (Figure 7.5).

As Lord Roskill said, 'Why is it conceded that the appellants owed a duty of care to others **not** to construct the flooring so that those others were in peril of suffering loss or damage to their persons or their property, (but) that a duty of care should not be equally owed to the respondents who though not in direct contractual relationship with the appellants, were as nominated subcontractors **in almost as close a commercial relationship** with the appellants as it is possible to envisage short of privity of contract; so as not to expose the respondents to a possible financial loss for repairing the flooring should it prove that the flooring had been negligently constructed.'

In *Simaan General Contracting Co. Ltd* v. *Pilkington Glass Ltd* 1988, however, another claim for economic loss, *Junior Books Ltd* v. *Veitchi Ltd*, was distinguished but not overruled and referred to by Dillon LJ as no longer being a 'useful pointer to

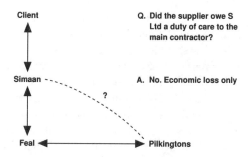

Figure 7.6 *Simaan General Contracting Co. Ltd* v. *Pilkington Glass Ltd*

Q. Was a duty owed when making the will to warn the testator not to have the will witnessed by a beneficiary or the spouse of a beneficiary, thereby showing that a duty was owed to such named beneficiaries?

A. Yes. The solicitors should have contemplated the problem. Therefore proximity proven.

Figure 7.7 *Ross* v. *Caunters*

any development in the law'. In this case S Ltd had a contract to build in Abu Dhabi. A green glass wall was to be built by an Italian firm of subcontractors. They bought the glass from Pilkingtons. The glass was of varying shades of green and red. The client refused to pay S Ltd and the company decided to sue Pilkingtons in tort and not the Italian company in contract, which they could have done under the Supply of Goods and Services Act 1982. The Court of Appeal found that, as this was a claim for economic loss, Pilkingtons owed no duty of care to S Ltd (Figure 7.6). It therefore seems that, unless there is some physical damage, there has to be a negligent misstatement (see p. 173), thereby giving rise to the special relationship needed.

In *Ross* v. *Caunters* 1980 a firm of solicitors drew up a will. They failed to warn the testator that the will should not be witnessed by the husband of one of the beneficiaries (which nullifies the gift). When the testator died, the beneficiary sued the solicitors for the loss of her legacy. Although the solicitors agreed that they had been negligent, they said their only duty was owed to the testator. Applying the proximity test, however, there was a sufficient degree of neighbourhood or proximity between the parties because beneficiaries in such a position should be owed a duty of care as they would have been contemplated as being affected by the lack of advice to the testator (Figure 7.7). Thus, this was another example of pure monetary loss and professional negligence for which there are more restricted duties of care (see pp. 173–5).

What was interesting about this case was that despite being a negligent misstatement it could not be one to which *Hedley Byrne* v. *Heller* applied (see p. 173) as the statement had not been made to the beneficiary, so could not have been relied on by him. In a more recent case, *White & another* v. *Jones & another* 1995 solicitors failed to act quickly enough on receiving instructions in July to draft a new will for a testator following the reconcilation between him and his daughters. He had previously cut them out of his will. The testator died in September without having the new will. It was held that there should be an extension of the *Hedley Byrne* v. *Heller* principle allowing a duty of care to be owed to the daughters, despite there being no contract, no statements made or, indeed, a statement made to them. Lord Mustill in a dissenting judgment said that 'A new rule with principled reasoning could not be devised to encompass such an extensive new area of potential liability.' One can see that this is still a difficult and unsettled area of the law.

To sum up the proximity aspect, therefore, we find that a duty of care may exist in relationships which are contractual or just short of contractual, or so close that no one else could have been contemplated as being affected, e.g. solicitor/client, builder/client, doctor/patient, dentist/patient, teacher/pupil, instructor/student situations and in the examples given previously. Perhaps you could make another list. Such relationships may be so close that a duty of care should exist in negligence. As one moves away from such close proximity to the defendant, where there is no contractual or near contractual link the possibility of there being a duty of care gets less certain.

The reasonable foresight aspect

'You must take care to avoid acts or omissions which you can reasonably foresee would be likely to injure your neighbours'; 'that I ought reasonably to have them in contemplation', Lord Atkin in *Donoghue* v. *Stevenson*; 'there must be reasonable foreseeability of the relevant loss', *Caparo Industries* v. *Dickman* 1990.

The *Donoghue* v. *Stevenson* reasonable foresight approach is very wide, being merely another aspect of 'proximity'. For example, in cases such as *Ross* v. *Caunters*; *Junior Books Ltd* v. *Veitchi Ltd*, which have probably survived the overruling of *Anns* v. *Merton LBC*, the required degree of proximity between the parties is so clear that it is obvious that the defendant should have owed the plaintiffs a duty of care. The consequences of their negligent actions should have been plain.

In less obvious situations, where the relationships between the parties do not appear to be so clear cut, the plaintiff's solicitors will have to examine all the circumstances and try to prove to the court that the defendant placed in the situation of the case **ought** to have contemplated his client as being someone who would be affected by the defendant's negligent actions. If they can prove this to the court's satisfaction, then they will have proved the required proximity between the parties, and thus shown the existence of the duty of care.

Expert and other evidence and common sense should be applied to decide whether, using the reasonable foresight approach, certain people could be neighbours within the law, and thus be sufficiently 'proximate to the wrongdoer'.

Let us imagine a hypothetical builder contemplating insulating work using asbestos. We now know that asbestos can cause a number of different types of cancer. If he sat down and thought about what he was going to do, he would immediately see that the following people could be affected by this work: his workmen; occupiers of the building to be insulated; building control inspectors; subsequent occupiers of the building;

people living close to the building; people living further away; trespassers.

We know that his workmen are owed duties of care because of their contracts of employment and because of the pre-1932 common law rules which impose such duties on employers.

Others in the list, however, may be owed duties of care because they are so closely related in law that no other persons could be contemplated as being affected by the builder's negligence, e.g. the occupiers of the building (especially if they have a contract with the builder).

The rest of the list must probably be tested by using the reasonable foresight approach and this is construed quite narrowly. Thus, if they could only be harmed by a fantastic possibility and not a reasonable probability, then they will not be neighbours within the law (and thus not sufficiently 'proximate'). It is at this end that policy decisions may preclude the existence of a duty of care, and they will have no remedy in negligence.

The *Caparo Industries* v. *Dickman* approach to 'reasonable foresight' is narrower than that of *Donoghue* v. *Stevenson*. Under *Caparo Industries* v. *Dickman*, there must be reasonable foreseeability of the *relevant loss*. Remember the *Caparo Industries* v. *Dickman* approach is to be used in novel fact situations. Thus, if the actual damage would have been unforeseeable (had the defendant thought about the situation), then the court may find that there was no duty of care owed in that situation.

It must be just and reasonable that a duty should exist

There are certain alleged cases of negligence in which, although it appears that there are all the elements for a successful case of negligence, the courts either refuse to find that a duty of care is owed or they impose stricter pre-conditions on finding the existence of such a duty.

Thus, certain people cannot be sued for negligence **at all**. Lawyers, for example, when conducting court cases are immune from legal action (but not for pre-trial work). Judges cannot be sued for negligence in conducting a court case, nor can arbitrators when making an award in an arbitration. A valuer is not an arbitrator, however, if he is merely carrying out a valuation, and thus has no immunity.

Matters of public policy also enter into this question, e.g. *Ashton* v. *Turner & another* 1980. Partners in a crime should not be allowed to sue for negligence when it goes wrong (see p. 182).

Other important categories are concerned with liability for negligent statements and liability for pure monetary (pecuniary or economic) loss.

Negligent statements – liability for words

Is it possible for someone who gives wrong advice or gives incorrect information to be sued for negligence? The answer, as always, is that it all depends.

Lord Atkin in *Donoghue* v. *Stevenson* 1932 probably never contemplated there being liability for negligent words, unless they resulted in damage to person or property.

For example, in *Clay* v. *A.J. Crump* 1963, an architect allowed a wall to remain standing on a demolition site. This lack of advice was held to amount to negligence when the wall fell and injured a workman.

Nevertheless, through this century, and in some cases even before, there was a certain degree of liability for bad advice.

Thus, people in contract with each other, such as solicitor and client, could sue for breach of contract. If the solicitor gave negligent advice, he could be sued. Similarly, if someone in a fiduciary (of good faith) relationship to another, such as a doctor and patient, gave bad advice, then he could be sued.

If the maker of a false statement is trying to induce someone to enter into a contract with him, then a number of courses of action are open to the other party. If the maker of the statement is telling downright lies, he can be sued for the tort of deceit, also called fraud. Certain requirements of the tort must be complied with. As the statement is a misrepresentation, the misled party can proceed under the Misrepresentation Act 1967 (see p. 72) if the representation is made negligently or even innocently.

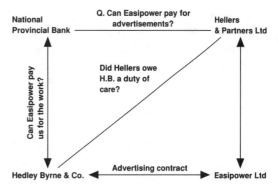

Figure 7.8 *Hedley Byrne & Co. Ltd* v. *Heller & Partners Ltd* 1964

The above causes of action, however, are and were only available to contracting parties, parties in a fiduciary relationship, or people who had been deceived by a liar.

The leading case of *Hedley Byrne & Co. Ltd* v. *Heller & Partners Ltd* 1964 has extended liability for the making of negligent statements to people who are not in such relationship.

H B & Co. advertising agents were preparing an advertising campaign on behalf of a client, Easipower Ltd. They therefore booked advertising space in newspapers and time on radio and television in their own name, and thus they would have to pay for this and recover their costs from Easipower.

In order to check whether Easipower would be able to pay the bill, H B & Co. asked their bankers to find out from Easipower's bank (Heller & Partners Ltd) if they were financially sound. Initially, they confirmed by telephone that the company was sound. After a further request as to their creditworthiness, Hellers wrote that the company was 'good for its ordinary business engagements'. However, the letter contained an exclusion clause, so that the statement was made 'without responsibility on the part of this bank'. Easipower Ltd then went into liquidation and H B & Co. lost a lot of money (Figure 7.8).

The question asked in this case was, did Hellers own H B & Co. a duty of care in the tort of negligence? The case went to the House of Lords.

The Lords in differing speeches agreed that the 'reasonable foresight' approach to the neighbour principle would not be enough to establish a duty of care. A sufficiently close relationship would be required. This was termed a 'special relationship'. How this relationship was decided upon, they failed to agree. But, in the present case they felt one did exist.

Unfortunately, because of the exemption clause, H B & Co. lost their case.

So, the important aspect of this case is that it allowed negligence cases to be brought for negligent words, be it advice or representations, in a wider range of situations than the courts had previously allowed.

Thus, solicitors can be sued in negligence or in contract by their clients for their advice.

Architects when giving certificates in building contracts can be sued in negligence by the employer if they negligently perform this task: *Sutcliffe* v. *Thackrah* 1974.

Liability can of course be excluded as it was in the Hedley Byrne case itself. However, such exclusion clauses must now be read in the light of the Unfair Contract Terms Act 1977 (see pp. 59–61) and the Unfair Terms in Contracts Regulations 1994.

Liability for economic loss (monetary or pecuniary loss)

For many years persons suffering from negligence which resulted in economic loss only, with no attendant physical damage to property or person, could not recover damages in negligence. Thus, if there was **only** economic loss, they could not sue in negligence at all unless the principles in *Hedley Byrne & Co. Ltd* v. *Heller and Partners Ltd* 1964 could be applied.

(Note: shoddy workmanship will result in immediate economic loss if the work has to be redone, e.g. wallpaper applied upside down. If actual damage is caused to the property, then there is no problem about suing in negligence.)

In *Junior Books Ltd* v. *Veitchi Ltd* 1982 (see p. 174), however, it was held that where there is a sufficiently close degree of proximity between the parties, damages for economic loss only could be recovered. To quote Lord Roskill: 'The appellants must be taken to have known that if they did

the work negligently (as it must be assumed that they did) the resulting defects would at some time require remedying by the respondents expending money upon the remedial measures, as a consequence of which, the respondents would suffer financial or economic loss.'

Lord Roskill, thus, encouraged 'the next logical step forward in the development of this branch of the law. I see no reason why what was called 'damage to the pocket' simpliciter should be disallowed, when 'damage to the pocket' coupled with physical damage has hitherto always been allowed.'

Thus, in *Junior Books Ltd* v. *Veitchi Ltd* the argument by the subcontractors that their duty of care was limited only to 'exercise reasonable care, so as to mix and lay the flooring as to ensure that it is not a danger to persons or property, excluding . . . the flooring itself' was rejected. Unfortunately, this case appeared to be the zenith in the development of these arguments. In *D & F Estates* v. *The Church Commissioners for England* 1989, a case involving a claim for damages for replacing defective plastering, no physical damage had been caused. It was merely shoddy workmanship and the claim was not allowed. The decision was followed in *Murphy* v. *Brentwood District Council* and *Department of the Environment* v. *Thomas Bates & Son Ltd* 1991. One argument put forward in D & F Estates was called the complex structure exception. This has been given limited approval in some cases. The exception allows for a duty to be found, where **only** economic loss has been caused, if the damage to one part of a complex structure thereby causes damage to another part. Whilst the initial damage is economic loss and thus not recoverable, the damage to the rest of the structure may be recoverable. Its application seems to be strictly limited, however.

To conclude therefore by pointing out some anomalies. If someone buys a bottle of ginger beer with a snail in it and suffers no injury, he cannot sue in negligence, probably because, whilst there is a close enough relationship of proximity for injury to the person, there isn't for economic loss and there is no damage, as reasoned in *Junior Books Ltd* v. *Veitchi Ltd*. The injured consumer has two courses of action, however. He can go back to the shop and ask for his money back under the Sale of Goods Act 1979 (see p. 114) and he can complain to the local authority who may prosecute whoever is responsible, be it manufacturer or shop owner, under various regulations.

If someone is injured or his property is damaged partly or wholly as a result of a defect in a product supplied to him and there was no negligence on the part of the supplier, then obviously one cannot sue in negligence. However, under the Consumer Protection Act 1987, in such cases the person suffering **can** sue under the Act. This does not cover damage to property where the property is the thing supplied. Nor does it apply to property incorporated into a sale of land. But it does apply to property incorporated into property on land (provided the land is not also included in the sale).

When is the duty owed?

A person can only be owed a duty of care **at the time the damage occurs**. This seems obvious if one thinks in simple terms of a car accident, for example. Only the person who has been run over is owed a duty of care. Such a person may be one of a class, e.g. members of a bus queue, who are hit by an out of control bus.

This is less easy to understand in relation to damage to buildings, because, if one is only owed a duty when damage **actually** occurs (and not when the damage is discovered), subsequent purchasers of houses could be excluded from having a right of action. *Pirelli General Cable Works Ltd* v. *Oscar Faber & Partners* 1983, however, confirms this point of view. Thus, the date on which the cause of action is founded in an action for damages caused by negligent construction or building design is the *date of the damage*. Purchasers of such property known as successors-in-title would still be able to sue (but subject to the limitation periods) (see p. 211).

Breaching the duty of care –
the second element

Having established that a duty of care was owed in the circumstances, the plaintiff must then prove to the court that there was a breach of that duty, i.e. that he was negligent in fact.

This is done in two stages. **First**, taking the circumstances of the case, the court decides on the **standard of care**, i.e. what should have been done by the defendant. **Second**, the defendant's actual behaviour is measured against the established standard. If it falls short, he is negligent; if it matches or is above the standard, then there is no breach.

The standard of care

The standard of care is not that of 'superbuilder' nor that of a fool. The court 'fixes' the standard by examining evidence which would show what the **average reasonable** person placed in the circumstances of the case should have done. Experts may be called to give evidence of trade practice.

So, the average bricklayer building a wall has the standard of care of the average bricklayer, i.e. to build carefully and safely in a workmanlike manner. The same bricklayer driving his car home has the standard of care of an average driver. If he decides to repair his television set, his standard of care is that of the average layman. Whether he is extremely intelligent or not, or normally lazy or careless, is not important, because this introduces a further subjective element, which has been rejected by the courts.

Furthermore, the knowledge which one is supposed to possess depends on the current state of knowledge. So, at present, we are probably doing many things which in years to come will prove to have been dangerous. Once we have that knowledge available (even if the actual defendant is ignorant of it) then ignoring it could amount to negligence.

Other factors which may raise or lower the basic standard of care

Is the risk reasonably likely?

The courts are not going to be impressed with a defendant who failed to avoid an obvious risk. Once again the reasonable foresight test is used. Thus, failure to see that safety helmets are being worn where necessary on building sites will amount to negligence because of the extreme likelihood of injuries caused by falling materials.

Even if the risk is not likely but is reasonably foreseeable then, when determining the standard of care, this should be taken into account. This follows dicta in the *Wagon Mound No 2* 1967, a Privy Council shipping case which involved an unusual chain of events. For example, in *Haley* v. *London Electricity Board* 1965, the Electricity Board's employees had dug a hole in a pavement, and although they erected adequate barriers to prevent a sighted person from falling down the hole, they were inadequate for a blind person. So, although the risk of harm being caused by their action was slight, the fact that a blind person could have fallen down it was reasonably foreseeable.

If the risk is a 'fantastic possibility' then probably there would be no need to do more than take the usual precautions.

Are the plaintiffs at particular risk?

Having decided that certain categories of people are owed a duty of care, then if they would be at particular risk because of age, infirmity or other special reason such as being women of child-bearing age, the standard of care is much higher than normal. So, school teachers owe a high standard of care towards their pupils. Employers of handicapped people must see that such people are safe, e.g. by making sure that fire exits can cope with wheelchairs etc. In the building industry, apprentices, or even people doing a new job, are at particular risk, and their work should be adequately supervised, and warnings and advice should be given and seen to be understood.

In *Paris* v. *Stepney Borough Council* 1957, the employing council was held to be negligent in failing to give goggles to a one-eyed workman, although it would not have been negligent had he been fully sighted. This seems an odd decision, but the court specifically mentioned the high degree of risk to a partially sighted person. By analogy, it is possible that failure to see that ear muffs are worn by a partially deaf person working in relatively noisy circumstances would amount to negligence, but not if he were of good hearing.

Could the risk of injury or harm have been eradicated by a small outlay?

Of course, no court will be impressed with the defendant arguing that he could only have avoided the risk of harm at great expense (unless this was coupled with the argument that the risk was a fantastic possibility). But, if the harm could have been avoided at a small cost, and the defendant failed to take avoiding action, then the courts are more likely to find that the standard of care was higher than normal, and there would be a case of negligence.

Is the defendant an expert?

If I fall down a pot hole smashing my leg, and the only way to free it is by amputation, then the standard of care to be shown by an unqualified rescuer is that of the average rescuer in an emergency. But if my leg is to be amputated in hospital, then the surgeon is expected to show such skill and expertise as is appropriate to his position, qualifications and experience.

Thus, in the first case, the standard is quite low and, in the second, the standard is extremely high.

Similarly, a builder is expected to perform construction work safely and of a high quality.

Amateurs doing work which could better be done by experts should show reasonable care, as their standard of care would be that of the average amateur doing a particular job. In determining such a standard, the courts assume that all people have a certain degree of common knowledge, such as that water is a good conductor of electricity.

Are there regulations, codes of practice etc. to follow?

If there are statutory regulations laying down conditions to be observed, then these should be complied with, and in many cases, it will probably be more appropriate to sue for breach of statutory duty (see p. 209). If for some reason it is impossible to sue in this way, due to a lack of one of the essential pre-conditions, then in the alternative, the plaintiff can sue in negligence and plead that the regulations established a particular standard of care which was not met.

Similarly, if there are statutory codes of practice, whilst they do not constitute absolute legal requirements, they will in most cases establish a set standard of care.

If there are house or company rules or regulations, then this will indicate a certain level of care, which may or may not be what would have been acceptable in the circumstances. For example, some company rules may be far too stringent, and whilst a breach would perhaps be grounds for dismissal, it would not amount to negligence in tort.

Is the defendant handicapped in some way?

If the defendant has some handicap such as extreme old age, infirmity, or is very young, then this does not necessarily establish a very low standard of care. The reasonable man test is still applied but, of course, the average reasonable man would have the handicap of the defendant placed in the circumstances involved.

Very old people often have impaired faculties and should not perhaps drive or be expected to look after children. Their age, however, does not excuse them if their alleged negligent act could in no way have been attributable to their longevity. Foolishness is not the prerogative of the young. Similarly, physically handicapped people could still, for example, telephone for help in many situations, even if they could not be expected to run for help. Mentally handicapped people would not generally be expected to be responsible for their acts, but this all depends on the degree of handicap and the circumstances involved.

What about the otherwise normal average man, who causes an accident as a result of having a heart attack or a fit?

If the attack was reasonably foreseeable because he had been warned by his doctor, or he had suffered one before, then his standard of care would be high, and ignoring the likelihood could amount to negligence. If, on the other hand, he had no inkling of the impending attack, then he could never be found to be negligent. This could have important consequences for an employer who is of course vicariously liable for an employee's torts. Should an employee work whilst ill and unfit and thereby cause injury to another,

then the employer will be liable as well. Thus, it is possible that more stringent supervision or regular medical examinations should be given in potentially dangerous work.

Was there an emergency?

Whether this lowers the standard of care depends on the circumstances of each case. If, for example, there is a fire, there should be rules laid down to follow to avoid worsening the situation. If these are not followed, then there may be a case of negligence. But if there are no regulations and the defendant does the best one could in the circumstances, then he will have acted as the average reasonable man should have done in the circumstances, and there can be no negligence.

Is the plaintiff an expert?

If the defendant is working with a plaintiff who is an expert, then this may reduce the standard of care, so as to negative any claim of negligence against the defendant, provided that the expert knew what was being done. If negligence is proved, however, the defendant could claim that the plaintiff was contributorily negligent, or even that there was consent (see pp. 182–3).

What did the defendant do?

Having determined the standard of care which should have been exercised in the given situation, the plaintiff must then prove that the defendant did not conform with that standard. So, if car drivers should normally be sober, then a defendant causing an accident whilst inebriated will be found to be negligent.

Normally, the burden of proving the breach will fall on the plaintiff's shoulders. It is not for the defendant to prove that he was not negligent. Evidence must be produced to satisfy the court that on a **balance of probabilities** the defendant was negligent in the circumstances. For this reason, all evidence must be properly produced and preserved. Obviously, the types of evidence will be different in each situation, but they could include photographs, pieces of equipment or machines, medical evidence, witnesses accounts,

RES IPSA LOQUITUR
('The facts speak for themselves')
People do not usually get hit by
falling barrels of flour unless there
has been negligence of some sort.

Figure 7.9 *Res ipsa loquitur*

experts' evidence, accident reports, letters, accounts, contracts, drawings and plans. The more evidence the plaintiff produces of a persuasive nature, the more likely the court will agree that the defendant was negligent.

A conviction in criminal proceedings may now be used as evidence in civil proceedings when establishing the existence of a tort.

If the evidence is overwhelmingly in favour of the plaintiff from the start, then it is highly unlikely that the case would ever actually go to a court hearing. The lawyers would then 'settle out of court'. Sometimes, however, despite the fact that the defendants have admitted their liability, the amount of damages, known as the quantum, is disputed. This is particularly so in relation to personal injuries or fatalities, as there is a highly complex approach by the courts to arrive at a final global amount. In such situations, the case will go to court, but merely for the judge to decide on the amount of damages to be awarded, and no evidence need be produced.

A special rule of evidence – res ipsa loquitur, *'The facts speak for themselves' (Figure 7.9)*
Occasionally, the facts of a particular situation point immediately to the existence of negligence. Decided cases have included barrels of flour falling out of warehouse windows and hitting someone, surgical swabs being left in a patient's body after an operation, and dangerous chemicals being used to 'finish' underpants.

In such cases, the plaintiff has the right to shift the burden of proof from his own shoulders onto the shoulders of the defendant, who then has either to admit liability or prove that there was no negligence. The plea of *res ipsa loquitur* is thus only a rule of evidence allowing the plaintiff to sit back whilst the defendant tries to extricate himself from the action brought against him.

Of course, such a rule is not as simple as that. There are a few conditions attached to its use.

First, the victim's injury must be one that would not actually have happened without there having been some sort of negligence. Discovering surgical instruments lodged in one's body after an operation is probably the best example of this. Negligence is the **only** explanation of their presence.

Second, there must be no explanation for the occurrence other than one of negligence. Thus, if you are alleging that someone was drunk when he caused the accident, then *res ipsa loquitur* cannot be successfully pleaded, and the plaintiff must use the evidence available to prove negligence.

Third, the facts (*res*) must indicate negligence on the part of the defendant. For example, a plaster found in the middle of a cake can only indicate that the bakery was responsible, as it must have been lodged before baking. No one else could be responsible. However, in the case of motor accidents, the plea of *res ipsa loquitur* can be used to indicate equal personal liability if it is unknown in which proportion each driver was to blame.

In *Kealey* v. *Heard* 1983, *res ipsa loquitur* was successfully pleaded against the owner of two adjoining houses which were being converted. Scaffolding erected by an independent contractor collapsed, injuring another independent contractor, a plasterer. He sued the owner of the houses for negligence and was successful. The court took the view that the house owner was in overall control, despite having instructed competent (or so he thought) contractors. As scaffolding does not collapse unless negligently erected, *res ipsa loquitur* applied.

Damage must result from the breach – the third element

Damage must be caused

First, there must be damage caused. If there is no damage, there can be no action, even if one can prove the first two elements of the tort. This is because negligence is *not* actionable *per se* as is, for example, trespass.

Damage can take any form

Second, damage can be either physical, i.e. damage to person or property, or it can be purely economic (monetary). However, to sue for pure economic loss, there must have been a negligent misstatement or it must fall into a special category.

The damage must not be too remote from the negligent act in fact

If there is damage, then it must not be too far removed from the negligent act in fact. So, if I fall over on a slippery floor suffering no apparent injury, and a year later suffer backache, which may or may not be due to the fall, it would be extremely difficult to sue successfully for the resultant damage. There has to be a **causal** link between cause and effect. If there is none, an action will fail.

A good example of a case failing through lack of causal factual connection between negligence and injury is found in *Barnett* v. *Chelsea Hospital Management Committee* 1961. Here, nightwatchmen drank tea containing arsenic which someone had

malevolently slipped into the pot. They visited the local hospital, where the casualty officer, who was rather off-colour, was in bed. The duty sister telephoned the doctor, describing their symptoms of vomiting and diarrhoea. The doctor told her to send them home and if they were no better in the morning, to see their own doctors. At no time did he examine them. The next day, one of them died. There had undoubtedly been a duty of care owed to these men and a breach of that duty. But it was proved that even if they had been examined, no test could possibly have given a result quickly enough to save the man's life. Thus, the widow's claim against the hospital committee failed.

Occasionally, the courts take a more liberal view of this aspect if it is difficult to say whether or not the defendant's negligence caused the injury to the plaintiff. In *McGhee* v. *National Coal Board* 1975, for example, a workman contracted dermatitis, probably as a result of not being able to shower after working in the defendant's brick kilns. Current medical knowledge made it impossible to determine whether the provision of showers could have prevented the dermatitis. Despite this, the House of Lords found for the plaintiff. It seems a very generous decision (for the plaintiff) as the plaintiff's dermatitis may have been caused in another way, totally unconnected with the lack of showers.

The damage must not be too remote in law – the reasonable foresight approach again

In order to limit the liability for **all** the direct **factual** consequences of negligence, the courts now recognise that one should only be liable for those consequences which could have been **reasonably foreseen** at the time of the negligence. This test was approved by the Privy Council in another *Wagon Mound* case, the *Wagon Mound No 1* 1967.

Because greater strides are always being made in science, it is this element which could defeat a claim by a worker for damages caused by cancer, for example, if the employer could not have known that what he was doing, which otherwise amounted to negligence, would cause cancer.

Similarly, there would probably be no responsibility for the deterioration in a building material if, at the time of its use, the industry thought it was safe (provided, of course, that the damage was only created by the deterioration and not by negligent building work).

It is **never** a defence in law to claim that you did not realise that what you were doing was legally wrong. Ignorance is no defence. Thus, employers or anyone in business have a particularly difficult task in taking note of all modern developments affecting their work. Very important problems are usually the subject of government department circulars which are sent around to all appropriate bodies or industries, and these are ignored at peril.

Injury caused by a number of different sets of circumstances

It often happens that a person suffers injury, not from **one** act, but from a chain of events, which started with the defendant's negligence. Should the defendant be responsible for **all** the final damage?

The factors considered previously still apply, i.e. first, is the injury a factual result of the negligence, and second, was the first injury reasonably foreseeable at the outset?

Novus actus interveniens – a new intervening act

If, however, a new happening occurs which breaks the chain of causation completely, then the defendant ceases to be responsible for any other injury. The new intervening act will only do this if the act or event was **not** reasonably foreseeable by the defendant.

So, if I fall on the proverbial slippery floor caused by negligent cleaning and am then hit on the head by a door being opened, this second event will probably not break the chain of causation. This is because the second occurrence is a likely and reasonably foreseeable consequence of the initial negligence. Nor will the person opening the door be liable in any way, because he would not expect to find lecturers lying around on floors behind closed doors.

In *Knightly* v. *Johns* 1982, for example, the plaintiff, a police motor cyclist, sued a number of defendants. The first crashed his car in a tunnel. The fourth, a police inspector, went to the scene of the accident instead of immediately closing the tunnel, as was more sensible and was required by standing orders. The inspector then ordered the cyclist to go into the tunnel in order to close it. Whilst travelling against the traffic, he collided with the second defendant. At first instance, the judge gave judgment against the first defendant only. In the Court of Appeal, an appeal by the first defendant was upheld and the judge gave judgment against the fourth and third defendant (who was the chief constable and therefore vicariously liable for the police inspector's negligence). This was done because, although the first defendant's negligence had created a situation which required a rescue of some sort, he should have only been liable for reasonably foreseeable consequences. As the inspector's behaviour was not reasonably foreseeable, this amounted to a break in the chain of causation, i.e. it was a *novus actus interveniens* which completely relieved the first defendant from liability.

What sort of later events will not break the chain of causation from the original negligence?

It all depends on whether these later events are reasonably foreseeable or not. The following are examples which could occur in the building industry.

Acting in the agony of the moment

Occasionally, the plaintiff in 'the agony of the moment' caused by the defendant's negligence chooses to do something which in fact worsens the situation, e.g. he may grab something for support, which collapses and falls onto him causing him severe injury. This will not amount to contributory negligence (see p. 183). In fact, it is a reasonably foreseeable deliberate act on the part of the plaintiff which will not break the chain of causation between the original negligence and final injury. If the defendant had thought about the negligent act which caused the plaintiff to fall in the first place, then it is reasonable to expect him to try and save himself. The fact that this effort worsened the situation does not alter the defendant's blameworthiness for all the resultant injury.

Rescue by plaintiffs

Here the defendant behaves negligently and the plaintiff attempts to rescue him from harm.

In the *Knightly* v. *Johns* 1982 situation, it is clear that a rescue would have been and should have been mounted. Thus, the first defendant would have realized that his negligence would require a rescue of some sort. However, the plaintiff was not injured as a result of the rescue attempt in itself. Had he been, he may have been able to sue the first defendant successfully as it would have been a reasonably foreseeable result of the negligence.

Similarly in *Harrison* v. *British Railways Board and another* 1981, the defendant passenger tried to get on a moving train. The plaintiff who was a guard gave an incorrect signal to the train driver in order to stop the train. At the same time he pulled the defendant onto the train which had not stopped and was now accelerating. Both fell from the train, the plaintiff badly injuring himself. Could the plaintiff sue the defendant for negligence?

The court ruled yes. If someone puts himself into a position which is dangerous and from which one could expect to be rescued, then such people do owe duties of care to their rescuers.

Without such liability, no-one would be keen on mounting rescues where there was negligence, as they would not be able to recover damages for any injuries. However, the rescue must fulfil three pre-conditions, otherwise the rescuer might be thought to be consenting to run the risk of the negligence and the defendant may successfully plead the defence *volenti non fit injuria* (see p. 182).

The three pre-conditions

1. The rescue must be reasonably foreseeable. This is self-evident.

2. The rescue must be necessary. The classic example of an unnecessary rescue is in the case of the runaway milk float horse *Cutler* v. *United Dairies* 1933. The horse had bolted into a field, where his milkman master was trying to calm him down. The milkman called for assistance to a passer-by. During his attempts to pacify the horse, the passer-by was kicked, and he sued the dairy. The dairy was found to have been negligent in using the horse, as it was known to have bolted on more than one occasion, and was not the ideal beast to pull a milk float. But the Court of Appeal stated that the rescue was unnecessary and the plaintiff lost his case.

   Necessary acts of rescue have been found where there was a risk to life and limb, and even, in some cases, a risk to property, e.g. where a house is on fire.

3. The rescue must be reasonable. What is reasonable would depend on the circumstances, but the courts take a very wide view of this and seldom make decisions like *Cutler* v. *United Dairies* 1933.

Miscellaneous actions taken by the plaintiff

If a plaintiff commits suicide as a direct result of depression caused by the defendant's negligence, then his widow will probably be successful, as in *Pigney* v. *Pointer's Transport* 1957. The plaintiff had suffered head injuries which caused his depression. As the injuries were thus still existing in the form of the depression, the suicide was therefore reasonably foreseeable.

In a similar case, however, *McKew* v. *Holland Hannen & Cubitts* 1969, the plaintiff aggravated the situation by attempting to go down some stairs without a handrail whilst still suffering from a leg injury caused by the defendant's negligence. This was held to be so unreasonable on the part of the plaintiff as to break the chain of causation *vis-à-vis* the final injury, but not for the original injury. Of course it could have been argued that he was not deliberately being silly, but rather he was being contributorily negligent (see p. 182).

In the above situations, the second event which worsened or even caused the final injury was perpetrated by the plaintiff. But often the second act or event is caused by someone else, or even is a chance happening. The same rules apply. Is the final injury a factual result of the initial negligence, and if so, was it reasonably foreseeable? If the answers to both questions are yes, the defendant will be liable for all the resultant loss.

Defences

The defendant in a court action does not sit back and cross his fingers, hoping that the plaintiff will not be able to prove his case. He will attempt to raise appropriate defences which could cancel out any negligence.

The burden of proving a defence obviously falls on the defendant and he must produce his own evidence in support.

Necessity

It is vaguely possible that the defendant may claim that the tort was committed necessarily to prevent another greater harm occurring. However, most of the cases involving necessity as a defence are concerned with other torts where the act was deliberate.

A more likely effect of doing something negligently to reduce a greater injury is that it will lower the standard of care required in the circumstances, so that a case of negligence will not be substantiated.

Volenti non fit injuria – consent or 'there is no injury to a willing person'

This is an absolute defence, and if proved, cancels out the negligence, thus defeating the plaintiff's claim. It arises where the plaintiff:

- knew that the defendant was being negligent,
- knew of the risk,
- continued to co-operate with the defendant, and
- suffered injury of the type one would expect would result from such negligence.

Nowadays the courts are very reluctant to find a case of consent existing in all but the most open and shut cases. For this reason, strict guidelines are adhered to for its application.

The plaintiff must know that the defendant is being negligent

There can be no valid consent if the plaintiff did not know there was any negligence. Thus, the defendant must prove there was either express or implied consent *to the negligence*. Express consent will be either oral (and therefore witnessed to stand up in court) or written, which may be witnessed (but see the effects of the Unfair Contract Terms Act 1977, p. 59). Consent may be implied where the plaintiff continues to co-operate despite the negligence.

(Note: consent concerns the negligence and not mere hazards of the job. People who do dangerous work may be paid danger money and they consent to run the risk of the ordinary hazards of the work (and possibly sign an agreement to that effect) but they are *not* consenting to actual negligence.)

The plaintiff must have known of the risks which could be caused by the negligence

Someone who consents to run the risk of a negligent act will not lose his action for negligence if he was injured in some other way.

Consent must be freely given

Sometimes, a worker is put under pressure to continue to work, knowing of the presence of danger caused by negligence. This usually takes the form of a threat of dismissal. In such cases, the defence will fail, as the consent was brought about under duress.

Because this is a complete defence which would cancel out all liability on the part of the defendant, the courts are more likely to agree to there being contributory negligence on the part of the plaintiff (see below).

However, the following case is a classic example in which *volenti* (consent) was successfully pleaded.

In *Ashton* v. *Turner and another* 1980, the plaintiff and the two defendants went on a drinking session, during which it was planned that the plaintiff and the first defendant would commit a burglary, using the second defendant's car.

The plaintiff was injured in a chase following the burglary. He sued both defendants for

Figure 7.10 Warnings like these may absolve an occupier of liability under the Acts

negligence: the first, for dangerous and drunken driving (for which he had already been convicted in the criminal courts), the second for allowing the driver to drive without insurance.

The High Court was asked whether the plaintiff was owed a duty of care. The judge decided that he was owed no duty of care as a matter of public policy, and also stated that the first defendant would be able to plead *volenti* successfully.

Exclusion of liability

Can the defendant plead consent if he puts up notices such as 'the public enter this property at their own risk'?

This sort of notice is an example of how people seek to exclude their liability (whether or not there has been any negligence) (Figure 7.10). They may also seek to do this by excluding their liability in clauses in their contracts with other people (see p. 59).

Until a few years ago, these exclusion clauses worked well in the defendant's favour. In contract, such clauses were pleaded to nullify liability as in *Hedley Byrne Ltd* v. *Heller & Partners* 1964 (see p. 173). In negligence, the notices gave plaintiffs the opportunity of not using the premises, and if ignored, the defendant could have tried to plead *volenti* as a defence (and could possibly succeed if the elements of the defence were proved).

In 1977, however, the Unfair Contract Terms Act was passed, severely limiting the scope of such exclusions. Under the Act, liability for what

amounts to negligence which causes death or personal injury cannot be excluded by contract or notice **at all** in the course of a **business**. Where negligence results in economic loss only, such an exclusion of liability is valid, but only if it is **reasonable**, taking into account all the circumstances, such as the balance of bargaining power between the parties.

It should be noted that exclusions between parties **not in a business relationship** are still valid. Also, the Act does not cover unenforceable contracts with minors (see p. 67) and effective exclusions could be inserted into these.

Employers cannot exclude their liability for vicarious liability for employee's torts causing injury or loss to fellow employees under the Law Reform (Personal Injuries) Act 1948. Nor can car drivers exclude their liability in tort towards passengers under the Road Traffic Act 1988. Thus, notices which were common at one time when hitch hiking was less hazardous, such as 'the driver takes no responsibility for any damage or loss done to passenger's property or person', are no longer valid, nor can they be used to prove *volenti*, e.g. if the driver is drunk. The circumstances of the case must therefore be examined to find whether the elements of the defence can be found.

Contributory negligence

This is not a complete defence, but it acknowledges the fact that, whilst negligence may have been proved against the defendant, the plaintiff may, at the same time, by his own fault have contributed to his injuries or loss, e.g. by failing to wear his safety helmet or harness.

The Law Reform (Contributory Negligence) Act 1945 introduced this defence and by S.1(1) in such cases, 'the damages recoverable . . . shall be reduced to such extent as the court thinks fit and equitable, having regard to the claimants' share in the responsibility for the damage'.

The wording of the Act allows the defendant to plead this, without having to prove *negligence* on the part of the plaintiff. All he needs to do is to show that the plaintiff was at fault in some way, which contributed to the damage. The plaintiff should have reasonably foreseen that his behaviour would contribute to his injury; otherwise the defence will fail.

As a result of this, the court will work out the damages in the usual way, and then reduce them proportionately, taking account of the blameworthiness of the plaintiff, e.g. that the plaintiff by failing to wear his seat belt was 50 per cent to blame, and therefore gets only 50 per cent damages.

(Note: there is no set amount attributed to the non-wearing of seat belts; it will all depend on the circumstances.)

Once again it is clear that it is in the worker's own interests to do all he can to prevent greater injury to himself by complying with all safety requirements, such as wearing safety clothing.

Liability for dangerous premises

This can be dealt with under three separate heads:
1. liability of the occupier to lawful visitors;
2. liability of the occupier to other people; and
3. liability of non-occupiers for premises in a dangerous state.

An occupier's liability for the state of his premises – Occupiers' Liability Act 1957: the duty towards lawful visitors

As far as liability for premises is concerned, the tort of negligence is only applicable to **activities being** carried on there, and is probably not applicable to premises being dangerous **because of their state**. Thus, if someone is injured on site by a brick being dropped on his head, then the proper course is to sue either for negligence or for breach of a statutory duty, such as under the construction regulations.

If, on the other hand, the brick had become dislodged in a storm from an improperly maintained wall, then the injured man could sue under the Occupiers' Liability Act 1957.

In some cases, an injured person would probably sue for both negligence and breach of the Act, just to be on the safe side if one or the other causes of action failed, e.g. as in *Wheeler* v. *Copas* 1981 on p. 185.

The 1957 Act was introduced to improve the common law situation, where occupiers of premises owed varying degrees of responsibility towards entrants. What it did not change, however, was the concept that, as the occupier is the person in control of the premises, he can therefore make sure that the premises are safe to entrants. This area of the law pre-dated *Donoghue* v. *Stevenson*, and therefore the two areas have developed slightly differently.

The common duty of care owed under the 1957 Act

All **lawful visitors** to premises are now owed a **common duty of care** by the occupier. This is defined by S.2 as a duty to take such care as in all the circumstances of the case is reasonable to see that the visitor will be reasonably safe when using the premises for all the purposes for which he is visiting.

This is a significant difference from negligence. In negligence, the plaintiff must first prove he *was* owed a duty of care by applying the proximity or neighbourhood tests. Under the Act, the plaintiff merely has to prove that he was a **lawful visitor** in order for a statutory duty to be automatically owed to him.

Who are lawful visitors?

Before the 1957 Act, there were a number of different types of visitors, who were each owed a differing degree of care by the occupier. The categories are worth noting for two reasons. They indicate first the types of visitor who may enter premises, and second, certain visitors who, whilst entering the land quite lawfully, are **not** owed the statutory duty and would have to rely on the Occupiers' Liability Act 1984, ordinary negligence or the old pre-1957 law (see below).

The categories of visitors were as follows:

Category 1 Invitees, i.e. people who have been allowed to enter for a special purpose, and whose presence is in the interest of the occupier, such as a private guest or customer in a public house.

Category 2   Licensees were people who merely had permission to enter the premises, but who held no interest for the occupier, e.g. neighbours' children allowed to enter to recover a ball.

Category 3   Contractors were people who had entered the land by virtue of a contract, e.g. a window cleaner or builder.

Category 4   Persons entering under a statutory or common law right, e.g. health and safety inspectors, policemen, building control inspectors.

Category 5   People in general exercising a public or private right such as a right of way.

Category 6   Trespassers.

Under the 1957 Act, the common duty of care is only owed to lawful visitors, being those in categories 1–4.

Unless the 1984 Act applies, category 5 entrants are still owed only a duty not to **actively** do things which could make the entry dangerous, e.g. dig up the road without warning. This is so if they could not be regarded as licensees. Thus, in the case of *Holden* v. *White & another* 1982, a milkman fell over a manhole cover which was in a state of disrepair in the defendant's pathway. At the time of the incident, he was in fact delivering milk to a third party, who merely had the right of access over the pathway. The Court of Appeal agreed that the defendant was an occupier but did **not** owe a duty to the milkman **under the 1957 Act**. So to sue the defendant, he would have had to sue in negligence and prove fault on her part. Since May 1984, however, such situations may now be covered by the 1984 Act.

Category 6, i.e. trespassers, are not owed any duty under the Act, but they may be owed duties under the 1984 Act (see p. 188).

Who are occupiers?

There is no statutory definition of occupier, although most authorities use Lord Denning's concept of the occupier, being the one in **control** of the premises: *Wheat* v. *Laçon & Co. Ltd* 1966.

There is no need for the occupier to have a legal or equitable interest in the property. So a contractor by virtue of Cl. 23 of the JCT form of contract has a **legal** licence to be in possession of the site. But, in other situations, a building contractor may be an occupier by virtue of his contract to build, say, a garage or extension, which otherwise gives him no legal or equitable interest in the land. Nor is there a need for the occupier to have **exclusive** possession of the property. Thus, the owner of a building site and the building contractor may both be in occupation of the site. Such people may still be in occupation or control even if there are independent contractors working there. Thus, the building contractor may be in occupation and so may a roofing subcontractor. Each separate set of circumstances may give a different result.

By S.2(4)(b), the occupier may in certain situations delegate his responsibility for technical aspects of his occupancy, by employing an independent contractor to deal with that particular aspect. For example, it is common to entrust building maintenance to outside contractors. If such contractors fail to maintain part of the building properly, then the contractors cannot be sued under the Act as they are not occupiers. The plaintiff will have to sue the contractors in negligence. However, the transfer of responsibility is only sound if the occupier appointed properly qualified or experienced contractors to do the work. An occupier may still therefore find himself liable (see p. 186).

What are premises?

The word premises includes land, buildings, ships, diving boards, cranes, ladders, scaffolding, boarding, grandstands, platforms, sea promenades etc. The list is not closed. In *Wheeler* v. *Copas* 1981, a farmer contracted with a bricklayer to help him build a house. At one point, the bricklayer required a long ladder. When he used it, it broke under his weight. Expert witnesses gave evidence that the ladder was unsuitable for building work, although it was suitable for farm work. The bricklayer successfully sued for injuries caused, either by the farmer being negligent in lending the ladder, or as an occupier failing to discharge the common duty of care towards his lawful visitor. Because of the bricklayer's expertise, however, his damages were reduced by 50 per cent due to contributory negligence (see p. 183).

The common duty of care – what levels of care must be attained

The levels of care to be taken are really the same as those found in negligence, and reference should be made to p. 175.

Children: S.2(3)(a)

However, the 1957 Act specifically mentions the special duties owed to children, in that they are expected to behave less carefully than grown-ups. The standard of care owed to child visitors is very high, and gets higher as the child is younger. Nevertheless, in certain cases the occupier will be excused liability, because it would be reasonable to expect a very young child to be accompanied by an older person, as in *Phipps* v. *Rochester Corporation* 1955, where a boy aged 5 fell into a trench when accompanied by his 7 year old sister. The occupier was held not to be liable.

Also, the occupier should not have any **allurements** on his land which attract children. Despite the fact that the Act does not apply to trespassers, if something is kept on the land which is tempting and dangerous, i.e. an allurement, then the occupier may nevertheless be liable for injury. Presumably, this is because the child probably would not have gone on the land but for the allurement. 'Allurements', thus, in this context have included poisonous berries (such as on laburnum) and railway trucks. Building sites may inadvertently provide allurements to children.

It is also possible that a child trespasser may become a licensee to enter land if the occupiers know of their presence and do nothing to stop them from entering, nor forbid them from doing so. If a child trespasser crossed the thin 'dividing line' between trespasser and becoming a licensee, then he is owed the statutory common duty of care. If he remains a trespasser, then he may be owed duties under the 1984 Act or at common law (see p. 188).

Specialists: S.2(3)(b)

The 1957 Act furthermore recognises also that the occupier's standard of care will be lower in the case of visitors who are carrying out their normal work. Such visitors should take care to protect themselves from injury caused by the normal risks of their work, and therefore should not be the responsibility of the occupier. Thus, if demolition contractors are carrying out demolition work of an old house, they should take all reasonable steps to prevent injury to themselves, and cannot sue the occupier under the Act for failing to maintain the house properly, as they should have inspected to view the state of disrepair.

Delegation to a competent independent contractor: S.2(4)(b)

If the occupier delegates certain duties of a specialist nature to a competent contractor, then he also discharges his liability towards persons injured through the fault of the contractor. The injured party would then have to sue the contractor for negligence or breach of statutory duty (see p. 209).

Extent of duties owed to contractual visitors

Contractual visitors may be either those entering by virtue of a contract **with** the occupier, e.g. a building contractor and a subcontractor (where the contractor has possession of the site), or those entering by virtue of a contract but not made with the occupier, e.g. a consulting engineer or architect who is in contract with the employer out of occupation and is permitted to be there because of the main contract. The first situation is governed by S.5, the second by S.3.

Extent of duty owed to visitor in contract with occupier: S.5

For example window cleaner, maintenance worker, subcontractor. Where there is a contract, its terms may regulate behaviour between parties, and obviously the contract must first be examined to see the extent of liability owed to the contracting visitor.

But, if there is no term in the contract, then by S.5(1) the **common duty of care** will be owed. Because there is a contract, the law of contract is

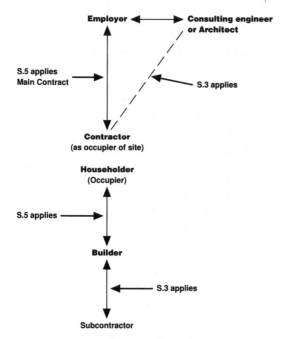

Figure 7.11 The application of S.3 and S.5 of the Occupiers' Liability Act 1957

primarily the basis of an action. But, if the common duty of care is implied (because of the lack of an express term), then the plaintiff can choose to sue for breach of contract or for breach of the Act (see Figure 7.11).

Extent of duty owed to a visitor who is empowered by a contract to enter the occupier's land: S.3
Section 3 entrants are visitors under the 1957 Act, and are thus owed the common duty of care by the occupier. They are not in contract with the occupier, and thus have no choice of action. They can only proceed in tort under the Act.

Excluding or restricting liability: S.2

Pre-1977
The occupier was free to exclude or restrict his liability under the 1957 Act under S.2(1) by 'agreement or otherwise'. Thus, until the Unfair Contract Terms Act 1977, occupiers of property frequently put up notices or inserted clauses in contracts excluding their liability altogether. This **was** perfectly legal and often rendered the 1957 Act of little use to the injured visitor.

(Note: such notices were valid to exclude liability for negligence as well; see p. 183).

After 1977
Occupiers of **business** premises, however, are now subject to the 1977 Act which makes such notices or contractual terms **void**, if they try to exclude liability for **death** or **personal injury** caused by the occupier's breach of any duties, including the common duty of care. By an amendment under the 1984 Act, recreational or educational reasons for entry are not to come within the definition of business purposes unless the business is concerned with providing education or recreation.

It should also be noted that people occupying premises for non-business purposes, e.g. a private house, can still exclude or restrict their liability.

Business occupants may exclude their liability for damage to property providing the exclusion is reasonable in the circumstances.

The 1977 Act does not apply to trespassers, so exclusions of liability or notices excluding liability will be valid against such entrants.

Defences

Warnings: S.2(4), a special defence
If a warning is given of the danger which, if observed by the visitor, would make him safe, then this will absolve the occupier from liability. The examples in Figure 7.10 would be perfectly sound, provided that they would protect the type of visitor expected. So a written warning is useless if blind people are to be expected at a hospital for example (see p. 176).

Volenti non fit injuria: S.2(5) consent and willingness to run the risk
This should be read in conjunction with the rules relating to warnings above. A warning plus the defendant's proven willingness to enter the dangerous premises would therefore be a good defence. Knowledge of the danger or risk from

another source, however, may or may not be grounds for a good defence if the visitor then enters the dangerous premises. The plaintiff would have to prove that the visitor knew of the danger, e.g. because he had visited the site on a previous occasion. Consent can then be pleaded as a defence, but the court would ask the question, why was there not a warning if the occupier knew of the danger to visitors? The duty to warn seems to be implicit in the Act. What is more likely is that the court will use the partial defence of contributory negligence, e.g. *Wheeler* v. *Copas* 1982 (see p. 185). Contributory negligence may occasionally cancel out damages altogether. But this will only be so if the knowledge of the danger would have made the visitor absolutely safe.

Liability of an occupier to 'non-visitors' to his property

The Occupiers' Liability Act 1957 does not apply to trespassers. However, the Occupiers' Liability Act 1984 may, as it imposes a duty of care on an occupier in relation to other people, not being visitors under the 1957 Act. If neither of the Acts apply, then the rules of common law must be used.

The Occupiers' Liability Act 1984

The 1984 Act replaces the common law rules in determining:

- whether any duty is owed by an occupier of premises to persons who are not his visitors in respect of any risk of their suffering injury by reason of any danger due to the state of the premises or to things done or omitted to be done on them;
- if such a duty exists, what that duty is.

The words 'occupier' and 'visitor' are used in the same sense as under the 1957 Act.

To whom is the duty owed?
By S.1(3) of the 1984 Act, an occupier owes a duty to **another** (not being his visitor), in respect of the risks mentioned above if:

1. he is **aware** of the danger or has **reasonable** grounds to believe that it exists;
2. he **knows** or has **reasonable** grounds to believe that the other is in the vicinity of the danger or that he may come into the vicinity – this is so, whether or not the other person had lawful authority to be there; *and*
3. the risk is one against which, in **all** the circumstances of the case, he may be reasonably expected to offer the other some protection.

Thus, in *Pannett* v. *Mcguinness & Co.* 1972 a child was injured on a demolition site. The demolition contractors had been burning rubbish and, knowing of the allurement of bonfires, had employed a watchman to chase the children away. The watchman went absent one day and a child got burnt. The contractors were held vicariously liable for the workman's tort. The children's presence would have been reasonably foreseeable, and they should have made sure that they were kept out.

Similarly in *Herrington* v. *British Railways Board* 1972, a child was electrocuted on the live rail on British Railway property. Evidence was brought to show that the fence through which the child had crawled was in a bad state of repair. Also, it was clear from the path worn in the ground that there was a well-used 'short cut' to a park used by local children. British Railways Board was held liable.

In both those cases, the result would have been the same had they been dealt with under the 1984 Act. In each case, the occupier was aware or had reasonable grounds to believe that a danger existed; they knew or had reasonable grounds to realise that people were in the vicinity of the danger; and lastly, the risk was of the kind which they could have guarded against and protected those people. Such a duty may also have been owed in *Holden* v. *White and another* 1982, had the new Act been enacted earlier.

The extent of the duty
If a duty exists under the 1984 Act, by virtue of applying the pre-conditions, then the duty is 'to take such care as is reasonable in all the circumstances of the case to see that (the other person)

does not suffer injury on the premises by reason of the danger concerned'.

It must, therefore, be noted that the 1984 Act does **not** impose the same duty as under the 1957 Act. The 1984 Act only requires protection from 'the danger concerned'. The 1957 Act requires that the visitor be reasonably safe when using the premises for **all** purposes for which he is visiting. The 1957 duty is thus considerably wider than the 1984 Act.

Warnings

By S.1(5) of the 1984 Act, any duty owed under the Act may **in an appropriate case** be discharged by taking such steps as are reasonable, in all the circumstances, to give warning of the danger concerned or to **discourage** persons from incurring the risk.

It will be interesting to see how the courts interpret this section in the future.

Acceptance of risk

By S.1(6) no duty will be owed to someone who willingly accepted the risks of a dangerous situation (see p. 182).

Not applicable to highways

The 1984 Act does not apply to highways. The common law rules are therefore applied.

Finally, it should be noted that the 1984 Act does not extend the duty to cover damage to property as does the 1957 Act.

At common law

In situations in which neither of the Acts apply, then only minimal duties are owed at common law to persons entering premises occupied by another.

Persons who enter premises by reason of exercising a public or private right of way are only owed a duty in tort not to actively make the exercise of that right dangerous. This would probably also cover highways. However, there are many other duties owed under the Highways Act 1980 which impose much more stringent duties on many categories of people, not just occupiers.

As far as other situations are concerned, it is possible that we have now returned to the old pre-Herrington situation based on a House of Lords decision of the 1920s, *Addie & Sons* v. *Dumbreck* 1929. That case said that an occupier should not intentionally injure uninvited entrants nor should they recklessly disregard their presence on land. Herrington's case in 1972, however, regarded that decision as socially unacceptable in the twentieth century and substituted a test of humanity. Whether this test has survived the 1984 Act is open to conjecture. In the writer's opinion, it probably has, as there are bound to be cases which are unable to satisfy the pre-conditions of S.1 of the 1984 Act.

The test of humanity takes into account the likelihood of the presence of the non-visitor; the type of entrant (children obviously being more favoured than vandals and burglars); and also whether an occupier has the finances to eradicate a source of harm. Fortunately, with the passing of the 1984 Act, few cases will fall into this category.

Non-occupiers responsibility for dangerous premises

On the basis of the ordinary duty of care in negligence, people who create dangers should guard against injury. So a builder who is not in either sole or joint occupation of a site, but perhaps is doing some repair or maintenance work, must still discharge his duty of care by preventing accidents.

He can do this by warnings, guarding dangerous premises or preventing people from entering. In *Billings (AC) & Sons Ltd* v. *Riden* 1958, such a duty of care was said to be owed by a building contractor to persons he could reasonably foresee would be affected by his work. A warning may be insufficient if this does not make the premises safe.

Non-occupiers who create reasonably foreseeable risks may also be liable to **trespassers** in negligence if there was a likelihood of a trespasser being injured.

The above situations are concerned with people who are working on the property, even if not in technical occupation.

Amazingly, however, before the 1970s, people who merely sold or leased premises which they had worked on causing them to be unsafe were virtually immune from action, as they were no longer occupiers. It is a true case of caveat emptor, i.e. let the buyer/lessee beware. This was so unless there was some other basis for a claim, such as a breach of contract. This is still the case where someone merely sells premises and fails to point out dangerous defects in the property. It is only where they tell lies or actively do something to make the place unsafe, or misrepresent facts to induce a contract, that the misled buyer has any course of action (see p. 72).

Thus, builders who owned the land on which they built were in a strong protected position compared with builders who built for other people. Landowner/builders were immune from actions in **negligence** (but not in contract) and so were lessor/builders. Non-owner or non-lessor builders were not immune. The law was obviously wrong and inequitable and a number of things occurred to change this position.

First, the Defective Premises Act 1972 imposes liability on landowner/builders to build in a workmanlike manner with proper materials, and to make a dwelling fit for habitation. Although houses covered by the National House Building Council Scheme were excluded from the provisions of the Act, being an 'approved' scheme, this is no longer the case as the scheme is not now submitted for approval.

If there has been a breach of the Building Regulations, then there is also statutory liability for damage caused under S.38 Building Act 1984. This section is still not in force.

As far as lessors are concerned, the Defective Premises Act 1972 imposes liability in negligence on lessors under a statutory obligation to repair, or where the lease or tenancy agreement contains a lessor's repairing covenant, and provided the lessor is aware of the defect or should be aware of it. Thus, lessor/builders are now covered by the Act. But, the Act is not all embracing, and there are some areas which are still uncertain, and liability may not be imposed.

The rule in *Rylands* v. *Fletcher* 1868: occupier's liability for dangerous things escaping

This is an example of a case creating a new area of liability in tort. Once again it was formulated before *Donoghue* v. *Stevenson* 1932 and thus has developed in a different direction to that of negligence. It is so restricted in its application that it would be an unusual occurrence in the building industry. Indeed, the Pearson Commission 1978 thought it would be better if it was abolished in relation to death and personal injury and a new Act be passed imposing strict liability in certain specified situations. Nevertheless, it is of importance because it is a tort of strict liability. The defendant is strictly liable for damage caused in circumstances covered by the rule, whether or not there has been negligence or whether or not a nuisance has been committed. The harm and the escape thus caused does not have to be foreseeable. The case concerned a mineowner and a millowner. The mineowner plaintiff had leased land for mining purposes which were adjacent to the defendant millowner's land under which there were some disused mineshafts (see Figure 7.12). The millowner needed more water for his mill and contracted with independent contractors to excavate a reservoir on this land to increase the water supply. The contractors failed to seal up the disused mine shafts which they did not realise connected with the mines next door. On filling the reservoir the plaintiff's mine became flooded causing severe damage. The plaintiff sued and won the case in the House of Lords, thereby establishing a new area of liability. Previous cases of a similar nature had been based entirely on precedent. The dicta in *Rylands* v. *Fletcher* established requirements for liability which could be found in certain situations. Thus the defendant must take onto land in his occupation, something which is not naturally there for his own use, which is likely to cause mischief and then escapes. If this occurs then the defendant is responsible for all the natural consequences of the escape, even if there is no negligence. Modern cases, however, are now

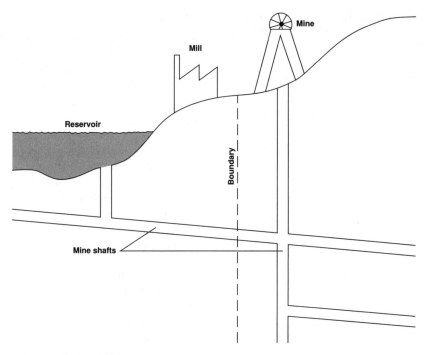

Figure 7.12 *Rylands* v. *Fletcher* 1868

closely aligning the rule with private nuisance and most cases will involve claims under both heads. Despite the occasional granting of damages for personal injury, it is primarily a land oriented tort.

A recent House of Lords decision, *Cambridge Water Co.* v. *Eastern Counties Leather Plc* 1994 2WLR 53 has introduced more doubts concerning the modern relevance of the rule. This case was concerned with a tannery belonging to the defendants. There they used a chlorinated solvent called PCE. Just over a mile away the plaintiffs' water company had a borehole from which they extracted water for domestic use. The solvent from the tannery seeped into the ground beneath, thereby adulterating the aquifer which supplied the water in the borehole. The plaintiffs sued for damages under the rule (as well as nuisance). The case finally went to the House of Lords. The Law Lords contended that the contamination of the water would not have been foreseeable nor was the damage which actually occurred foreseeable and they did not find the defendants liable.

The elements of the rule

Occupation of land
The rule applies to persons in ordinary occupation of land. They do not need to actually own it. Thus, a building contractor in possession of a site will come within the scope of the rule.

The 'thing' must not be naturally present on the land
Otherwise, the defendant could not have taken it there in the first place. But if he fails to control natural occurrences such as rainwater or wild rats, then he may be liable for nuisance or negligence (see p. 193).

The 'thing' does not have to benefit the defendant
He merely has to bring it there for his own use. It does not have to be for his own good. Thus, it has been suggested that local authorities could possibly be responsible for the escape of sewage from their sewage works without proof of negligence.

The 'thing' must be dangerous if it escapes

This does not mean that the thing itself must be inherently dangerous, only that it will cause mischief if it escapes, as in the facts of the case. The water was inherently harmless. It was only when the water flooded the mineshaft that it caused the damage. Since the Cambridge Water case, however, if it is not foreseeable that a particular type of damage will occur, then there will be no liability.

There must be an escape of the 'thing'

In *Read* v. *J Lyons & Co.* 1947 a ministry factory inspector was injured in the defendant's shell-filling shop. She sued only under *Rylands* v. *Fletcher*. As there was no escape from the defendant's property, the plaintiff lost her case.

The escape and damage must be foreseeable

In the Cambridge Water case it was not known nor foreseen that the chemicals would escape and, even if they had foreseen the escape, the damage that in fact occurred was also unforeseeable.

Examples of the rule being applied

The rule has been applied to the following: water – *Rylands* v. *Fletcher* 1868; petrol in car tank causing a fire which spreads – *Musgrove* v. *Pandelis* 1919; explosives – *Rainham Chemical Works* v. *Belvedere Fish Guanoco Ltd* 1921; poisonous tree planted whose leaves fall elsewhere – *Crowhurst* v. *Amersham Burial Board* 1878; gas escaping – *Batcheller* v. *Tunbridge Wells Gas Co.* 1901; electricity – *National Telephone Co.* v. *Baker* 1893.

The defendants must use the land in a non-natural way

This is defined quite normally and means that a special use which increases the possibility of harm to others must be shown. So ordinary use is *not* 'non-natural'. Nor must the use be for the general benefit of the community. It appears therefore that *Rylands* v. *Fletcher* 1868 will thus be applied in exceptional circumstances.

Remedies

The plaintiff may recover damages for physical damage to property, possibly for personal injury although this is predominantly a land protecting tort, but probably not for mere pecuniary loss.

Defences

Act of God

This covers uncontrollable events such as earthquakes and lightning. Such extremes, with modern knowledge and methods of forecasting and dealing with such eventualities, have meant that this defence has not been used in English law cases in recent years.

Plaintiff's consent

Consent to the thing being taken onto the defendant's land may be expressed or implied. If express consent has been given there will definitely be a good defence. Even if it is only implied from acquiescent behaviour this may well be a good defence.

Statutory authority

Where bodies such as public limited companies act under statutory **duties** to supply gas, electricity, water etc., the rule is excluded and the plaintiff would have to prove negligence. If the organisation only has a statutory power to do something, i.e. they may do something if they wish, then they can be sued under the rule.

Unforeseeable act of a stranger

The rule only imposes strict liability for things within the occupier's control. An unforeseeable event which is caused by a stranger is a good defence. Note that a stranger is someone such as a trespasser who was not reasonably expected to be there, e.g. a vandal. Also this defence probably does not apply where the stranger does something negligent (which is reasonably foreseeable).

Nuisance

There are three types of nuisance. **Public** nuisance is a crime created by the common law. **Private** nuisance is a tort, and **statutory** nuisance is nuisance created by Act of Parliament. All three types may possibly be committed during building work.

Public nuisance

A public nuisance has been defined as 'an act or omission which materially affects the reasonable comfort and convenience of life of a class of Her Majesty's subjects': *AG* v. *PYA Quarries Ltd* 1957.

Thus, for a nuisance to be public, it must affect a large enough class of people, such as a village. Because of its ancient beginnings, it has embraced different types of behaviour, which one would not normally consider to be nuisances. Thus, obstructing a highway or making a highway dangerous are public nuisances, as are keeping a brothel and polluting a public water supply, holding a pop festival causing noise, smells etc.

Public nuisances are indictable offences, and therefore they can be tried in the Crown Court in the usual way. They can also be dealt with in a civil **relator** action by the Attorney General or the local authority under S.222 of the Local Government Act 1972, who act on behalf of the affected public.

Once the criminal or relator case has been proved, if an individual member of the public, not necessarily a landowner, particularly suffers as a result of the nuisance, then that individual can sue in tort and get damages, even for personal injury, e.g. *Halsey* v. *Esso Petroleum Co. Ltd* 1961, where there was an excess of noise, smell and acid smuts caused by the defendant's business. Otherwise there is usually no right to sue for damage in respect of public nuisance.

It should be noted that, as public nuisances are essentially crimes, it is **no** defence to say that there has been consent to the nuisance, as one can never consent to a crime. But one can always consent to a tort being committed.

There does not have to be any indirect invasion of private land, as there has to be in private nuisance.

Builders are less likely to be involved in a public nuisance action than in private nuisance, unless the operation of their building yard or a long-term building project causes a large number of people to be affected by their activities.

Public nuisance on highways

The special dangers that users of highways face are reflected in this aspect of public nuisance.

A public nuisance on a highway may be committed by someone obstructing the highway. The obstruction must either be permanent, such as a wall, or be of sufficient duration to be a nuisance. Thus, an obstruction for a reasonably short period may not necessarily be a nuisance.

A public nuisance may also be committed by someone who occupies premises close to a highway. If he fails to keep them in proper repair and if they then collapse and injure someone, then that person will be able to sue, whether or not the owner or occupier knew of the danger: *Wringe* v. *Cohen* 1940. The law takes the view that he should have known of the danger.

It should be noted that such liability is only concerned with damage caused by disrepair, and not by natural causes or dangers caused by trespassers. This, of course, is of great concern to builders, who should maintain property close to highways, not just to avoid prosecution under the Health and Safety at Work etc. Act 1974, nor to avoid claims for negligence (which requires proof of a degree of carelessness) but to avoid this special liability.

Private nuisance – a tort

Definition

A private nuisance has been defined by Professor Winfield as an unlawful interference with a person's use or enjoyment of land or some right over or in connection with enjoyment of the neighbour's land or some right over or in connection with it.

Such interferences may be caused by noise, smoke, smell, water, gas, fumes, roots or any type of behaviour which causes the neighbour to be unable to use or enjoy his property. The courts agree that the list of nuisances is by no means closed. Thus, it can be seen that the building site operator is always in danger of committing a nuisance. It must be noted that this tort protects interest in **land** only; such interests include interference with easements or profits à prendre (see p. 97).

Criteria for suing in nuisance
Unlike negligence, there is only one essential element to be found. But many other factors have to be taken into account when deciding whether a nuisance has, or has not, been committed.

Damage or harm must have been caused – the essential element
Nuisance is like negligence. There *must* be some damage or harm suffered on the part of the plaintiff in relation to his land.

This may take the form of physical damage to the property which could be caused by vibrations, roots undermining walls, or water. If there is physical damage, the court looks no further at whether damage has been caused and is less likely to investigate its triviality. Indeed the court may even presume it.

If, on the other hand, the damage is an interference with the use of land or its enjoyment, then it must not be trivial. The courts take the view that there should be 'give and take' between neighbours. So an occasional smoky, smelly fire or a noisy do-it-yourself session would not be actionable. But weekly bonfires or continuous noisy drilling and sawing probably would amount to a nuisance. The damage that results in these situations would be that the plaintiff is unable to use a room or his garden in the normal way because of the nuisance. Damage may also be presumed in nuisance involving interference with easements or profits à prendre.

The behaviour must be unusual, excessive, unreasonable
This follows on from the previous section. Normal, reasonable and moderate use of property cannot be actionable. Building work, which has to be done almost continuously for a particular period, could thus amount to being excessive and possibly unreasonable.

In *Andreae* v. *Selfridge & Co. Ltd* 1938, demolition contractors caused an unreasonable amount of dust and noise, which affected the plaintiff's hotel trade to its detriment.

The behaviour must have gone on for some time
Generally, an isolated act will not be actionable. But, the isolated incident arising out of a continuing state of affairs may indicate an interference with the use of the property. Thus, if builders habitually throw down materials close to another's land, causing the occupier to fear being hit, then the one occasion that, for example, a tool lands on that property may amount to evidence of the existence of the nuisance.

The nuisance must be caused by another person on neighbouring property and not on the plaintiff's own premises

Character of the neighbourhood
If the plaintiff's land is in a built up area, the defendant is not allowed to claim that his behaviour causes no nuisance, merely because of that. But it is reasonable to expect less personal comfort in towns and industrial areas compared with, say, the tranquillity of the Hebrides. It will thus affect the question 'is the damage trivial?'

In *Gillingham Borough Council* v. *Medway (Chatham) Dock Co. Ltd* 1993 the grant of planning permission to operate a commercial dockyard on part of the former Chatham Royal Dockyard Ltd undoubtedly changed the nature of the neighbourhood from relatively quiet and residential to extremely noisy and commercial. An application by the Council (yes, the same) for an injunction for public nuisance was turned down. The court specifically refused to equate planning permission with statutory authority (see p. 197). But from that case and subsequent decisions such as *Wheeler* v. *JJ Saunders Ltd* 1995 regarding the keeping of pigs and *Hunter* v. *Canary Wharf Ltd* 1997, this approach seems to have been followed in respect of applications for injunctions but not necessarily in relation to applications for damages or the grant of damages in lieu of an injunction. The building of Canary Wharf in London caused interference with television reception and the building of the related Limehouse Link Road caused a large deposit of dust. The Court of Appeal and subsequently, on appeal, the House of Lords held that the interference with reception could not be an actionable nuisance. The Court of Appeal agreed that in granting planning permission no immunity from action in nuisance was being given. (This point was not appealed against.)

Is it necessary to prove that the defendant behaved maliciously?

No. Malice is not an element of the tort. It is in torts such as malicious prosecution. But malicious behaviour may sometimes tip the balance in favour of the plaintiff.

Could the nuisance have been prevented easily?

If it could have been prevented for a relatively small cost, then the nuisance should have been stopped, and the court will take this into account.

This is important when considering nuisances caused by natural things on land. In *Leakey* v. *National Trust* 1980, houseowners, close to the defendants' land in which there was a large natural mound called 'Burrow Mump', were caused annoyance by earth and rubble which fell naturally from the 'Mump' onto their land. After adverse weather conditions, a large crack appeared in the mound. The houseowners reported this to the defendants, who said they were not liable for natural earth movements. Later, earth and tree stumps fell against a house. The houseowners sued the defendants in nuisance. In the Court of Appeal, the judges said that the defendants were under a duty to remove or eradicate the nuisance, even if it was naturally caused.

Abnormally sensitive plaintiffs

The tort only protects normal use of property. If, however, the plaintiff proves that there is a nuisance anyway, the plaintiff with a special use of the property or an abnormal sensitivity to such nuisances can claim greater damages than the average plaintiff.

Reasonable foresight

In the House of Lords decision *Cambridge Water Co.* v. *Eastern Counties Leather Plc* 1994 it is now clear that reasonable foresight of the type of harm caused is necessary in order to sue successfully in nuisance (and under the rule in *Rylands* v. *Fletcher*). (For the facts in the Cambridge case and greater discussion see p. 191.)

What sort of plaintiff may sue?

As the tort only protects interests in land normally, the plaintiff must prove that he has a right to enjoy the land itself or an easement over that land. Thus, he must normally be in possession, unless the damage is physical and affects the property in some permanent way, in which case persons out of possession may sue.

So, owner/occupiers, tenants in possession, licensees in possession and lessors (if their reversion has been damaged) can sue. In *Khorasandjihan* v. *Bush* 1993 some doubt was cast on the traditional approach that the plaintiff must have some interest in the land. The case involved the daughter of the house receiving nuisance telephone calls. Her case requesting an injunction against the maker was successful despite the fact that she had no interest in the property. Unlike a wife who has a right to reside in the matrimonial home, children have no rights in relation to the family home. The Court of Appeal went against its longstanding decision in *Malone* v. *Lasky* 1907 and was influenced by the persuasive Canadian precedent of *Motherwell* v. *Motherwell* 1976, a decision of the Appellate Division of the Alberta Supreme Court. However, in the appeal to the House of Lords in *Hunter* v. *Canary Wharf Ltd* 1997, one of the grounds for appeal was in relation to whether a mere licensee could initiate an action for private nuisance. Their Lordships specifically dealt with the case of *Khorasandjihan* v. *Bush* above. They felt that to allow mere licensees to sue was to face the problem of defining the category. Where does one stop? In the Court of Appeal, following Khorasandjihan, the category included those with a 'substantial link' with the land. Those who called a particular house their home had that substantial link. The Lords, however, said that whilst such a categorisation obviously included spouses or partners, their children and possibly other relatives, did this also include lodgers, au pairs and resident nurses? The way in which this interpretation was going was unacceptable in relation to a 'land' based tort and was encouraging the idea that it could protect the person. As nuisance requires a less rigorous approach than negligence, requiring there to be a balance struck between neighbours, *Khorasandjihan* v. *Bush* had to be overruled in this connection. Therefore only those residents with rights of occupation would be able to claim for the dust nuisance from the road. Mere licensees therefore have no rights in relation to private nuisance.

Who can be sued?

The occupier

Obviously he should be primarily liable as he is in control of the premises. Vicarious liability is also a factor (see p. 163). He will be responsible for his own and his family's acts, those of his employees, independent contractors (possibly, depending on the degree of control), his invited guests and persons who have licences to enter his property and whom he fails to control.

If one looks at the building contractor who is in prime control of his site, this is a heavy burden.

The occupier is even responsible for nuisances which were created before he came to the property, e.g. by previous owners blocking the neighbour's light, which the new owner fails to remedy. If he makes use of something which is a nuisance, he is said to adopt the nuisance, and he is more likely to be liable than if he merely failed to remedy the nuisance.

Creator of the nuisance

He is always responsible for a nuisance, even if he has left the land on which he created the nuisance and can no longer abate the nuisance.

Persons authorising a nuisance

If a person lets land for a purpose, which by its very nature entails the commission of a nuisance, then that person may be liable in tort. This applies even if it is only foreseeable that the nuisance may be committed by the occupiers. By including a clause in the letting agreement prohibiting behaviour which could amount to a nuisance, the lessor cannot be said to authorise the nuisance.

Liability for the nuisance of a trespasser only occurs if the occupier has been negligent in discovering the existence of the nuisance.

An occupier who carries on or adopts a nuisance created by a former occupier will obviously be liable for nuisance: *Sedleigh Denfield* v. *O'Callaghan* 1940.

In *Sampson* v. *Hodson-Pressinger* 1981 a new landlord of a building previously converted into flats was liable in nuisance as he knew, at the time of the assignment of the freehold reversion to him, that a roof terrace allowed noise to penetrate to the flat below. He was effectively authorising the nuisance (see below). The Cambridge Water Co. case confirmed the position that, once one knows or should know of the nuisance, then the occupier will be responsible for it if he fails to take reasonable steps to remove it.

Defences

Consent

See p. 182. Even if there has been consent, the courts may still take the view that the complainants did not know to what extent they had consented and may still allow them to proceed in nuisance – or they may grant a limited injunction.

Prescription

See p. 99. In order to gain the legal right to continue to commit a nuisance by prescription, the owner of the land must prove that he has been committing the nuisance for 20 years at least. He must not have done this with the other landowner's permission, nor must he have done it secretly, nor used force to exercise it (*nec vi, nec clam, nec precario* – without force, secrecy or permission). But, the behaviour amounting to a nuisance should be capable of existing as an easement (see p. 97). So, in practice, prescription may be of little use, especially for builders, who are generally not going to be in occupation of land for 20 years.

Statutory authority

If an Act of Parliament or a relevant rule or regulation imposes a duty on someone or somebody to do something which inevitably will cause a nuisance, then the authority will be a defence, provided that no negligence was attached to the action. It will be up to the defendant to prove that he took care to avoid negligence. This is so even if the Act expressly made the defendant liable for nuisances.

There are slightly different rules in relation to **powers**. (Duties **must** be carried out but powers only **may** be.) Once again there must be no negligence on the part of the defendant but this time the statute must either expressly exclude liability for nuisance or it must make no mention of the matter. In *Allen* v. *Gulf Oil Refinery Ltd* 1981, under powers given by the Gulf Oil Refinery Act 1965, an oil refinery was built. Local villagers complained about noxious odours, vibrations and excessive noise. The House of Lords held that the section of the Act relied on gave the company immunity from any action for nuisance which might be the inevitable result of building such a refinery.

Remedies

At common law

Damages. On proving his case, the plaintiff is entitled to compensation for the actual damage caused to his land, such as depreciation or loss of business. Entitlement is as of right, as it is a *legal* remedy, and even if the plaintiff has behaved inequitably (see below) the court cannot refuse damages if he satisfactorily proved his case and there are no adequate defences. The court will take into account and offset the loss caused by reasonable behaviour, when the job which caused the nuisance was necessary. For example, in *Andreae* v. *Selfridge & Co. Ltd* 1938, there would have been loss of custom due to the work anyway, but not as much as was actually caused.

Where the nuisance is a continuing nuisance, the plaintiff can recover damages for remedying it, even though he has acquired the interest in the property after the nuisance had commenced. For example, in *Masters* v. *Brent LBC* 1978, the council

planted trees outside a house whose roots caused subsidence, thereby undermining the foundations. The plaintiff purchased the house later and was able to sue and recover the cost of remedying the damage.

In equity

1. Injunction. This is a court order which orders someone to do or not to do some specific thing. Usually, in relation to noise or other building work nuisance, it orders the defendant to stop the work or behaviour which amounts to a nuisance. Such injunctions are called **prohibitory** or **restrictive** injunctions. They may be awarded before the case actually comes to court for trial (**interlocutory** or **interim**) or if they are awarded at the trial they are called **perpetual**. However, this does not mean that the injunction necessarily carries on in perpetuity.

Because an injunction is an equitable remedy, it is **discretionary** and will only be awarded if damages are an inadequate remedy on their own. If there is no likelihood of a repeat of the nuisance then an injunction would not be the appropriate remedy.

In using the court's discretion the equitable maxims such as 'He who seeks equity must do equity' and 'Delay defeats equity' will be taken into account by the court. The first means that the plaintiff must have behaved equitably or fairly if he is going to avail himself of an equitable remedy, e.g. if he had told his neighbour that he did not mind the Senior Citizens tap dancing class practising every week next door then he may find that he will be refused an injunction. (But if he proves a nuisance exists he can still get damages which is his by right.)

The second maxim means that if one requires an equitable remedy the plaintiff must not delay and must go to court as soon as he can.

In *Cambridge Water Co.* v. *Eastern Counties Leather Plc* 1994 it was pointed out that an injunction should be awarded to prevent future harm and that those cases where injunctions are sought should be carefully distinguished from those where the main remedy sought is damages.

Often, a request for an injunction accompanies one for damages. The injunction may be permanent or it may request modification of behaviour, e.g. operations to be carried out during specified hours only. An injunction may also be granted but suspended in order to give the defendant the opportunity of modifying his behaviour.

Quia timet injunctions may be granted even though there has been no actual nuisance, only threatened. However, in order to obtain such an injunction one would have to show that there was imminent danger of irreparable damage (in nuisance, to land).

Historically, injunctions were only granted by the Court of Chancery (the court applying Equity) and the court had no power to grant damages as this was a legal remedy only available in the common law courts. However, the Chancery Amendment Act 1858 (Lord Cairns Act) changed the law and permitted the award of damages in lieu of or in addition to an injunction, e.g. where only an injunction had been asked for. The remedy is now found in S.50 Supreme Court Act 1981. (Damages under this heading could be granted even in lieu of a quia timet injunction but this would be extremely unusual as it would allow the award of damages even though there had been no tort committed and thus no possibility of damages being awarded at common law!)

The leading case *Shelfer* v. *City of London Electric Lighting Co.* 1895 which involved nuisance caused by noise and vibration stated that damages in lieu of injunctions would only be awarded if there was

- only a small injury
- which could be estimated in money terms,
- which could be adequately compensated by a small amount of damages *and*
- the award of an injunction would be oppressive. But each case must be looked at individually and in the light of the equitable maxims.

Consequences of breach of an injunction. If someone carries on with acts in breach of the terms of the injunction then they will be in contempt of court and could be fined or sent to prison. Anyone knowingly aiding them in breach could

also be dealt with for contempt even if not party to the original action. If a company or other type of corporation is in breach then its property may be sequestrated and its officers committed for contempt.

2. Non-judicial action – abatement of nuisance. Self-help is not usually advisable. The person suffering from the nuisance, after giving the other party notice, enters the land of the person committing the nuisance and does whatever necessary to abate it. He must not do anything which causes unnecessary damage and, if there are alternative methods of abatement, then he must choose one that causes least harm. However, in so doing he is committing the torts of trespass to land and property. Fortunately, abatement is a defence to trespass. If someone does exercise his right to abate he cannot then sue for damages in nuisance. (There are often better remedies under legislation.)

Statutory nuisances

The law relating to statutory nuisances has been consolidated and updated in recent years. The main legislation affecting builders is the Control of Pollution Act 1974 and the Environmental Protection Act 1990 (the 1990 Act) Part III Ss. 79–82 as amended and supplemented by the Noise and Statutory Nuisance Act 1993 (the 1993 Act). Part III of the 1990 Act is concerned with statutory nuisances in general as amended by the 1993 Act. (Note that the Noise Act 1996 will not really be of relevance to building work as it is primarily concerned with night-time noise and few councils appear, at present, to be adopting its provisions.) The following is merely an outline of some statutory nuisances which could be committed during building work.

Section 79 defines certain matters which constitute 'statutory nuisances'.

Section 79(1)(a): any premises in such a state as to be prejudicial to health or a nuisance
Whilst it is not envisaged that builders would usually be affected by this section, if they are

landlords they could find themselves liable. Poor insulation, poor heating and ventilation have been the subject of cases against council landlords.

Section 79(1)(b): smoke emitted from premises so as to be prejudicial to health or a nuisance
This replaced the equivalent law found in the Clean Air Act 1956. Exceptions to this section include smoke from a chimney of a private dwelling in a smoke control area; dark smoke emitted from a chimney or a chimney serving the furnace of a boiler or industrial plant attached to a building or for the time being installed on any land; smoke from a railway locomotive; and dark smoke emitted, other than above, from industrial or trade premises. (But see p. 204).

Section 79(1)(c): fumes or gases emitted from premises so as to be prejudicial to health or a nuisance

Section 79(1)(d): any dust, steam, smell or other effluvia arising on industrial, trade or business premises and being prejudicial to health or a nuisance
Dust is probably the biggest problem for the builder in this respect.

Section 79(1)(e): any accumulation or deposit which is prejudicial to health or a nuisance

Section 79(1)(f): any animal kept in such a place or manner as to be prejudicial to health or a nuisance
This could include guard dogs.

Section 79(1)(g): noise emitted from premises so as to be prejudicial to health or a nuisance
Note that by S.79(7) noise includes vibration.

Section 79(1)(ga): noise that is prejudicial to health or a nuisance and is emitted from or caused by a vehicle, machinery or equipment in a street
This was introduced by the 1993 Act. This subsection does not apply to noise made by traffic.

Section 79(1)(h): any other matter declared by any enactment to be a statutory nuisance
It can be seen therefore that building work could easily cause these types of nuisance and the site operator could then be subject to delay if he ignores an abatement notice and incurs prosecution.

Under S.79 of the 1990 Act the local authority must inspect its area from time to time in order to detect any statutory nuisances which ought to be dealt with under S.80 below and, where a complaint of a statutory nuisance is made to it by a person living within its area, to take such steps as are reasonably practicable to investigate the complaint. If a local authority is satisfied that a noise amounting to a nuisance exists or is likely to occur or recur in the area, then it shall serve a notice imposing all or any of the following requirements:

- it may require the nuisance to be abated;
- it may prohibit or restrict the occurrence or recurrence of the nuisance;
- it may require the execution of such works and the taking of such steps as may be necessary to give effect to the requirements or as may be specified within the notice: S.80.

The notice must specify the period in which the requirements must be carried out. It must also contain the time or times within which the requirements must be complied with. Normally this should be a reasonable time but in cases under S.58 1974 Act (the equivalent of S.80) it has been held that no time need be specified and indeed in certain cases a very short time would be sufficient, depending on the facts. In *Strathclyde Regional Council* v. *Tudhope* 1983 the City of Glasgow Council served a S.58 notice on the Strathclyde Authority in respect of noisy roadworking operations, requiring that all pneumatic drills should be fitted with effective exhaust silencers and dampened tool bits. No time was stated for compliance and the Court held that the notice should come into effect at midnight following the date of service and that this was not unreasonable.

Should the abatement notice contain a prohibition on the occurrence or recurrence of the nuisance then this prohibition continues indefinitely.

It must be served on the person responsible for the nuisance. The 'person responsible' is defined in S.79(7) as the person to whose **act**, **default** or **sufferance** the nuisance is attributable. This shows that the person may be responsible in a positive way, by actually doing the act, or in a negative way, by not doing something he should be doing or by allowing a state of affairs to continue. If he cannot be found, it must be served on the owner or the occupier of the premises from where the nuisance is or would be emitted: S.80(2). Thus, the notice can be served on landlords even though it is the tenant who is creating the nuisance.

Section 80 thus gives the local authority wide-ranging powers. Not only may it give notice to stop or reduce existing nuisance, but it may also restrict the emission during certain parts of the day or week. Furthermore, a notice may be served where a nuisance is *likely* to occur, but has not yet started, e.g. where site work is going to commence. (This must be compared with a civil action for nuisance where the sufferer can only sue where there has been nuisance behaviour on more than one occasion (see p. 194). Normally civil action can never be used for behaviour which may occur in the future.) The carrying out of works necessary to give effect to the notice is a newer concept and may entail modifying machinery, making a building 'soundproof' or even moving the cause of the nuisance to a place where it is less likely to create a nuisance. The local authority no longer needs to specify what works must be carried out to comply with the notice. Now the authority must only 'require' the execution of such works and the taking of such steps as may be necessary to fulfil the terms of the abatement notice. These are vaguer words which allow the authority to show that work must be done without having to specify what exactly has to be done. It has happened in the past that the specified works have been carried out and yet the nuisance remained.

Should the work not be carried out as requested, the local authority can execute the works and recover the cost from the recipient of the notice: S.81(3).

Appeals

By S.80(3) appeals against the notice may be made to the magistrates' court within 21 days of receiving the notice. The Secretary of State has made regulations in relation to such appeals. These are the Statutory Nuisance (Appeals) Regulations 1990 made under this Act and the Control of Pollution Act 1974. There are various grounds for appeals. Perhaps the most important for builders, where the nuisance to which the notice relates is noise caused in the course of a trade or business, is that the best practicable means have been used for preventing or for counteracting the effect of the noise.

On hearing the appeal, the court may quash the notice, vary it in favour of the appellant or dismiss the appeal.

Furthermore, the regulations specify two situations in which a notice will be suspended:

- when the notice requires expenditure on works before the appeal has been heard; and
- in the case of category (g) or (ga) when the noise to which the notice relates is caused in the course of the performance of a legal duty, e.g. by a statutory undertaker such as a gas company.

But the notice shall not be suspended in the above situations if the noise is injurious to health or is of such limited duration that suspension would render the notice of no practical effect or the expenditure incurred on the works before the hearing would not be disproportionate to the public benefit. In such cases the notice must include a statement that it shall not be suspended pending appeal. Such a statement was included in the leading case, *Hammersmith LBC* v. *Magnum Automated Forecourts Ltd* 1978 (see below).

Breach of notice

By S.80(4) if the recipient of a notice contravenes any requirement contained in the notice without reasonable excuse then he commits an offence against this part of the Act. Summary proceedings will then proceed in the magistrates' court. If the offence is committed on **industrial, trade** or **business premises** then the guilty person will be liable to a fine on summary conviction not exceeding £20,000: S.81(6). But if the offence is committed elsewhere, then the fine will be not more than level 5 on the standard scale with a further fine equal to one-tenth of that level for each day the offence continues.

Defences

There are a number of defences. First, it is a defence if the recipient had 'reasonable excuse' to breach the abatement notice. This will depend on the facts of the situation but would probably cover emergency situations.

Second, if the nuisance was caused in the course of a trade or business, then it is a defence to prove that the 'best practicable means' have been used for preventing or counteracting the effects of the nuisance: S.80(3). Section 79(9) defines 'best practicable means': 'practicable' means reasonably practicable, having regard amongst other things to local conditions and circumstances, to the current state of technical knowledge and to the financial implications. In *Wivenhoe Port* v. *Colchester Borough Council* 1985, a case concerned with dust nuisance, the court held that the defence was not maintained by proving that the business would require extra money to operate or even would become unprofitable. It is therefore possible that a factory, vital to the community, may have the sympathy of the court if vast sums of money are required to eradicate the nuisance. 'The means to be employed' include the design, installation, maintenance and manner of operation of plant and machinery and the design, construction and maintenance of buildings and structures. The test, according to the section, is to apply only so far as is compatible with any duty imposed by law. Thus, if there are legal duties imposed at common law in nuisance or in negligence or in statute in the same situation, then this test must be compatible with those duties.

Finally the test is to apply only so far as compatible with safety and safe working conditions, and with the exigencies of any emergency or unforeseeable circumstances.

In addition to the 'best practicable means' defence, which *only* may be used in relation to trade or business circumstances, S.80(6) sets out other possible defences.

It will be a defence if the alleged offence was in relation to a notice served under S.60 Control of Pollution Act 1974 or consent had been granted under S.61 Control of Pollution Act 1974. These sections relate to construction sites (see below).

Section 80(4): the right to take High Court action

This section allows the local authority to take proceedings in the High Court, if it feels that proceedings for an offence under S.80(4) would be an inadequate remedy. In this way a statutory nuisance may by injunction be restricted, abated or prohibited. This is so even though the local authority has suffered no damage from the nuisance.

This may be instead of, or as is possibly more usual, following failure of summary proceedings in the magistrates' court. This occurred in *Hammersmith LBC* v. *Magnum Automated Forecourts Ltd* 1978. In 1976, the defendant company erected a new building in a quiet street and used it for a 'taxi care centre'. This provided 24-hour service for taxis selling fuel, washing facilities and vending machines. People living nearby complained about the noise, particularly in the early hours of the morning. On being satisfied that a noise amounting to a nuisance existed, the local authority served notice under the old Control of Pollution Act 1974 S.58(1) requiring Magnum Ltd to stop all activities on the premises between 11 pm and 7 am. In the notice, the authority had stated that the expenditure necessary to comply with the notice was not disproportionate to the public benefit. Thus, the notice was not suspended pending any appeal. The defendants continued to operate as before. They also appealed against the notice. The local authority therefore decided to apply to the High Court for an injunction. When the appeal came before the magistrates' court, the magistrates said that the High Court must now deal with the matter. When the case was heard in the High Court, it decided that the magistrates should determine the case. Before the magistrates' appeal was heard, the appeal from the High Court was heard by the Court of Appeal. It decided that the present situation was adequately covered by S.58(8) and that recourse to the High Court for civil remedies was thus available. The notice had specifically stated, as required, that it would not be suspended pending an appeal. When the nuisance continued in contravention of the notice, action had to be taken, and if the authority was of the opinion that proceedings under S.58(4) would be inadequate, then there could be no alternative but to seek an injunction. The Court of Appeal thus granted an interlocutory (i.e. pending the outcome of the action) injunction, pending determination of the matter in the magistrates' court.

This section can only be relied on if the nuisance complained of is a statutory nuisance. If the nuisance is only a private nuisance then the local authority or the person suffering only have the right to proceed in the High Court if they have suffered interference with their own land. If the nuisance is a public nuisance then the local authority would have to use S.222 Local Government Act 1972 (see p. 193).

If the local authority felt at the outset that there would be no response at all to an abatement notice, then the authority could proceed immediately under S.81(5) in the High Court and the whole matter would be dealt with in the High Court, without any reference to the magistrates' court.

A person aggrieved's rights: S.82

This section is a considerable improvement on its old equivalent S.59 1974 Act which gave only occupiers of property rights to complain to a magistrates' court. Furthermore, the 1993 Act has greatly extended this section by covering category (ga) cases as well.

Now a magistrates' court may act on a complaint made by **any person** that he is **aggrieved by the existence of a statutory nuisance**. Who is a person aggrieved? It appears from cases that he should not be 'a mere busybody who is interfering in things that do not concern him' per Lord Denning in *Att.-Gen. (Gambia)* v. *N'Jie* 1961. There should be some connection between the aggrieved person and the statutory nuisance complained

of. Thus, if someone's health is being affected or he is suffering from a nuisance directly then he should be able to make a complaint. Similarly, if his family is being affected he will be able to complain.

The magistrates may, if satisfied that the alleged nuisance exists or that, although abated, it is likely to recur on the same premises (or, in the case of a category (ga) nuisance, the same street), make an order for either or both of the following purposes:

- requiring the defendant to abate the nuisance within a time specified in the order and to execute any works necessary for that purpose;
- prohibiting a recurrence of the nuisance and requiring the defendant, within a time specified in the order, to execute any works necessary to prevent the recurrence.

The court may also impose a fine.

It should be noted that this section does not allow action to be taken if there is only a likelihood of a statutory nuisance occurring. One must either *exist* or there must be a likelihood of its recurrence. (Cf. the local authority's duties.)

The local authority's rights are different from those of the aggrieved person, in that magistrates' court action would only occur following the service of an abatement notice by the authority. However, before instituting proceedings under S.82, the aggrieved person must now give notice in writing of his intention to bring proceedings and the notice must specify the matter complained of. In relation to category (g) noise nuisances at least three days' notice must be given. In the case of all other categories a minimum of 21 days' notice is necessary.

The order must be unambiguous and does not necessarily have to specify what works must be done to abate the nuisance unless it would be unclear what is required otherwise.

If after receiving the court order the defendant, without reasonable excuse, contravenes any requirement or prohibition imposed by the order, then he will be guilty of an offence and liable to summary conviction in the magistrates' courts to a fine not exceeding level 5 on the standard scale together with a further fine of an amount equal to one-tenth for each day the offence continues after the conviction: S.82(8).

It is not available if the nuisance is such as to render the premises unfit for human habitation: Ss. 82(9) and 82(10).

Compensation: S.82(12)

This subsection gives the complainant an opportunity to be compensated by the defendant. Thus, if it is proved that the nuisance existed at the date of the complaint, whether or not it is shown that the nuisance still exists or is likely to recur, then the courts will order the defendant(s) to pay the complainant a reasonably sufficient sum to compensate him for any expenses properly incurred by him in the proceedings. Note that this is not the same as damages in a civil case where the plaintiff would be compensated for the interference with the use of his land. Nevertheless, this is an improvement on the previous law (although there was a similar provision under the Public Health Act 1936 for non-noise statutory nuisances).

Construction sites

One of the most difficult areas involving noise control is that relating to construction sites which, by virtue of the type of work involved, often create noise nuisance. The Act has therefore formulated special rules peculiar to construction site work. Such work is defined by S.60(1) as works for erecting, constructing, altering, repairing and maintaining buildings, structures or roads; and the breaking up, opening or boring under any road or adjacent land in connection with building, demolition or dredging work and any work of engineering construction.

Section 60 notices

If it appears to a local authority that any of the above work is to be carried out or is being carried out, then it may serve a notice imposing requirements as to the way the work is to be carried out and, if necessary, publish the notice of the requirements in such a way as the authority thinks appropriate. Such a notice may specify:

- plant or machinery which is or is not to be used;
- times of operation;
- levels of noise for
  (i) emission from the premises,
  (ii) emission from any particular part of the premises during specified hours.

It may also provide for any change in circumstances.

The local authority must have regard to the Control of Noise (Code of Practice for Construction and Open Sites) 1984. It must also ensure that the 'best practicable means' are employed to minimise noise, including suggestions for alternative machinery.

The local authority must have regard to the need to protect persons from the effects of noise in the locality in which the premises are situated.

The notice must be served on the person who is going to carry out the works and on such persons as have control over the operations, as the local authority thinks fit.

Appeals may be made to the magistrates' court within 21 days and the Control of Noise (Appeals) Regulations 1975 apply. The following are the grounds of appeal:

- that the notice is not justified by the terms of S.60;
- that there has been some informality, defect or error in, or in connection with, the notice;
- that the authority has refused unreasonably to accept compliance with alternative requirements, or that the requirements of the notice are otherwise unreasonable in character or extent, or are unnecessary;
- that the time, or, where more than one time is specified, any of the times, within which the requirements of the notice are to be complied with is not reasonably sufficient for the purpose;
- that the notice should have been served on some person instead of the appellant, being a person who is carrying out, or going to carry out, the works, or is responsible for, or has control over, the carrying out of the works;

- that the notice might lawfully have been served on some person in addition to the appellant, being a person who is carrying out, or going to carry out, the works, or is responsible for, or has control over, the carrying out of the works, and that it would have been equitable for it to have been so served;
- that the authority has not had regard to some or all of the provisions of S.60(4).

If the person served with a notice fails to do anything required within the specified time and without reasonable excuse, then he commits an automatic offence. But, if consent has been given under S.61, it will be a defence.

Prior consent on construction sites: S.61
This section gives those people responsible for construction work, and the local authorities, an opportunity to settle any problems relating to the potential noise before the work starts. Indeed, the Department of the Environment circular suggests that there might be some sort of 'early warning system' by the local planning authority giving attention to work with potentially serious noise problems. Furthermore, advice is given in the Codes of Practice that the noise requirements of a local authority should be ascertained *before* the tender documents are sent out, so that they can be incorporated into those documents.

Application may be made before the work begins and consent may be given to the applicant. It has been suggested that the consent should be sought before the tender is made in order to take account of any requirements made by the local authority.

If building regulation approval is required, then the application must be made at the same time as the request for approval. Perhaps greater use of the consent procedure would reduce the number of S.80 notices or High Court injunctions served on builders, which can contribute to the lowering of profits on a contract. The application must contain the following particulars: (i) the works and the method by which they are to be carried out; (ii) noise minimisation steps.

If the local authority considers that sufficient information has been given and that if the works were carried out in accordance with the application it would not serve a notice under the preceding section in respect of those works, then the authority shall grant its consent within 28 days and must not serve a notice under S.60, but in doing so may:

- attach conditions;
- limit consent where there is a change in circumstances;
- limit duration of consent.

If there is a contravention of any of these, then there will be an offence. The consent may be published if the authority thinks fit. Where the local authority does not give its consent or gives its consent but subject to conditions, then the applicant may appeal to a magistrates' court within 21 days. Should proceedings be brought under S.60(8), then it would be a good defence to prove that the alleged contravention amounted to the carrying out of the works in accordance with consent given under this section.

Any consent given does not in itself constitute a defence to proceedings under S.82. Where consent has been obtained by someone other than the site worker, e.g. employer/architect/consulting engineer, then those people *must* bring the consent to the notice of the site worker; otherwise the applicant will be guilty of an offence.

Smoke
Building work may be affected by smoke control laws in various ways.

First, ordinary smoke which amounts to a nuisance to the neighbourhood is a statutory nuisance under the Environmental Protection Act 1990 (see above). These provisions do not apply to smoke from private dwellings, which may nevertheless be covered by bye-laws. Thus, builders should ensure that any burning of rubbish is done without causing a private nuisance and without it becoming a hazard which could cause accidents and thereby liability for negligence.

Second, under the Building Regulations, there is strict building control on the height of industrial chimneys so that they prevent smoke, gases etc. from becoming prejudicial to health, or a nuisance.

Third, under the Clean Air Acts of 1956 and 1968 the emission of 'dark' smoke (which is defined by reference to a special chart) from industrial or trade premises is an offence, for which **no nuisance** need be proved. This is a form of strict liability and allowing dark smoke to occur on only one occasion can result in prosecution.

Fourth, there is local authority control of the building of new non-domestic furnaces.

Builders may of course also be affected when building in smoke control areas, in which only smokeless fuel can be used and **no** smoke may be emitted from buildings.

Generally, as the occasional bonfire is not a nuisance, most builders will be able to burn waste material providing they do so in a controlled and reasonable way. (See also smoke on or near highways, p. 222.)

Miscellaneous

Part II of the Environmental Protection Act 1990 is of interest to the builder as it imposes both criminal and civil liability on persons dumping waste not in accordance with the requirements of the Act, particularly if it is deposited 'in a manner likely to cause the pollution of the environment or harm to human health': S.29. Waste management licences are required and it does not matter that it is known whether the waste is hazardous or not. The word waste is widely defined and includes many types of building waste from mere rubble to asbestos. By S.33(1), where any damage is caused by waste deposited in land by any person who deposited it knowingly, or caused or permitted it to be deposited, the person is liable for that damage provided that the deposit constituted an offence under the Act. There are defences such as taking all reasonable precautions and exercising due diligence to prevent commission of an offence. Thus, turning a blind eye to where one's lorry drivers are tipping waste is commercial suicide, incurring both criminal and, more economically damaging, civil liability.

Trespass to land

Definition

Trespass to land is committed by someone who intentionally goes onto land in the possession of another, or, having been asked to leave, remains on that land after permission to be there has ceased (Figure 7.13). Putting something onto the land would also amount to trespass. It should be noted that some types of trespass to land have now been criminalised (see p. 208).

Wrong
As trespass is a **tort** the occupier can only sue the trespasser unless there is an Act of Parliament or other legislation making the trespass a crime, e.g. as on railways.

Figure 7.13 Typical signs erected to deter trespassers

Examples

Thus, a person merely crossing another's land is committing trespass. Building contractors could commit trespass by dumping rubbish or materials, or digging a tunnel on another's piece of land, erecting a sign in another's air space, or swinging a crane through that air space, as in *Woollerton & Wilson Ltd* v. *Richard Costain Ltd* 1970. When working very close to the boundary of the land, it is quite easy to commit a trespass, and therefore prior consent should be obtained to site scaffolding and ladders, e.g. *Westripp* v. *Baldock* 1939 (but see p. 207).

It should be noted that trespass is a **direct** interference. Thus, roots and branches of a tree spreading into another's land is a nuisance, and not trespass: *Lemmon* v. *Webb* 1894.

No damage necessary

There is no need to prove that the trespasser did any real damage, because the tort is **actionable** *per se* (for itself). Merely committing the tort is enough, and ignorance is no defence.

If damage is committed, however, compensation will be awarded, to reflect that damage.

If the damage is serious, then the matter may best be dealt with in criminal law, under the Criminal Damage Act 1971, and the police called. In criminal law, however, the first priority is to punish the offender and not to compensate the sufferer. Thus, a civil action may still be necessary to recover full damages.

Trespass protects possessory rights to land

As we have already seen, English law often protects possession of property. Thus, if a plaintiff wishes to sue someone successfully for trespass, he must show that he was in actual possession of the land at the time of the alleged trespass. The owner, if out of possession, would not therefore be able to sue for trespass (although if necessary he has another remedy for recovery of land, see p. 208). So, those building contractors who are in possession of a site would be the proper persons to sue for an injunction to prevent trespassers repeating their trespass, and not the site owner.

Similarly, lessees can sue for trespass where they, and not the lessors, are in possession of the leased premises. Lodgers do not have sufficient possession to be able to sue.

Trespass actions used for boundary disputes

It should be noted that an action for trespass is often used for the court to decide who actually owns or possesses a particular piece of land. Whilst in such cases the damages would be very low, the end result is that the legal position of the owners and the person in dispute will have been sorted out. Also, many such disputes do not concern large tracts, but may be mere boundary disputes. For this reason, when marking out sites, the boundaries should be clearly shown to correspond exactly with the plans and the plans on the title deeds. Also, if a description of the land is used, this should be accurate.

What amounts to land?

See p. 86, but note that land includes the surface of the ground, fixtures, buildings, rooms, growing plants, the earth below and the air space above, subject to modern limitations. Merely flying over someone's land is not a trespass: *Bernstein* v. *Skyviews Ltd* 1977.

The trespass must be intentional

This means that the trespasser must have meant to enter on the land, *even if he was mistaken* about some aspect, e.g. because he thought he was allowed to enter the land, or he did not realise he was on private land, but not if he had been thrown onto the land by another.

Defences

Consent

If the occupier has given permission for the 'trespasser' to be on his land, then this would be a good defence. Written permission is preferable as this is better evidence in any court proceedings at a later date. In this context consent may sometimes be expressed as in a licence. Builders, under building contracts, have licences to be present on the employer's land.

Statutory rights to enter

Certain public officials have the right to enter another's land, e.g. police, electricity board and tax officials (see p. 89).

Public rights of way

If land is subject to a **public** right of way, then those exercising that right are not trespassers. But they must not do anything beyond the exercise of that right; otherwise it could amount to trespass.

Access to Neighbouring Land Act 1992

Normally builders, householders and other landowners who need to go onto someone else's land in order to be able to work on their own property have to get permission from the other landowner. If they do not and then enter the other land they commit trespass, even if the work is absolutely necessary. Problems arise where the servient landowner refuses to give his consent (or withholds consent until a hefty payment is made). Now under the Access to Neighbouring Land Act 1992 an originating application may be made for an access order against that person to the appropriate county court in which the land is situated. Such an order will only be made if the court is satisfied that the works are reasonably necessary for the preservation of the dominant land, i.e. the land upon which the work is required, and that they cannot be carried or would be substantially more difficult to carry out without entry on the other land. The court will not make such an order if any person would suffer interference with or disturbance of his use or enjoyment of the servient land or any occupier would suffer hardship because of the entry on the land. The access order is given on terms as to how the work is to be carried out and as to any compensation for loss, damage or injury and disturbance. If the owner of the adjoining land is unknown, a search can be made at the Land Registry who will give the necessary details. If the land is unregistered, however, discovery of the name may prove more difficult and require some detective work. Asking the oldest inhabitants of an area who the land used to belong to is usually helpful.

Party Wall etc. Act 1996

This Act covers work on party structures, i.e. walls or fences built along a boundary line and which may actually be part of a building. Obviously, problems can occur if one needs to do anything to one's wall which is also a party wall. If someone wishes to build a new party wall from scratch, then by S.1 he must give his neighbour at least one month's notice. If the neighbour agrees within 14 days, then the wall must be built halfway across each one's land or as they agree, sharing the costs. If he does not agree, the wall may only be built on the builder's land, although in certain circumstances the footings may be built below the dissenting neighbour's land. Where work is to be done on existing structures, then S.2 gives wide powers to the builder, subject to many safeguards. If it is necessary to excavate or construct work within 3 or 6 metres of his neighbour's building, the builder may be required to strengthen the neighbour's foundations at his own expense in certain situations. Rights are given under the Act to enter and remain on the neighbour's land in order to carry out the above works and an occupier would be guilty of an offence if he refused admittance.

Easements

These are private rights, e.g. a right of way or of light. The owner of an easement has the right to enter the land to exercise that easement.

Necessity

It may be that the defendant committed the tort of trespass in order to prevent a greater wrong occurring, e.g. entering land to put out a fire. If this is so, then it could be a defence.

Abatement of nuisance

See p. 198.

Recaption of chattels

Here, the owner of chattels enters another's land to recover them. He can only do this if the occupier of the land, or someone else, wrongfully took them there in the first place, or they got there by accident. This is an unusual right, and may prove a defence to an action for trespass.

Re-entering land by person entitled to have possession

It must be remembered that, if the person entitled to possess the land is out of possession, then he has **no** remedy in trespass. He can, of course, take other types of action, as set out in the note below. Does this mean that, if he re-enters the land to regain possession, he could be sued for trespass by the occupier?

The answer is yes, but he can raise a number of different defences, provided he used only reasonable force to eject the trespassers and provided the occupiers were not, or had not been, tenants, in which case he would need a court order to evict them.

Remedies

If the plaintiff **already** has possession of the land he may apply for the following remedies.

Damages

Once the plaintiff has proved his case, he is entitled to damages **as of right**, as this is a legal remedy. Because the tort is actionable *per se*, no actual damage is needed. But if damage has been caused, damages must be awarded to compensate for the loss he has suffered. If the trespasser did no more than tip-toe across your front lawn, the damages awarded could be **nominal** (from Latin, nomen-nominis-name), e.g. lp, because the judge may be unimpressed with your case and feel it was wasting the court's time. If the defendant had actually occupied the land, by living there, or leaving rubbish behind, then the plaintiff is entitled to ask for rent.

Injunction

This is an equitable and therefore discretionary remedy. It will only be awarded where it is 'just and convenient'. A prohibitory injunction is awarded, to stop the defendant from continuing the trespass. If the injunction is ignored, the plaintiff can ask for the defendant to be committed to prison. Occasionally, the injunction will be suspended as in *Woollertons & Wilson Ltd* v. *Richard Costain Ltd* 1970 above, thus allowing the work to be finished.

Declaration by the judge

A declaration may be made by the judge concerning the rights of the disputing parties, and may be used in boundary disputes, especially if there has been no actual trespass proved.

(Note: if the owner of the land wishes to *remove* people in unlawful possession of his land, then trespass is not the appropriate action to take. He must, therefore, bring an action for recovery of land to get possession, and may claim damages for the period of occupation by the trespassers. Such damages are called **mesne** (pronounced mean) **profits**.)

Repossession *only* can be claimed by more modern methods under RSC (Rules of the Supreme Court) Ord 113 (White Book) or in the CCR (County Court Rules) Ord 26 (Green Book). Here, an originating application (and not a writ) can be issued, supported by an affidavit, setting out a claim against the persons in possession. This is useful when recovering land from 'squatters' and can be used even if the names of the squatters are unknown.

Criminal trespass

There have always been some Acts of Parliament which have criminalised particular trespasses on land, usually in relation to security, central and local government, railways and other dangerous statutory undertakers.

The Criminal Justice and Public Order Act 1994, however, has criminalised some aspects of trespass and now, in specified situations, trespassers will be dealt with in the Criminal Courts and not privately in tort. The relevant sections are designed to deal with trespasses by squatters and travellers, and whilst it is unlikely that builders would *commit* breaches of public order, some mention ought to be made of the new aspects to trespass, as it is possible that a building site might become occupied by people who have no right to do so.

By S.61, if a police officer present at the scene *reasonably* believes that two or more persons are trespassing on land with the common purpose of residing there for **any** period, that reasonable steps have been taken by or on behalf of the occupier to ask them to leave AND

(a) any of these has caused damage to the land or property or used threatening, abusive or insulting words or behaviour or

(b) that those persons have between them six or more vehicles on the land,

then the police officer can direct them to leave the land and to remove their vehicles and/or property: S.61(1). By S.61(4) if such a person fails to leave the land as soon as it is reasonably practicable or, having left, returns within three months, he commits an offence. Defences to proceedings taken against offenders could include that there was no trespass at all or that the trespasser had a reasonable excuse to remain on the land.

Breach of a statutory duty – a tort

Certain Acts of Parliament and regulations impose duties on many people, but especially employers and employees, to do or not to do specific things, e.g. the Construction (Working Places) Regulations 1996.

A breach of such duties is primarily punished by criminal proceedings.

Whether the relevant Act or regulations allow someone to be able to sue in **civil** law for damages caused as a result of the breach of statutory duty depends, first, on what is intended by the criminal Act (see p. 18). In order to sue there must be a clear Parliamentary intention in the legislation to confer private rights of action on members of a specific class, or, in the absence of a specific mention, then the absence of any other remedy and a clear intention to protect such a class would probably indicate a wish to confer such a right. See X (*Minors*) v. *Bedford County Council etc.* 1995. Having said that, in the majority of cases it is perfectly valid to sue, following a prosecution for breach of a statutory duty, whether or not there has been a successful prosecution.

Probably in the absence of an express bar, such a civil action will be permitted as under S.38 Building Act 1984 when it comes into effect. Thus there is **no** right to sue for breach of the general duties laid down by the Health and Safety at Work

etc. Act 1974, as this is specifically stated in the Act. But if there has been a breach of one of the regulations or relevant Acts which come under the umbrella of the Act, then there can be a civil action.

The most common examples of breaches of statutory duty are, for our purpose, found in the sphere of health and safety at work.

What must be proved to be able to sue?

There are five things which must be proved:

1. That the regulation or Act imposes a duty to do something.
2. That the particular regulation or Act has been broken.
3. That the plaintiff is within the class the regulation was designed to protect. Thus visitors to a site are probably not owed statutory duties under the Construction Regulations (but, of course, they are owed duties in negligence and under the Occupiers' Liability Act 1957).
4. That the Act or regulation conferred a right to sue on the plaintiff.
5. That the plaintiff is suffering from an injury which is not too remote from which the regulation was designed to protect against, even if the injury did not occur in the precise way anticipated. Thus in *Donaghey* v. *Boulton & Paul Ltd* 1968 a man fell through a hole in a roof and not through fragile roof material as anticipated by the Construction Regulations designed to protect people from falls from roofs. He was able to recover damages.

Some of the statutory duties are absolute. Thus, failing to do the precise act required amounts to a breach. In other cases, the duty is not specified and principles of the same sort found in negligence will be used to determine criminal (and civil) liability. The existence of phrases such as 'by the best practical means' or 'so far as is reasonably practicable' will indicate a 'reasonable man' approach as in negligence. There are, of course, statutory definitions of such phrases in some Acts.

Defences

Contributory negligence
See p. 183.

Volenti non fit injuria
See p. 182. An employer cannot plead this as a defence in a claim by his employee against him for his own personal liability. It may be pleaded, however, in a claim for vicarious liability. But note a plea of volenti is not often successful.

A claim based on a delegation of the statutory duty by the employer to the plaintiff
Where the plaintiff clearly knew of the statutory obligations which had been validly transferred to himself, this may be raised as a proper defence. There is, of course, a duty imposed on the employer to see that the employee understands the extent of this duty and the consequences of failing to carry out its requirements properly.

Specified under the Act
By S.38 Building Act 1984, for example, a prescribed defence may be provided for under the Building Regulations. This section is still not yet in force.

Liability for spread of fire

Historically, this has always been treated in a separate way from other branches of tort. However, it now seems to have been absorbed into the mainstream of tort except in some areas mentioned below.

Strict liability in *Rylands* v. *Fletcher*

See p. 190.

Strict liability in public nuisance

See p. 193.

Strict liability for breach of statutory duty

See p. 209. It should be noted that by S.71 of the Health and Safety at Work etc. Act 1974, there can

be civil liability for a breach of a statutory duty imposed under the Building Regulations. Part B of the regulations is concerned with safety in fire of buildings and it is possible that someone could be sued for breach of such regulations, but see above.

Negligence

The tort of negligence will, of course, apply to the spread of fire, despite the existence of the Fire Prevention (Metropolis) Act 1774 which toned down an earlier type of liability for the spread of fire, which amounted to almost strict liability. The Act says that it is a defence for an occupier, if the fire began 'accidentally' on his land. Of course, this is now in line with modern thinking in negligence, that where there is no fault, there can be no liability. Nevertheless, if a fire is started and spreads and negligence is proved, the occupier is responsible not only for his own actions but also for those of his employees or independent contractors. This is of particular importance in demolition work, where burning rubbish is an essential and continuous part of the job. In *H & N Emanuel* v. *GLC* 1971 the council gave permission to the Ministry of Works to clear a GLC site of houses. The Ministry contracted with a demolition contractor who negligently allowed a fire to spread and cause damage. The GLC was found to be liable, as an occupier, for the contractor's negligence.

In *Alcock* v. *Wraith* 1991 it was held that generally, in trespass, nuisance and negligence, an employer is not liable for the torts of an independent contractor. But if there is a special risk involved or the work itself is likely to cause danger or damage, then the employer may well be liable. This case involved the repair of a terraced house roof by an independent contractor, who in carrying out the re-roofing had removed slates from the adjoining roof so that there was no overlap, had inadequately carried the join between the two roofs and encroached 200 mm beyond the employer's roof. The court also held that, although the employer had a right to interfere with the neighbour's roof in order to carry out the work, the employer was under a duty to see that reasonable skill and care were used and this duty could not be delegated to the independent contractor. (For definitions of independent contractors see p. 164.)

Limitation Act 1980

It is unfair that plaintiffs do not take advantage of their right to sue within a reasonable period. The 1980 Act, which consolidated previous legislation, lays down various periods in which the action must be brought. For most torts this is a period of six years with the exception of actions brought for damages for negligence, nuisance or breach of duty, when the damages claimed include damages for personal injuries, when the period is three years: S.11. The limitation period begins to run from the date on which the cause of action accrued. This can be a problem in cases concerning personal injury and thus in such cases the time runs from either the injury or the plaintiff's knowledge of the injury: S.11(4). This knowledge is that the injury is significant, that the injury is due to the alleged negligence of the defendant and that the plaintiff knows of the identity of the defendant: S.14(1). In any event the courts have the discretionary right to override the time limits where this is fair subject to certain statutory criteria.

Similarly, there was also a problem with latent damage in the case of building claims (see p. 175). However, the Latent Damage Act 1986 amended the Limitation Act 1980 in respect of such damage. It does not apply to actions for personal injury. Thus, by S.14(A)(4) an action may be brought either six years from the date on which the action accrued or three years from the 'starting date', whichever is later. The starting date as defined in S.14(A)(5) is the earliest date on which the plaintiff had both the knowledge required to bring such an action for damages and had the right to do so. The knowledge required is concerned with both the material facts concerning the damage and that the damage was attributable to the alleged breach, the identity of the defendant and anyone else additionally responsible. No action can be brought after the expiration of a 15 year period from the date of the alleged negligence.

8 The law relating to highways

Introduction

The word highway has a rather old-fashioned ring to it, conjuring up visions of Dick Turpin on Black Bess, galloping furiously to York.

In fact, highway is an all-embracing word used to cover many different types of routes which may be used by the **public** in general, without permission or limitation, except perhaps as to the manner of use. Thus, a highway may be a road, footpath, street, bridlepath, metalled road or a mere track. But, it must be used by the public, as of right.

Builders are affected by the laws relating to highways, as they travel across them to transport materials to site, they construct roads, they deposit things such as skips there, they park vehicles on them and erect hoardings and scaffolding close by.

Relevant law

The law is found principally in the Highways Act 1980 as amended, which consolidated previous Highways Acts, and at common law, particularly in tort.

Definition

A highway may be defined as a particular route across a piece of land which is open to the public, over which they may pass or repass **as of right**. Section 328 of the Highways Act 1980 specifically includes bridges and tunnels if they form part of a highway, but excludes ferries and waterways.

1. They must be open to the public in general. Otherwise, if a right of way exists over a piece of land, it may only be an easement, for which there must be a dominant and servient tenement (see p. 97) or it will be a private right of way for the benefit of a small community such as a village, which may be created by custom.

2. There must be a definite route, which may be shown on a map. Thus, a right to cross a field in any direction cannot be a highway.

   Whilst the public must be allowed to pass and repass over the land, it is not necessary for the highway to be a thoroughfare (i.e. a through road or path). It could, for example, be a cul-de-sac.

3. If a road or any other purported highway can only be used by special permission, e.g. by licence, as on a toll road, or if the landowner merely puts up with the public using the land, then the public will not have the use **as of right**.

4. The route will still be a highway if its use is restricted in some way. For example, it may be restricted to horse riders or pedestrians. Highways are usually maintained at public expense by the highway authorities and appropriate government departments. In exceptional circumstances, highways may be maintained privately by the owner of the land over which the highway passes. Other routes exist which are not the liability of anyone, known as **private streets** (see p. 225).

   It should be noted that private streets may also be highways, but if the right of the public to use them is restricted, then they will not be highways. Thus, 'private roads' on exclusive housing estates are usually private streets in legal terms.

As a highway is a route over land, all the usual rules relating to land are applicable (see Ch. 4).

Types of highway – Part II Highways Act 1980

The following types of highway are roads maintainable at public expense.

Trunk roads

These are main roads which form the basis of the national system of routes for through traffic in England and Wales. Local roads may be designated trunk roads. The Department of Transport is the highway authority for trunk roads, but county councils or metropolitan district councils may become agents for the department and either they, or even district councils, may carry out work for them. Long distance routes are the responsibility of the Department of the Environment.

Special roads: S.16

These are roads reserved for particular classes of traffic, most of which are motorways.

Any other publicly maintained highways

Generally the responsibility of the county council. Publicly maintained roads are those roads which:

(i)   were publicly maintained before the Highways Act 1835, *or*
(ii) have been **adopted** for maintenance at public expense under statutory rules contained in the various provisions of the Highways Act. This may be under the following sections.

Section 37 Highways Act
Where someone dedicates part of his land for use as a highway by the public, it will become publicly maintainable if the highway authority certifies to its dedication and to the fact that the road has been made up to their satisfaction.

Section 38
A highway authority may adopt an existing highway by agreement. This is the usual method when a developer proposes to build a new road for later adoption by the highways authority. In such cases the builder need not make a deposit or give security as he has to do under the Advance Payments Code (see p. 226). However, it is sensible for the highways authority to request a bond with a good surety so that, if the builder defaults on the agreement and fails to make up the road, the surety will pay the authority. The advantage of this method is that the actual cost to the builder may be far less than the amount needed to be deposited under the Advance Payments Code. Only the parties to the agreement, i.e. the highway authority, the builder and the surety, are bound and subsequent builders finishing work off after the liquidation of the original building could not be made to pay towards the cost of completing the roads: see *National Employers' Mutual General Insurance Association* v. *Herne Bay Urban District Council* 1972.

Section 53
The owner of a privately maintained highway may apply to a magistrates' court for an order to extinguish his liability to maintain. He must then deposit a requisite sum with the highway authority, who will then adopt it for public maintenance.

Sections 54, 116
Where a privately maintained highway is diverted by order of the magistrates under S.116, the new highway is maintainable at public expense.

Section 228
A private street may become publicly maintained if private street works have been executed.

Section 229
A private new street must become publicly maintained if payment has been made under the Advance Payments Code S.229 (see p. 226).

Section 69 Housing Act 1988

If land has been designated as a housing action trust area and street works have been carried out there on a road which was or has become a private street (see p. 225) the housing trust may serve a notice on the street works authority asking it to make a declaration that the street be a highway maintainable at public expense.

Creating a highway

A highway may be created in three ways, at common law by dedication and acceptance, under the Highways Act or by statute.

Dedication and acceptance

1. Dedication by landowner. A landowner may dedicate, i.e. set aside for a special purpose, a right over his land to the public, allowing them to pass and repass. He can do this **expressly** by executing a deed. This may occur when a builder buys a piece of land and builds a housing estate on it with new roads, intending those roads to be used by the householders and the public. The fact that the new road may be used by the public will be written into the deeds transferring the land to the new owners.

The landowner can also dedicate the highway in an **implied** way. This would mean that the dedication is obvious from the landowner's actions. This may occur where he allows the public to pass over a particular road without restriction.

Whether express or implied, the dedication must have been **intended** by the landowner. There can be no dedication either if the landowner restricts the right in some way, e.g. so that only certain people such as the local villagers can use the path. To prevent dedication, many landowners close the route by a gate or chain every now and then to show that they have no intention of allowing it to become a highway.

A landowner who does not own the fee simple cannot dedicate the land for the public's use, as he only has a restricted interest himself.

There can be no dedication if the public pay for permission to use the highway, e.g. on a toll road.

As we know that bridlepaths and footpaths can also be highways, any restriction as to the **type** of use will not alter the fact that it is a highway, provided all the other requirements are fulfilled.

2. Acceptance by the public. In addition to dedication there must **also** be acceptance of the dedication by the public. This is indicated by the public's use of the highway **as of right**.

Acceptance may be examined by the courts in much the same way as other customary rights in terms of how long it has been going on etc. (see p. 99).

Highways Act

In most cases S.31(a) of the Highways Act 1980 provides a simpler method of declaring whether there has been a valid dedication and acceptance. This states that 'where a way over any land ... has been actually enjoyed by the public, as of right and without interruption for a full period of 20 years, the way shall be deemed to have been dedicated as a highway unless there is sufficient evidence that there was no intention during that period to dedicate it'.

By S.31(2) the period of 20 years is calculated retrospectively from the date the highway's dedication is first called into question.

In order to negative the landowner's apparent intention to dedicate the way as a highway, the landowner may erect a notice declaring this to be so, and this is sufficient to prevent the dedication: S.31(3).

If the land is leased, the person owning the freehold reversion may exhibit such a notice on the land: S.31(4). Such a notice must be visible to the highway users. But if the notice is torn down or defaced, e.g. by hostile road users, the owner can give notice to the appropriate highway authority, informing them of the fate of the notice, and that he has not dedicated the way as a highway: S.31(5).

In order to clarify his position, e.g. if there seems to be a dispute brewing, a landowner may deposit a map with the authority and a statement indicating which routes (if any) have been dedicated and those which are not. These are available for public inspection: S.31(6).

So to give an example of how the dedication and acceptance test can be used in practice, imagine that Big C Construction plc have started to build houses on Lord Avarice's estate. Lord Avarice told the builders that the highway through the estate was not a public highway, as he had always maintained a notice stating this fact as required by the Act. When the company started to rip up the road, a lobby of motorists from the surrounding area forced them to stop work by obtaining an injunction. From this date, the 20 year period will be calculated retrospectively. Thus, if in the ensuing court case, the motorist's lawyers can prove that there has been 20 years' use of the road without interruption, as of right, and there has never been a notice or plans deposited, then the road will be declared a highway.

(Note: if Lord Avarice had barred the way occasionally, and this can be proved, this will indicate there was no use **as of right** and that there never had been an **intention** to dedicate).

Creation by statute

Highways can be created by Acts in many other ways apart from dedication and acceptance. There may be a local Act covering a particular area or roadway or a general Act such as the Enclosure Acts of the past. But the most important Act is the Highways Act 1980 Part III Ss. 24–35.

The Act makes provision for the highway authorities and the minister to create new highways in general such as motorways and to provide road ferries as well as new footpaths and bridleways. Highways created in this way will be maintained at public expense. They can also create walkways, now popular in shopping precincts, which have certain characteristics of highways, but whose use is regulated by byelaws and departmental regulations. The land on which the highways are to be constructed may be bought by agreement or by compulsory purchase. Landowners who are affected by the new highways may be granted

compensation under the Land Compensation Act 1973 as amended, for the depreciation in the value of their property. The Act also provides for acquiring rights over someone's land, but not buying the land itself.

Under S.34 a new private street, say on a housing estate, may be declared to be a highway by the highway authority. This also applies to a road which has been designated on a development plan under planning legislation.

Under S.35, walkways may be created by **dedication agreement** between the owner of the land and the local highway authority or district council. The agreement will contain strict provisions as to the walkway's use and maintenance.

Ownership of highways

As has already been shown, a highway does not have to belong to the highway authority, even if it is maintainable at public expense.

But, if the road is maintained at public expense, the Act vests a sufficient proportion of the road surface in the highway authority, so as to allow it to carry out required maintenance: S.263. The underlying subsoil thus remains the property of the landowner, but subject to such restrictions on use as are contained in the Act and other legislation, e.g. the Town and Country Planning Act 1990 (see Figure 8.1).

The rights of the public over the highway

It is always important to differentiate between a highway and private property, as certain acts may be quite legal on private land but may amount to an offence if committed on a highway, e.g. driving a motor vehicle without a licence.

The extent of the rights of the public over a particular highway depends on what was specified by dedication and acceptance or in any legislation. Thus, for example, a motorist has no right to drive his motor car along a bridlepath. Parking may be permitted on roads, provided it is not

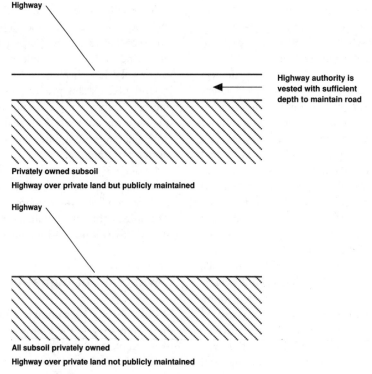

Figure 8.1 Ownership of highways

prohibited and that no obstruction is caused. At common law, obstruction of a highway may amount to a public nuisance (see p. 193), as may allowing hedges to encroach onto the highway, depositing rubbish or dumping something on the highway, or allowing the road to fall into disrepair. Direct interferences with the highway may amount to trespass (see p. 205) even if the behaviour merely exceeds the normal and reasonable use expected of a member of the public.

Protection of public rights: S.130 Highways Act 1980

It is the **duty** of the highway authority to assert and protect the rights of the public in relation to their use and enjoyment of any highway for which it is responsible: S.130(1).

Any local authority has a **power** to protect the rights of the public in their use and enjoyment of any highway within their area, but for which they are not responsible as highway authority: S.130(2).

Highway authorities must prevent as far as possible the **stopping up** or **obstruction** of highways as well as the unlawful encroachment of any roadside waste. To this end, they may take or defend legal proceedings and take such steps as they think expedient: S.130(3–5).

Stopping up and diverting highways and stopping up of means of access to highways

As we have seen in conveyancing (see pp. 104–110), it is not enough to buy land on which to build. Searches and enquiries must be made to check that the land is not adversely affected by matters which will prevent building or restrict the use of the land in some way. One particular aspect is important in relation to highways. If there is a public (or private) right of way over the land, the builder must respect this right or face the consequences. If he builds on a public right of way, he will commit a criminal offence. If he builds on a private right of way, he faces civil action by those benefitting from the right (such as the owner of an easement).

What he must therefore do is to arrange for the diversion or the stopping up of the highway **before** the work is started.

However, in *Ashby* v. *Secretary of State for the Environment* 1980, the Court of Appeal ruled that the Secretary of State could confirm a diversion order made under the Town and Country Planning Act 1971 *after* the work had begun. Reliance on such a possibility is not good commercial practice, especially in relation to breaches of the Town and Country Planning Act 1990 under which orders may be made to demolish houses built in contravention of planning permission.

Part VIII Highways Act Ss. 116–130: Section 116: the magistrates' power to stop up or divert highways

If the magistrates, after viewing the highway in question, are of the opinion that the highway is unnecessary or can be diverted so as to make it **nearer** or more **commodious** to the public, then the court may, on application by a highway authority, authorise its stopping up or diversion. The section does not apply to trunk roads or special roads (as they are usually by definition necessary and beneficial).

By S.117, a person who wishes to have the highway stopped up or diverted, such as an estate developer, may request the highway authority to make an application on his behalf. He may be required to pay for the application, but he has no right of appeal against a refusal. He may need to make such an application if he wishes to build on a particular piece of land which is affected by the presence of a highway. It must be noted that the highway will only be diverted if it is no longer needed or would be commodious to the public.

If such an order is made, it only extinguishes public rights over the highway. Private rights of way remain (and could thus still be a problem in practice). By S.128 anyone who uses an access to a stopped up highway is guilty of an offence unless it is a public right of way.

Section 116 is not the only statutory means of closing or diverting highways. The following are also important.

Acquisition of Land Act 1981 as amended
S.32 authorises the Secretary of State for the Environment or authorised authority to make an order **extinguishing** a non-vehicular public right of way over land that has been compulsorily purchased or acquired by agreement as an alternative to compulsory purchase. If the order is made by an acquiring authority, the order must be confirmed by the Secretary of State who will not confirm it unless a suitable alternative right of way has been or will be provided. If the order is unopposed, no confirmation by the Secretary of State is necessary.

Housing Act 1957
By S.4 the Secretary of State may give his approval to a local authority to make an order **extinguishing** a public right of way over land purchased by the authority as part of a clearance area.

Town and Country Planning Act 1990
By S.247, the Secretary of State may, by order, authorise the **stopping up** or **diversion** of a highway if he is satisfied that it is necessary to enable development to be carried out in accordance with any planning permission or by a government department. The order thus made may also require the provision or improvement of other highways to compensate for the loss of the highway.

Also, by S 251, the Secretary of State may make an order **extinguishing any public right of way** acquired by a local authority for planning reasons.

In practice, therefore, the builder will use Ss. 116, 117 of the Highways Act 1980 or S.247 of the Town and Country Planning Act 1990.

Section 118: stopping up of footpaths or bridleways

Public path extinguishment order

A council may by order confirmed by the Secretary of State extinguish a public right of way over a footpath or bridleway if the way is not needed for **public use**.

Section 119: diversion of footpaths and bridleways

Public path diversion orders

If an owner, lessee or occupier of land which is crossed by a footpath or bridleway satisfies the council that to enable his land to be used efficiently or to provide a shorter or more commodious path or way, the footpath or bridleway should be diverted, then the council may make an order to be confirmed by the Secretary of State for the Environment. The order may

- **create**, from a particular date, a **new** footpath or bridleway, so as to effect the diversion **and**
- **extinguish** the public right of way over all or part of the path as seems necessary to the council.

The landowner/lessee/occupier who requested the diversion order may be required to defray or make a contribution towards the cost of compensation which may be awarded to someone whose land has depreciated in value or who has been disturbed in his enjoyment of the land or who has suffered damage as a result of the creation of a new footpath: Ss. 28, 119(5). Also, a contribution to expenses may be required.

Section 122: power to make temporary diversions where highway is about to be repaired or widened

Where a highway authority or a person who is liable to maintain a highway is about to repair or widen a highway, they may construct on adjoining land a temporary highway for use while the work is in progress. The owner or occupier of the land on which the temporary highway is constructed may recover compensation from the highway authority or other person constructing the road, if he sustains any damage as a consequence of the work.

The section does not authorise interference with land which is part of a house-site, garden, lawn, yard, court, park, paddock, plantation, planted walk or tree nursery.

Stopping up of means of access to highways

Under Ss. 124–129 of the Highways Act 1980, there are provisions for the stopping of private accesses to highways by order, S.124; private accesses to premises, S.125; and agreements to stop up private accesses, S.127.

Building and improvement lines: Ss. 72–75 Highways Act 1980

Improvement lines: S.73

A highway authority may wish to widen certain streets, and therefore the appropriate local authority may prescribe a line to which the street may be widened called an **improvement line**. This is a local land charge and must therefore be registered on the register of local land charges (see p. 107).

Once an improvement line has been made **no new building** may be erected and **no permanent** excavations below street level shall be made beyond the line, without first obtaining consent.

Breach of this amounts to an offence and a fine is levied for each day the contravention continues.

Building lines: S.74

The highway authority may also prescribe a building line similar to the improvement lines. Once prescribed, no new buildings (or permanent excavations) other than boundary walls or fences may be erected without consent beyond the building line, which is a specified distance from the centre-line of the highway. It is thus a frontage line beyond which no building may project. Contravention of this amounts to an offence.

It should be noted that statutory undertakers such as water or electricity boards are not affected by the above.

General restrictions on damaging highways

By S.130 the highway authority is under a duty to assert and protect the rights of the public in their use and enjoyment of the highways under their authority and also to prevent the stopping up or obstruction of those highways or even highways *outside* their authority, if it would be prejudicial to the interests of their area. Therefore, under S.130(4), they are authorised to take or defend actions in this respect. (This is additional to the rights mentioned under S.222 Local Government Act 1972; see p. 193.)

Penalty for damaging highways

By S.131, anyone without lawful authority who causes damage to a highway, by digging a ditch, excavating, removing soil or turf, depositing anything or lighting a fire on the highway, is guilty of a criminal offence.

It is also an offence to remove or obliterate road signs, milestones or direction posts.

Section 131(A) added by Rights of Way Act 1990

Anyone who, without lawful authority or excuse, disturbs the surface of a footpath, bridleway or carriageway of a highway so as to make it inconvenient for the exercise of a public right of way is guilty of an offence. By S.131(A)(2) the appropriate highways authorities are under a duty to ensure that such proceedings are brought if it is desirable in the public interest. Whilst this section was introduced mainly to prevent farmers ploughing up edges of fields and thus interfering with rights of way, nevertheless it is obvious that builders could be adversely affected if they do not take care when working near highways.

Unauthorised marks on highways (graffiti)

Under S.132, anyone painting or fixing a picture, letter, mark or sign upon a highway or tree, structure or works without the highway authority's consent, statutory authority or reasonable excuse is guilty of an offence. The authority is empowered to remove the offending mark etc.

Damage to footways of streets by excavations

Under S.133, if a street footway (pavement or similar) is damaged by a person excavating or doing other work on land adjoining the street, the highway authority may make good the damage and recover the expense from either the landowner or the person causing the damage.

Obstructing highways and streets

Section 137: obstruction

If someone without lawful authority or excuse in **any way** wilfully obstructs the free passage along a highway, then he is guilty of an offence.

In addition to this section under the Highways Act 1980, there are a number of other legal methods of restricting similar behaviour, e.g. by Reg. 122 Road Vehicles (Construction and Use) Regulations 1986, trailers must not be allowed to stand on roads so as to cause unnecessary obstruction. Also, the rules relating to public nuisance apply (see p. 193).

Section 138: offences for erecting buildings in carriageways

Anyone erecting a building or fence or planting a hedge on a carriageway without lawful authority or excuse, is guilty of an offence. A carriageway is part of a highway used by vehicles: S.329.

Sections 139, 140: builders' skips

Skips must not be left on highways without the consent of the highway authority. Such consent may be granted unconditionally or subject to conditions. Conditions may include restricting the dimensions of the skip, the manner of making visible to traffic, determining its siting, lighting, guarding, care and disposal of contents, and removal at the end of the period.

Anyone depositing a skip without permission is guilty of an offence.

Once permission has been granted, the owner must clearly mark details of his name, address and telephone number on the skip in accordance with the Builders' Skips (Markings) Regulations 1984. He must keep it properly lighted at night, so that it is visible to traffic and pedestrians. After it has been filled, it must be moved as soon as is practicable. All conditions of the consent must be complied with. Even if permission has been given, the skip may be removed or repositioned at the wish of the highway authority or a uniformed police constable: S.140.

It is a defence to prove that the offence was due to the act or default of another person and that the person charged took all reasonable precautions and exercised all due diligence to avoid the commission of such an offence by himself or anyone under his control. Such a defence can only be relied on if, within seven days before the hearing, the defendant serves a notice on the prosecution giving such information, identifying or assisting in the identification of that other person, e.g. a subcontractor, as he has in his possession.

Section 139 does not authorise the creation of a nuisance or of a danger to users of the highway. Nor does it impose liability on the highway authority for any injury or damage caused to someone resulting from the presence of a permitted skip.

Under S.140(A) inserted by New Roads and Street Works Act 1991 Sch. 8, should the skip remain on the highway longer than was permitted and the skip has not been removed within a reasonable period, then a charge may be made.

Section 141: restrictions on planting of trees etc. in or near a carriageway

It is an offence to plant trees or shrubs in a carriageway or within 15 ft from the centre of a carriageway.

However, under S.142, the highway authority may grant a licence to the occupier or owner of premises adjoining a highway to plant and maintain, or to retain and maintain, trees, shrubs, plants or grass in such parts of a highway as may be specified in the licence.

Section 143: structures erected on a highway

Anyone erecting a structure, which includes anything capable of causing an obstruction, without permission or statutory authority may be required by notice to remove it within a specified period.

'Structure' includes any machine, pump, post or other object capable of causing obstruction and it may nevertheless be a structure if it has wheels. Thus a caravan can be a 'structure'.

Section 148: penalty for depositing things on a highway

It is an offence to deposit on a made-up carriageway dung, compost, material or rubbish on or within 15 ft from the centre of the carriageway, or by such deposit **interrupt** the use of the highway without lawful authority or excuse.

Section 149: removal of things so deposited as to be a nuisance

Anything deposited on a highway which is a nuisance may be required to be removed by the highway authority by notice to the person who deposited it there. If he then fails to remove it, the authority may make a complaint to the magistrates and ask for a **disposal order**.

If the authority considers that the unlawful deposit constitutes a danger (even by merely obstructing the view), and that the deposit ought to be removed without the delay involved in the above procedure, the authority may remove it forthwith and charge the cost to the person who made the deposit.

Section 150: duty to remove snow, soil or other obstructions from highways

A highway authority is under a duty to remove obstructions on highways caused by an accumulation of snow, falling down of banks on the side of the highway **or from any other cause**. If the authority fails to remove the obstruction, **any person** may make a complaint to the magistrates' court, who may then order the removal of the obstruction within a specified period.

The authority may take any reasonable steps to warn highway users of the obstruction, recover the cost and, if necessary, **sell** the obstruction.

The sort of obstruction that comes within this section must be construed to be in the same categories as snow and falling banks, i.e. natural obstructions which occurred suddenly, such as falling trees, and not lorries which are dealt with elsewhere in the Act.

Section 151: prevention of soil being washed onto the street

A local or highway authority may, by notice to the owner or occupier of any land adjoining a street maintainable at public expense, require him to execute such works as will prevent soil or refuse from that land falling, being washed or carried onto the street or into any sewer or gully, so as to obstruct the street or choke the sewer or gully.

Failure to comply amounts to an offence, and if convicted, fines are incurred for each day the offence continues.

Scaffoldings and hoardings and other projections

At common law

Public nuisance

The principles of *Wringe* v. *Cohen* (see p. 193) will also apply to building constructions which project across highways. Liability is strict, but only in relation to **disrepair**. Also, these rules only appear to apply to **artificial** projections and not to things such as trees, in which case **negligence** would have to be proved.

Private nuisance

See pp. 193–8.

Statute

Section 152 Highways Act 1980: powers as to removal of projections from buildings

A highway or local authority may, by notice to the occupier of any building, require him to **remove** or **alter** any porch, shed, projecting window, step, cellar, cellar door, cellar window, sign, signpost, showboard, window, shelter, wall, gate, fence or other obstruction or projection which has been erected or placed against or in front of the building **and** is an obstruction to safe or convenient passage along a street. If the recipient of the notice fails to remove the obstruction, he is guilty of an offence.

Section 153: doors etc. not to open outwards

Doors opening outwards into the street are prohibited unless the highway authority gives its consent for a public building. A notice to alter the door may be served on the occupier and failure to do so amounts to an offence.

Section 154: cutting or felling etc. trees that overhang, or are a danger to roads or footpaths

A local or highway authority may serve a notice on the occupier on whose land is growing a hedge, tree or shrub overhanging a highway, so as to endanger or obstruct the view of drivers of vehicles or the light from public street lamps. The notice may require the lopping or cutting of the obstruction.

If it appears that the tree etc. is diseased or dead, and is likely to cause a danger, a notice may be served to require its removal.

Section 169 as amended by, inter alia, *New Roads and Street Works Act 1991: control of scaffolding on highways – licences*

No person is permitted, in relation to any building, demolition, alteration, repair, cleaning etc. works, to **erect** or **retain** on or over a highway, any scaffolding or other structure which obstructs the highway, without a **licence** issued by the highway authority. The terms of the licence must be complied with.

Provided an applicant for a licence furnishes the authority with such particulars as are required, the authority must issue a licence in respect of the structure, unless:

- the structure would cause unreasonable obstruction of a highway, or
- an alternative structure could be erected to that proposed by the applicant, which would cause less obstruction and could be used for the work in question.

An appeal against a refusal to grant a licence or particular terms contained in a licence may be made to a magistrates' court. Once a licence is granted, subject to exceptions, a licensee must ensure that the structure is lit at all times between half an hour after sunset and half an hour before sunrise. He must also comply with all written directions concerning the erection and maintenance of traffic signs.

However, if no part of the structure is less than 18 ins horizontally from a carriageway and no part is less than 8 ft above a footway, then these two provisions do not have to be complied with (see Figure 8.2). He must give access to any apparatus belonging to or used by statutory undertakers, e.g. the electricity companies.

Failure to comply with the above amounts to an offence.

Civil and criminal immunity under S.169(7)

Provided there is a licence granted for the structure and the terms of the licence are complied with, then **no civil** or **criminal** proceedings lie in respect of any **obstruction** of a highway, in relation to either the licensee or the highway authority.

Danger or annoyance to users of highways and streets

Section 161: deposits on highways endangering or injuring users

A person who, without lawful authority, deposits **anything** on a highway which causes injury or endangers a user commits an offence: S.161(1).

A person lighting a fire on or over a carriageway of a highway, which injures or endangers a user or interrupts the use of the highway without lawful authority, is guilty of an offence: S.161(2).

Section 161(A) was introduced by the Highways (Amendment) Act 1986 and was intended to avoid stubble burning by farmers. However, it could equally apply to fires on demolition sites. Thus, if a person lights a fire on any land not forming part of a highway, being a carriageway, or directs or permits a fire to be lit on any such land and as a result a user of the highway is injured, interrupted or endangered by that fire, or by smoke from that fire (or any fire caused by the fire) then he is guilty of an offence. It is a good defence if he could prove that there were reasonable grounds to believe that such dangers were unlikely or that he took all reasonable precautions to avoid them or he had a reasonable excuse not to.

A person is also guilty of an offence if he allows any filth, dirt, lime or other offensive matter to run on to a highway from any adjoining premises without lawful authority: S.161(4).

Section 162: placing ropes etc. across highways

The placing of a rope, wire or other apparatus across a highway so as to be a likely cause of danger to users is an offence, unless adequate warnings were given of the danger.

Section 163: prevention of water falling on or flowing onto a highway

A local or highway authority can require steps to be taken by the occupier to prevent water from his premises falling upon users of a highway or, so

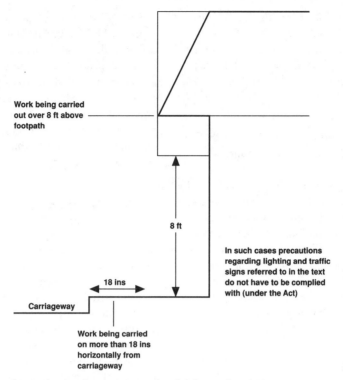

Figure 8.2 The exception to the requirements regarding lighting and maintenance

far as is reasonably practicable, surface water from flowing onto or over the footway of the highway.

Section 164: removal of barbed wire

Similarly, an authority may require the occupier of land adjoining a highway, on which there is a fence made with barbed wire and which amounts to a nuisance to the highway, to abate the nuisance within a specified time. By S.164(1)(b), such barbed wire is deemed to be a nuisance if it is likely to be injurious to persons or animals **lawfully** using the highway. Thus, barbed wire used to discourage entry to a site would generally not fall under this section.

Section 165: dangerous land adjoining a street

A local authority may serve a notice on the owner or occupier of land which adjoins a street if in or on such land there is an **unfenced or inadequately fenced source of danger** to persons using the street. The notice may require work to be carried out, to repair, protect, remove or enclose the land, to obviate the danger.

Section 167: retaining walls near streets

This section imposes conditions on the construction and maintenance of retaining walls which are within 4 yards of a street and over 4 ft 6 ins high. It requires that their erection shall be in accordance with plans, sections and specifications approved by the local authority.

Precautions to be taken when doing work on or near streets or highways

Section 168: building operations affecting public safety

If, in the course of carrying out building work in or near a street, an accident occurs which

- gives rise to the **risk** of serious bodily injury to a person in the street or
- would have done so if the local authority or highway authority had not prevented this happening by exercising their powers under S.78 Building Act 1984 (emergency measures to deal with dangerous buildings),

then the land or building owner on which the operation is being carried out is guilty of an offence.

It is a defence to prove that the person charged took all reasonable precautions to avoid endangering passers-by, or that the offence was committed by some other person and that he took all reasonable precautions to avoid committing the offence by himself or any person under his control.

'Building operation' includes, *inter alia*, construction, repair, maintenance etc., demolition, laying foundations and erecting or dismantling cranes or scaffolding.

Scaffolding: S.169

See p. 222.

Section 170: control of mixing of mortar etc. on highway

It is an offence to mix or deposit on a highway any **mortar** or **cement** or any other substance which is likely to stick to the surface of the highway or to solidify in drains or sewers.

Section 170(2) allows the mixing of such substances in receptacles or on plates, thereby preventing it from touching the highway. A licensee with a street works licence is permitted to carry out such work, provided the work cannot reasonably be done elsewhere than on the highway.

Section 171: control of deposit of building materials and the making of excavations in streets

Permission may be given by a highway authority to someone to **temporarily** deposit building materials, rubbish etc. or make a temporary excavation in a street maintainable at public expense.

Such consent may be subject to conditions. Appeals against the refusal of consent may be made to the magistrates' court.

The deposit or excavation shall be properly fenced and lighted in the dark. It must not remain there longer than necessary.

Failure to comply with any of the above amounts to an offence. Charges for occupation of the highway for work under S.169 (scaffolding) or S.171 will be made.

Section 172: hoardings to be set up during building

Before a building is erected or demolished, its outside repaired or altered, in a street or court, then **prior** to the commencement of the work, a close boarded hoarding or fence must be erected to the county, metropolitan district or London Borough authority's satisfaction, so as to separate the building from the street or court.

This obligation may be dispensed with by the consent of the authority. Once erected, it must be properly maintained, lighted if necessary, and be removed when required. Contravention amounts to an offence.

Section 173: hoardings to be securely erected

A hoarding adjoining a street must be securely fixed, to the council's satisfaction; otherwise an offence is committed.

Section 174: precautions to be taken by persons executing work in streets

Any person executing work in a street other than works under the New Roads and Street Works Act 1991 must:

- erect such barriers and traffic signs to prevent danger to traffic, regulate and warn traffic, and remove them when no longer necessary;
- light and guard the works in the hours of darkness;

- when the work so requires, cause any building adjoining the street to be shored up, or otherwise protected.

Failure to do so amounts to an offence. Any person removing any such warning sign, barriers etc., or extinguishing a light, is also guilty of an offence.

Section 175: liability of servants of highway authorities or of persons responsible for maintenance of highway

If a servant of a highway authority or district council or a person who is himself responsible for the maintenance of a highway **causes** any object or heap of material to be put onto a highway so as to become a danger to traffic, without taking all reasonable precautions, he is guilty of an offence.

(Note: this is a statutory duty, which could lay an *individual* open to civil action for breach of statutory duty or negligence.)

Restrictions on construction of bridges or buildings or placing of rails, beams etc. over highways

Section 176: restriction of construction of bridges over highways

No bridge may be built across a highway without a licence from the highway authority.

Section 177: restriction on building over highways

No building may be constructed over a highway maintainable at public expense, unless it is in the exercise of statutory powers and a licence has been obtained from the highway authority.

Section 178: restriction on placing rails etc. over highways

No overhead beam, rail, pipe, cable, wire or similar shall be placed over, along or across a highway without the consent of the highway authority. Contravention amounts to an offence.

Cellars. Construction under streets or carriageways

Section 179: cellars

Consent must be obtained from the appropriate authority before constructing a vault, arch or cellar under any street. The appropriate authority outside London is the highway authority if the street is a highway. Otherwise it is the local authority in whose area the street is situated.

Failure to do this amounts to an offence. This section does not apply to street works under Part III of the New Roads and Street Works Act 1991.

Section 180: openings into cellars etc. under streets

Consent must be obtained before making an opening in the footway of a street, as an entrance to a cellar or vault. Such consent may be given with conditions attached as to the manner of construction of the door or covering.

Section 181: breaking open highways

Anyone who without lawful authority or excuse places apparatus, e.g. pipes, cables etc., in or under a highway, or breaks open a highway in order to place or repair such an apparatus, is guilty of an offence. A licence may be granted by the highway authority for the purposes of such a work.

Part XI Highways Act 1980: Ss. 203–237

Private streets

The Act defines a street as any 'highway, road, lane, footpath, square, court, alley or passage' whether a thoroughfare or not and which may be a part of a street.

A private street is not maintainable at public expense, but it may become so if the relevant statutory requirements are complied with.

Generally, no-one can be forced to maintain such streets. However, the frontagers (the people abutting onto the private street) may be forced by the highway authority to carry out emergency repairs, say, for example, to a hole in the road. If they refuse to do the repairs, the authority may execute the works and recover the costs from them.

Adoption procedures

No-one, unless they are very wealthy, really wishes to maintain a private street for their own convenience, especially as they cannot be forced to do so. Builders developing a housing estate will, of course, be in this category. Neither they nor the prospective purchasers would wish to pay for road maintenance. Adoption by the highway authority is therefore necessary and it can be achieved in two main ways. (Note: the general outline for adoption is on p. 213.)

1. The Private Street Works Code: Ss. 204–218. This is concerned with **existing** private streets, which are not made up to the satisfaction of the highway authority. The authority may therefore pass a resolution to that effect and make up the street, bearing the whole of the cost itself or sharing the cost with the frontagers. The work required may cover total making up or it may merely be some aspect, e.g. lighting or sewerage.

Notice of the job specification and estimated cost and apportionment of costs between frontagers must be displayed in a local newspaper or in the relevant street for certain specified periods. A copy of the notice must also be served on those frontagers who are to share the costs. Copies of all the documents must be available for inspection at the highway authority's offices (district and county).

The affected frontagers may object against the charge. The local authority must then take the matter for a decision before the magistrates.

Once the work is finished, a final apportionment is made of the costs and notice is served on the frontagers who may appeal. They may be allowed to pay in instalments. If they fail to pay, the costs may be recovered as a civil debt in the magistrates' court (see p. 28). They can also recover the money in the county court or High Court. Alternatively the authority may charge the premises with a **local land charge** (see p. 107). This gives the authority the powers of a mortgagee and the property could be sold by them to cover their costs.

Once the work has been completed, the street may then be adopted for maintenance at public expense under S.228. The frontagers may object to this and, if the objections are in the majority, the street may remain a private street, now made up, but not maintainable at public expense, and not maintainable by the frontagers, except for emergency repairs.

2. Advance Payments Code: S.204, Ss. 219–225. This is more familiar to builders who are developing a new estate with new roads and who **at the outset** intend that the roads should be maintainable at public expense.

Within six weeks of the builder's plans having received Building Regulation approval, the authority gives the builder six weeks' notice of the required sum needed by the authority to make up the road. This must be paid to the authority or security must be given for it.

Not such deposit need be paid if the builder builds the road at his own expense or if the new buildings are being built abutting into an existing publicly maintained highway.

After the work is completed, the sum is adjusted and more money may be required or some may be repaid. Failure to pay is punishable by a fine.

If the builder has merely secured the payment of the sums (and not passed on the making up costs as an element of the sale price), the new houseowner must pay the proportionate sum required. Even if only one houseowner has paid the sum, provided at least half the street/road is built up, then the frontagers may call on the council to require the making up of the road under the Private Street Works Code and then adopt it as maintainable at public expense under S.228 (see p. 213).

9 Employment law

The contract of employment

Like all other areas of contract law, the basic principles apply. Thus, without the essential elements of a contract, the contract would be void and there would be no binding rights and duties. In addition to the rules of common law, there are a wide-ranging number of Acts of Parliament which govern different aspects of employment. The law has recently been consolidated in the Employment Rights Act 1996 (the Act).

For many of these statutory rights, and indeed, for much of the common law, it is important that the contract of employment between the employer and the employee is one of **service** and not **for services**, as **only** employees have protection under these Acts (see pp. 164–5). It must be remembered that whatever the parties call the relationship existing between them, in a dispute, all aspects must be examined to discover its true nature.

Form of contract

Like most contracts, the contract may be in any form. Thus, it may be oral, in writing, or implied from the parties' behaviour, or by reference to trade union agreements or practice within a trade or to employer's working rules or a combination of any form. Terms may also be implied by the common law or by statute.

For many people, the contract will be oral. For example, if a person goes to an interview for a job, he may be offered the job at the interview. If he accepts, the contract will be oral. If he accepts by letter, it is a combination of a written and an oral agreement. If he was offered the job by letter, which he accepts by letter, then it will be in writing. He may be given a formal contract to sign and this is usually desirable where there are unusual terms. If he is not supplied with all the terms in a written contract, then the employer must supply written particulars of the terms of employment (see below).

Employer's duties towards employees

The employer's duties are a combination of both common law and statute. Any unusual duty should be an express term of the contract. The parties may make such terms as they wish, provided they comply with basic contract rules and statutory restrictions.

He must provide written particulars of the terms of employment

This is a statutory duty under S.1 Employment Rights Act 1996. Because of the problems associated with oral contracts which may have been made when the employee was in no fit state to take in the terms of his new employment, S.1 requires employers to provide written particulars of the terms of the employment. These particulars must be given to the employee within two months from commencing the job (unless he works for less than one month).

The particulars must give details of the following (S.1):

● Names of parties to the contract (note that companies, partnerships and sole traders often trade under different names to their true ones).

- Job title.
- Date employment commenced.
- Date on which the employee's period of continuous employment began.
- Whether previous employment counts as part of the employee's continuous employment.
- Amount of pay or method of calculating this.
- Whether pay is weekly or monthly or at another interval if applicable.
- Hours of work and conditions relating to them.
- Holiday entitlement.
- Details of sick pay and incapacity for work.
- Details of pension and pension schemes.
- Length of notice for employees to give and receive on terminating employment.
- If the job is for a fixed period, the date on which it ends.
- Disciplinary and grievance procedures. (There is an ACAS code of practice dealing with disciplinary procedures – Disciplinary Practices and Procedures in Employment. The person to whom the employee should complain about a grievance should be clearly stated (see p. 234): S.3.(1).
- Place(s) of work and address of employer.
- Collective agreements which may affect his work (see below).
- Where employees are required to work outside the United Kingdom for more than one month, the period of work outside the United Kingdom, any additional remuneration or benefit payable as a result, the currency of pay and any terms or conditions.

Particulars provided for information only, if given after the job has started, cannot form the contract, and reference, for other purposes, would have to be made to the contractual terms contained in the (possibly oral) contract made earlier.

But, if the particulars are given *before* the contract was entered into, then they could form part of the contract. Once the particulars have been delivered to the employee, generally neither can then deny that any of the particulars do not form part of the working conditions.

The written particulars may refer the employee to an agreement such as a collective agreement, e.g. the Working Rules Agreement in the construction industry, which is issued by the Building and Allied Trades Joint Industrial Council: S.1(4)(j).

Alternatively, the agreement referred to could be a company master document. Access to copies of such agreements must be freely available: S.6.

Changes in the terms of employment

Should any change be made, these must be notified within **one** month. A statement should then be attached to the original document or written particulars, or reference may be made to an amended document to which access is available. However, changes must have been approved by both parties unless the contract contains a clause allowing one party to change terms such as those relating to working conditions.

Breach of duty to supply particulars

If the employer fails to provide the particulars, the employee may go to an industrial tribunal (see p. 39). The tribunal may then decide on any terms of employment which might reasonably have been imposed.

To provide statements of pay

Under the Act, every employee is entitled to a proper statement of pay, setting out his gross wages, all the appropriate deductions and net wages.

Anyone not receiving such a statement can apply to an industrial tribunal. The tribunal can then declare the employee's rights. It cannot order the employer to repay money which was underpaid. Action would have to be taken in the county court or Queen's Bench Division. But, if the deductions were not mentioned at all, the tribunal can award compensation for the 13 weeks prior to application whether the deductions were in breach of contract or not.

To make guarantee payments

This is a statutory duty under S.28 of the Act 1980. If the employer has no work for the employee, then the employee is entitled to a guarantee payment unless the lack of work is due to an industrial dispute involving another employee or another employer, or if he refuses an alternative reasonable offer of work: S.29. This provides protection for those laid off without pay. There is a maximum of five days' payment during each quarter with a maximum wage of £14.50 per day in 1996.

If the contract states that he will be paid for workless days in any case, then any money paid discharges the employer's duty to make guarantee payments: S.32.

In many cases, if employees are paid weekly or monthly, then the employer has **no right** in law to lay off staff without pay unless this is **expressly** incorporated into the contract of employment, or if the employee agrees to be laid off, or if it is implied by custom or past dealings.

Failure to pay any of the guarantee payments gives the employee the right to complain within three months to an industrial tribunal: S.34.

Should there be a **collective agreement** which deals with guarantee payments, then the particular body responsible for the agreement can apply for an exemption order from the Secretary of State for Employment, if such payments would be inappropriate in the circumstances. In the building industry, the Guarantee Payments (Exemption) No 29 Order, SI 1994/417 excludes from S.28 all employees who work under the National Working Rules Agreement of the Joint Industrial Council for the Building Industry.

Duty to provide work at common law

There is no definite duty to provide work for the employees, unless by failing to do so it would lead to loss of earnings. Also, it might indicate that the employer is trying to avoid the contract and thus there should be some remedy for the employee, e.g. for breach of contract.

Duty to pay workers at common law

If the employee does any work, then he must be paid. Details of the *method* of calculation should be inserted in the written particulars. If there is no **express** agreement as to the amount, then reference may be made to a collective agreement.

There is no duty imposed to provide references for an employee

If he does provide them, however, he should make sure that they are true as he may face a court action in tort for deceit or perhaps in negligence for a negligent misstatement by a prospective employer of his former employee or a libel action if the references are defamatory.

Also, in *Spring* v. *Guardian Assurance Plc* 1994, the House of Lords held that there **can** be liability for a negligently written reference in the tort of negligence. Until then, this had been a moot point for many years. As of course there will only be economic loss which, if you re-read the section on negligence, cannot normally be the basis for a successful action, this has to be by an extension of the principles in *Hedley Byrne* v. *Heller* (see p. 173). Thus, because the reference is a **negligent statement** there can be a claim. (Following Hedley Byrne principles, a suitable disclaimer may therefore exclude the writer from liability. However, such a disclaimer should probably be given to the former employee to have any effect in this instance.) The Law Lords also held that there was an implied term in contract that the employer would exercise due care and skill when preparing a reference.

Duty to see that the employee is safe at work under common law and statute

This duty has a number of different bases. First, at common law in tort, an employer owes a personal duty of care to his employees to provide safe plant and equipment, a safe place of work, a safe system of work and competent staff. This duty pre-dated *Donoghue* v. *Stevenson* 1932, but is really the same as the duty of care in the tort of negligence.

Second, it could be a breach of contract, if he failed to provide the above.

Third, under the Employers' Liability (Defective Equipment) Act 1969, an employer is strictly liable to employees who are injured in the course of their employment by equipment provided by the employer, providing that the fault in the equipment is that of a third party such as the manufacturer. No fault has to be proved on the part of the employer in such cases. But, if the equipment was unsafe because of poor maintenance, then the employer will have to be sued under the first head above, and fault would have to be proved.

Fourth, an employer is vicariously liable for torts committed by his employees in the course of their employment, and this would cover acts which cause harm to fellow employees.

Fifth, the Health and Safety at Work etc. Act 1974 encompasses a number of different Acts which cover the employee's safety and health whilst at work. Regulations such as the Construction Regulations are now under the ambit of this Act (see Ch. 10).

Sixth, S.44 Employment Rights Act states that an employee has the right not to be subjected to any detriment by any act or deliberate failure to act by his employer in the following five circumstances: first, where the employee has been designated by the employer to carry out health and safety prevention or reduction measures and the employee carried out such measure; second, where the employee, having been appointed to be a health and safety representative or committee member in accordance with statutory requirements or acknowledged to be such by the employer, performed or proposed to perform any functions in relation to those positions; third, where the employee was employed where there was no safety committee or safety representaives or, if there was, it was not reasonably practicable for the employee to raise a matter by those means and he notified his employer of harmful or potentially harmful circumstances; fourth, where there were dangerous circumstances which the employee reasonably believed to be serious and imminent and which he could not reasonably be expected to avert and he left or proposed to leave the workplace; fifth, where there were dangerous circumstances which the employee reasonably believed to be serious and imminent and he took steps to protect himself or others from the danger.

Duty to pay statutory sick pay – statute

Under the Social Security, Contribution and Housing Benefits Act 1992, it is the duty of the employer to pay statutory sick pay to an employee for the first 28 weeks of incapacity to work, caused by sickness or injury. It is an offence to avoid paying it.

In many jobs, the employer pays a greater amount in sick pay than the statutory minimum. But there is a Percentage Threshold Scheme which repays the employer's costs in any month if this exceeds 13 per cent of the employer's gross national insurance contributions, employer's and employee's, which the employer must deduct from the employee's wages on behalf of the DSS.

Sex and race discrimination

Under the Race Relations Act 1976, Equal Pay Act 1970, European Union Legislation and Sex Discrimination Act 1975, it is prohibited to discriminate against persons on racial or sexual grounds, or because of their marital status.

Discrimination can take various forms, e.g. restricting recruitment, giving less favourable conditions of employment, or dismissal.

Complaints about racial discrimination may be made to an industrial tribunal within three months of the alleged discriminatory act, where up to £11,000 (in 1996) can be awarded for loss (including injury to feelings).

The Commission for Racial Equality can investigate allegations of discrimination in companies and serve non-discrimination notices, which can be backed by a county court injunction.

Complaints about sex discrimination or discrimination against married people may be made to industrial tribunals, where compensation may be awarded as above.

The Equal Opportunities Commission can investigate allegations of sex discrimination and serve non-discrimination notices.

It should be noted that the definition of employment under the Act includes self-employment: *Hill Samuel Investment Group Ltd* v. *Nwanga* 1994.

Under the Employment Act 1989 Sikhs are exempted from wearing safety helmets on construction sites.

Duty of the employee

Duty of good faith towards employer – implied by common law

This is an all embracing duty and covers such matters as being honest, not accepting bribes or secret profits, not competing with the employer for business or soliciting the employer's clients, nor divulging confidential information learned during his employment and disclosing information which could affect his ability to do his job, such as a bad heart.

In some cases, certain aspects of this general duty will be specifically expressed in the contract. For example, it is very common to include in contracts a term restricting the employee from working within a certain distance of the employer, on leaving that employment. Such terms must not be so restrictive as to prevent the employee from working at all and they can be reviewed by the courts if necessary (see p. 69).

Obedience – implied by common law

The employer can expect that his employees should obey all reasonable instructions. What is reasonable will depend on the circumstances and it is unlikely that the isolated refusal would be grounds for lawful dismissal. But, of course, the degree of refusal is important. The use of obscene language when refusing to do the work (or at any time) could constitute grounds for dismissal.

Often, refusal to do work stems from lack of clarity in explaining the scope of the job in the first place. Only the title of the job is necessarily included in the written particulars. Perhaps a job description would be more useful. Trade unions, of course, often intervene in demarcation disputes, and this, sometimes in the end, leads to clarification of working conditions.

Duty to exercise skill and care and be generally competent for the job – implied by the common law

Obviously, the employee should be able to perform his job properly in accordance with the qualifications he holds, and his experience doing the job. The degree of competence will depend on what type of job he holds and there is a certain responsibility on the part of the employer to check that the employee is up to the job and that his qualifications are genuine. If they were not genuine or if he did not have the experience he professed to have, then this would undoubtedly be grounds for dismissing him. Where an employee has an easily testable skill, such as being able to do calculations or lay bricks, then it is wise to check this before offering him the job. If the skill is not easily testable, then probably a probationary period is a good idea. In either case, the hazards of sacking someone are lessened in such circumstances.

What is far more difficult, is to dismiss someone who becomes incompetent at a later date, i.e. he was competent to start with, but has grown lazy or careless.

Rights of the employee

Right to pay while suspended on medical grounds – statute

Under S.64 an employee is entitled to be paid for up to 26 weeks if he has been suspended on medical grounds under any statutory requirement or under recommended codes of practice issued by the Health and Safety Commission.

At present, such a suspension will have been for a good reason, to protect the health of the worker in relation to specific regulations concerning dangerous chemicals, radioactive substances or lead.

He is only entitled if he is still able to do the work and is not physically or mentally ill. He will lose this right if he refuses a reasonable offer of alternative work.

Right not to have action taken against him for trade union activity or membership

Under the Trade Union and Labour Relations (Consolidation) Act 1992, an employee may not be prevented from or penalised for becoming a trade unionist or taking part in trade union activities, **nor** may he be forced to become a trade unionist or a member of a particular union or category of trade union.

If a closed shop exists, an employee cannot be forced to belong to a trade union. Should pressure be brought on him, which led to his dismissal because he did not want to join the union, then this would amount to a case of **unfair dismissal** under the 1996 Act (see p. 237). Nor can he be forced to make a payment, e.g. to charity, instead of joining the union.

Examples of penalties against an employee could be blocking his promotion, pay rise or bonus.

(Note: in certain exceptional circumstances, trade union membership can be banned by government action, usually for security reasons.)

Right to have time off work: S.27 Employment Rights Act 1996

For trade union representatives: Ss. 61 and 62

Employees who are trade union representatives are entitled to **paid** time off work in order to carry out their union duties. Also, time off is given for them to undergo industrial relations training. Such matters are dealt with by an ACAS code of practice, Code 3 Time Off for Trade Union Duties and Activities.

For trade union members: S.61 Employment Rights Act 1996

This differs slightly to the above. Here, a member of an **independent** trade union is entitled to time off in order to take part in his own trade union activities or to activities where he is acting as the union's representative. Such activities do **not** cover industrial action. He is also entitled to remuneration at an appropriate hourly rate. The union must be one recognised by the employers. Failure to allow time off or to pay wages for such time would be grounds for complaint to an industrial tribunal.

For pregnant mothers

Any pregnant employee is entitled to paid time off to attend ante-natal appointments.

Right to have time off to perform public duties

Under S.50 certain employees are entitled to time off work to perform certain public duties, e.g. justices of the peace, local councillors, members of statutory tribunals, members of a health or police body, members of an educational, prison visitor or Environment Agency institution. They are not necessarily entitled to payment unless their contracts of employment specifically provide for this. Of course, in some cases such as justices of the peace, expenses may be claimed.

Under the Juries Act 1974, anyone employed or otherwise is under a **duty** to attend court for jury service and expenses will be paid.

Right to have 'equal pay' – Equal Pay Act 1970, Sex Discrimination Act 1986

Men and women in full or part-time employment have the right to be treated on equal terms as far as their pay and conditions of work are concerned. They must be in work which is of a similar nature or, if the business they work for operates a job evaluation scheme, they must have jobs with a similar grade.

Maternity rights

The Trade Union Reform and Employment Rights Act 1993 (TURERA) introduced new legislation to comply with EC Directive 92/85/EEC 'The Pregnant Workers Directive'. In addition to statutory maternity pay and the right to time off above, a pregnant worker is now entitled to 14 weeks' maternity leave during which she cannot be dismissed except in exceptional circumstances. Her contractual rights are also preserved during this period and on her return to work she must be given her old job back.

Termination of contracts of employment – common law

Contracts of employment may be terminated in the same way as other contracts (see p. 73). However, there is now a great deal of statutory law which gives the employee protection against summary dismissal without good reason or dismissal without the correct period of notice. Such statutory law is in addition to the usual common law remedies for breach of contract. This is still important, however, as statutory protection depends on the employee having worked for the employer for a particular length of time.

Methods of termination

Performance
Where the contract was for a particular task, the contract will end automatically on its completion.

Effluxion of time
If a contract is made for a fixed term then this ends automatically on the due date, without either party having to give notice.

Frustration
Its application in employment law has been shown in cases involving the death of an employee, an employee's serious illness, or injury or conscription. The frustrating event or circumstances must be outside the employee's control and thus imprisonment due to the employee's conviction after a trial is not a frustrating event (but would be grounds for the employer to repudiate the contract and dismiss the employee for misconduct). (See p. 234.)

Agreement
As in all contracts, the parties may agree to end the contract either by virtue of a contractual term or by agreement at a later stage (see p. 76). If consideration is paid by the employer for the employee's agreement to terminate, it is colloquially known as a 'golden handshake'.

Death of an employer (as a sole trader)
This will again terminate the employment unless there is a transfer of the business as a going concern (see below).

Liquidation of a registered company
This automatically ends the contract if the liquidation is as a result of a compulsory winding-up order.

If on the other hand the company is being wound up voluntarily, this does not necessarily terminate the contract unless the company is insolvent. Under the Transfer of Undertakings (Protection of Employment) Regulations 1981, SI No 1794, contracts of employment will not be terminated if the employees have been transferred from one company to another as a result of the transfer of the business as a going concern, even if the original company has been dissolved. The employee's rights are thus protected and preserved and the length of his employment will date back to the beginning of the original contract. This provision also applies to transfers of a business as a result of the sale or death of an owner and this includes individuals, partners and companies.

If the employee is dismissed by reason only of such a transfer, he will be regarded as having been unfairly dismissed (see p. 237). However, if there was an 'economic, technical or organisational reason' for a change in the workforce affecting the employee, then he may be fairly dismissed. It is up to the employer to prove that he acted reasonably in dismissing the employee: Reg. 8.

Dissolution of partnership
The rules above will more or less apply to an employee of a partnership. If the partnership ceases trading through insolvency, the employment will cease. If there is a transfer of business as a going concern, the regulations above will apply.

Bankruptcy of a sole trader
This will usually terminate the employee's contract, unless the trustee in bankruptcy or receiver obtains permission to continue the business (see p. 142).

Breach of contract

Only a serious breach will allow a party to repudiate the contract. However, if the wronged party affirms the contract, then after the elapse of a reasonable time, he will be unable to rely on the breach as a ground for terminating the contract.

Dismissal

Some contracts end by operation of law, e.g. on the liquidation of a registered company (see above). The rest will be summary dismissals, i.e. dismissal without notice or dismissals with notice.

Summary dismissal

The contract itself may contain terms giving either party the right to terminate without the giving of notice (summary dismissal).

Furthermore, the law recognises that there will be certain situations which allow the parties to terminate either where there has been a serious breach of contract or where a party has committed some act which amounts to repudiation of the contract. In both cases no notice need be given but, of course, the employer should be able to justify the dismissal; otherwise he may be sued for wrongful dismissal at common law or an action may be brought for unfair dismissal in an industrial tribunal. As a result of the Employment Rights Act 1996, claims up to £25,000 may now be brought in the industrial tribunal, if preferred. Otherwise, claims must be taken to the county courts or High Court. The employee could be sued for breach of contract.

Thus, the reasons for such a dismissal must be of a very serious nature. Most of the existing cases have concerned behaviour such as gross misconduct, serious dishonesty, immorality during working hours, serious disobedience, assault on the employer, competing for business with the employer and disclosing trade secrets.

Of course, the employer may be guilty of some serious breach or repudiating action and the employee may then regard the contract as terminated. This may amount to constructive dismissal, i.e. by virtue of the employer's behaviour, the employee regards himself as dismissed.

The ACAS Code of Practice No 1 (Disciplinary Practice and Procedures in Employment) sets out guidelines for the proper organisation of a disciplinary procedure. It emphasises that fair discipline means good industrial relations. Workers are entitled to know what standards of conduct are expected of them and the management, whose prime responsibility it is to maintain discipline, must operate within their system and operate fairly; otherwise they may face an unfair dismissal claim in an industrial tribunal.

The disciplinary machinery should be explained to employees in writing. It should set out the people who have power to discipline (and to dismiss).

An employee should be informed of the types of breaches of contract which would be grounds for summary dismissal.

Opportunities of answering complaints should be afforded to the employee and information of the complaint should be given. Unless there is a case of gross misconduct, no-one should be summarily dismissed for a first 'offence'. An appeal procedure should be operated.

The code suggests that normally a formal **oral** warning should be given first concerning the matter of complaint, unless the matter is very serious in which case a **written** warning may be given. If the worker fails to mend his ways, a **final** written warning should be given, informing him that dismissal or suspension or other penalty will be imposed. Only after ignoring the final warning should the employee be dismissed.

It should be noted that if the worker commits a crime outside his employment, this should not be a reason for his automatic dismissal. His whole position should be looked at to see whether he is now unsuitable for the job.

Records of the warning system should be kept (including, perhaps, written evidence of the oral warning), to provide the employer with a possible defence to a claim for unfair dismissal.

Dismissal with notice

Most cases, however, are concerned with the employee being given notice of his dismissal. Nevertheless, employees should comply with their contract and give the correct length of notice before leaving. If there is no such clause, they should give such notice as would be implied at common law.

It is unlikely that they would be sued for breach of contract if they gave too short a period of notice, unless their presence is vital to the employer. Whether the employee is dismissed by notice, without notice or as a result of the expiry of his fixed term contract, he is entitled to request a written statement giving the reasons for his dismissal: S.92 Employment Rights Act 1996.

Fixed term contracts

As these are usually intended to end on a particular date, no notice is required to end the contract.

But, the contract may contain a clause allowing termination **before the contract date** by a period of notice. If there is no such clause, then there is no such right.

Contracts for indefinite periods

Express terms of notice required by statute: S.1 Employment Rights Act 1996. Provided a new employee works for one month, written particulars of the employment must be given to the employee. One such particular is concerned with the length of notice required (see below).

Statutory minimum lengths of notice under S.86 ERA 1996. Provided the employee has worked for **one** month he is entitled to the following minimum periods of notice (in practice, he may get more under his contract).

Length of employment	Length of notice
4 weeks to 2 years	1 week
2 years to 12 years	1 week × years of employment
12 years +	Not less than 12 weeks

The employees on the other hand need give no more other than one week's notice if they have worked more than one month.

Redundancy

The law is found principally in the Employment Rights Act 1996.

An employee may be paid a redundancy payment if:

- his employer dismisses him by reason of redundancy or
- his employer lays him off or keeps the employee on short time in such a manner as is specified by the Act: S.135.

By S.134 an employee is dismissed by reason of redundancy:

- if the employer has stopped or intends to stop carrying on the business for the purposes of which the employee was employed, e.g. where a business is wound up;
- if the employer has stopped or intends to stop carrying on business in the place where the employee was employed;
- if the business no longer requires work of a particular kind which the employee was doing or such work would be stopping or diminishing, e.g. where mechanisation eliminates the need for manual labour such as riveting in the car industry.

Dismissal

For the purposes of the Act, an employee is treated as having been dismissed if:

- his contract is terminated by the employer, with or without notice, *or*
- his fixed term contract has expired without being renewed under that contract, *or*
- he terminates the contract with or without notice in circumstances such that he is entitled to terminate it without notice by reason of the employer's conduct (constructive dismissal, see p. 238): S.136.

Where there has been no actual dismissal under S.136

The above section applies if the employee has actually been dismissed (summarily or with notice). Often, however, an employee may be made redundant by force of circumstances and not by dismissal.

Thus by S.136(5), if an employee has his employment terminated by reason of:

- any act on the part of an employer (e.g. a 'moonlight flit'), *or*
- any event affecting an employer, e.g. death or frustration of contract,

then such acts or events are treated as if the employee has been made redundant, provided such matters would not otherwise have terminated the contract. Thus, if an employer goes bankrupt or the company is compulsorily wound up, which is treated as the fault of the employer, then S.136(5) does not apply.

Consultation

Before contemplating redundancy an employer should consult with the relevant unions to determine who should be selected for redundancy and for what reason the redundancies are required. Also, the Secretary of State for Employment should be notified in certain circumstances.

Time limits for redundancy payments

Within six months of the relevant date, the employee must have:

- claimed a redundancy payment by written notice to his employer (S.164);
- referred a question to the industrial tribunal concerning his right to a payment or amount of payment;
- presented a claim for unfair dismissal to an industrial tribunal; *or*
- agreed or accepted a payment (S.164).

The tribunal may, in certain circumstances, allow an additional six month period in which to claim, if there are good reasons why the employee failed to take such action. The relevant date is the date notice expires, expiration of the fixed term contract or when termination actually takes effect.

Qualifications for redundancy payments

Compensation for redundancy will only be paid to employees who have been *continuously employed for at least two years* and who are *over 18*. It will not be paid to employees over the normal retirement age. They must have been dismissed by

reason of redundancy and voluntary redundancies will count for payment. But, if the employees resign because of the threat of redundancies, then they have *not* been dismissed and will not get compensation.

An employee may be offered alternative employment by the employer. This must be 'suitable', i.e. comparable with the job he already held. If he unreasonably refuses the offer then he will probably lose his entitlement to a redundancy payment because he will not have been dismissed by reason of redundancy. However, there is a statutory trial period under which the employee can decide whether he would like the new job or not. Provided he terminates his job reasonably within that period, then he will still be entitled to a redundancy payment (see the case of the carpenters on p. 240).

Compensation for redundancy

A worker must have worked for the employer for at least two years prior to redundancy. Pay is calculated by reference to the number of complete years of employment with the same employer (sometimes previous employment counts) up to a maximum of 20 multiplied by a week's basic pay. The years worked before the age of 18 do not count (see Figure 9.1). Written particulars must be given.

Lay-off and short-time: S.147 ERA 1996

It should be noted that this is not concerned with the right to guarantee payments.

If an employee

- has been laid off or kept on short-time (with less than half a week's pay) for four consecutive weeks or more or for six weeks during a period of 13 weeks (three of which were consecutive), *and*
- has within four weeks following such a period of lay-off or short-time served a notice of intention to claim a redundancy payment on his employer, *and*
- has terminated his contract by giving the contractual length of notice within a specified period,

then he may claim a redundancy payment.

Age	Compensation
18–21	¹/₂ weeks' pay for each year in band
22–40	1 weeks' pay for each year in band
41–64	1¹/₂ weeks' pay for each year in band

Example

Thus, a 36-year-old person who has worked for the same employer since he left school at 16 will get:

For the years	16–18	nothing
	18–21	3 × ¹/₂ a week's pay
	21–36	15 × a week's pay

(A week's pay is calculated by reference to the salary paid when made redundant and does not include overtime or other extra payments, up to a statutory maximum.)

Figure 9.1 Calculation of redundancy payment

The employer may respond by actually dismissing the employee (who is then effectively dismissed by reasons of redundancy anyway) or by serving a counter notice giving details of work coming up, in which case a claim for a redundancy payment under these sections may be defeated. (If that is the case, however, he may be able to claim a guarantee payment.) One cannot claim for redundancy pay where lay-off or short-time was caused by a strike or lock-out.

The National Insurance fund

No tax need be paid on redundancy pay under the Act.

Should the employer not pay for some reason, the employee can make a claim to an industrial tribunal within six months of ending that employment (see p. 236).

If the employer cannot pay because of insolvency, the Department of Employment will pay the total amount from the National Insurance fund (S.166).

Remedies for dismissal of employee without good reason or insufficient notice

At common law – wrongful dismissal

As employment is a branch of contract law, the employee is always entitled to sue his employer (and vice versa) for breach of contract, if the other has breached his duties in any way.

If the breach is to dismiss the employee without good reason, or by giving too short a period of notice, the common law has developed a breach of contract action called **wrongful dismissal**.

Like all contractual breaches, the main remedy is for damages. It should be remembered that contractual damages are governed by the rule in *Hadley* v. *Baxendale* (see p. 79). Thus, damages awarded will not exceed the amount the employee should have been paid had he been given a proper period of notice before his dismissal. Also, because of the personal nature of the contract, no court will grant a decree of specific performance, as this would force the parties to enter into an employment situation again (cf. re-instatement and re-engagement below). This has been confirmed by S.236 Trade Union and Labour Relations (Consolidation) Act 1992.

Despite the introduction of proceedings for unfair dismissal, wrongful dismissal is still important for those who have not the requisite length of service and for those who wish to sue for sums higher than permitted under legislation. Claims for under £25,000 may now be heard in industrial tribunals.

By statute – unfair dismissal

The law relating to unfair dismissal is found primarily under the Employment Rights Act 1996.

An employee with at least two years' service who is dismissed with notice or without may take his employer to an industrial tribunal in a claim for **unfair dismissal** on various grounds.

The claim must normally be made within three months of the date of termination of the contract.

Claims regarding race, sex or trade union activity do not require the two year qualifying period.

Exceptions

By S.94 ERA 1996 every employee has the right not to be unfairly dismissed **except** in the following situations:

1. Where the employee has not been in continuous employment for at least two years. S.108(1).

2. Where the 'employee' has got a contract for services, i.e. he is **not** an employee: S.230(1) and (2) (see p. 164).
3. If the employee has reached the normal retiring age for that job or, if not, 65 in the case of men and women: S.109(1).
4. If there is a procedure agreed between the employer and trade union which deals with dismissals. However, such an agreement must be 'designated' by the Secretary of State for Employment: S.110.
5. Where the employee normally works outside Great Britain: S.196. Note: North Sea oil rig workers and workers within British territorial waters are regarded as working in Great Britain.
6. Where the employee has a fixed term contract of more than one year and the dismissal merely consists of the expiry of the contract without it being renewed *and* the employee has agreed *in writing* not to pursue a claim for unfair dismissal: S.197(1). (Thus contracts for fixed terms ought to be carefully checked before signing; otherwise the remedies for unfair dismissal will be lost, even if all that has happened is that the normal contractual date has been reached.)

   (Note: this is the only type of contract in which an employer can exclude his liability for unfair dismissal.)
7. Where employees are share fishermen, registered dock workers, policemen but not Crown employees unless in the services: Ss. 199 and 200.
8. If the contract is illegal either at its inception or in its performance (rule of common law), e.g. in *Tomlinson* v. *Dick Evans U Drive Ltd* 1978 an employee was paid without deduction of tax. The employer and employee had deliberately defrauded the Inland Revenue. The Employment Appeal Tribunal confirmed the industrial tribunal's decision that the contract was unenforceable and thus no rights accrued and the employee could not complain of unfair dismissal.

There must be a 'dismissal'
The employee must prove that he has been dismissed in fact and in law.

Section 95 ERA 1996 states the three circumstances in which there **will** be a dismissal.

1. Where the employer terminates the contract with or without notice. Thus, resignations or mutual agreements terminating the contract are **not** dismissals. Invitations to resign, which are often used to 'save face', may amount to a dismissal if it is implicit that if the employee does not resign then he will be dismissed.

   Where the words of dismissal are couched in the vernacular, e.g. '—— off' the industrial tribunal will have to determine the true intention of the employer.
2. Expiry of a fixed term contract without it being renewed under the terms of that contract. A fixed term contract may be for a particular period or may, as in some building contracts of employment, be for the duration of a project, provided it has a fixed date of completion. As a result of *Dixon* v. *BBC* 1978, a fixed term contract is still one even if either party can terminate by notice before the due date.
3. Where the employee terminates the contract with or without notice in circumstances such that he is entitled to terminate it without notice by reason of the employer's conduct. This is known as **constructive dismissal** and is concerned with circumstances in which the employer has behaved in such a way that the employee has decided to resign rather than continue to suffer the employer's behaviour. In *Western Excavating (ECC) Ltd* v. *Sharp* 1978 the Court of Appeal found that deciding whether or not someone has been constructively dismissed depends on whether the employer has committed a serious breach of contract or not. If he has not then there can be no grounds for the employee repudiating the contract.

What sort of dismissal is fair?
It is up to the employer to prove that the dismissal was fair. The following are grounds for dismissal to which the tribunal would agree with the employer as being capable of being fair: S.98 ERA 1996.

- Redundancy (see p. 235).
- Misconduct (see p. 234). Potentially fair.
- Incapability to do the work, whether through lack of qualification, experience or physical or mental incapacity (see p. 231). Potentially fair.
- If the employment is in contravention of statute law. Potentially fair.
- Any other substantial reason for justifying the dismissal of an employee in his position. Potentially fair.

Having proved that the dismissal was for one of the above reasons, the tribunal then has to decide whether the ground was a **reasonable** one for dismissing the employee in the particular circumstances.

Each case is then judged on its special merits. Whether the ACAS Code of Practice on Disciplinary Practice and Procedures and the Industrial Relations Code of Practice on Redundancy have been followed will also be taken into account (see pp. 228, 240).

The following are definitely **unfair** grounds for dismissal:

- where the employee was or intended to become a member of an independent trade union, *or*
- has taken or intended to take part in independent trade union activities, *or*
- was not a member of a trade union or a particular union or who had refused to join or remain a member of a union including a closed shop, *or*
- where the employee had been made redundant and had been the only one selected for redundancy, for one of the previous three reasons, out of a group equally affected by the threat of redundancy *or*
- where the employee had tried to assert a statutory right, by either bringing proceedings against the employer or otherwise alleging infringements of this right and had then been selected for redundancy (see p. 240).

The legislation thus provides protection for trade union members, activists and those who do not wish to become trade union members.

Strikes or other industrial action: S.239 TULRCA (see page 243)

If an employee is dismissed during an industrial action, an industrial tribunal shall not consider whether the dismissal was unfair or not **unless**

- one or more fellow employees have *not* been dismissed (implying that the employee had been victimised);
- one or more fellow employees have been offered re-employment and the claimant employee has not. (This is subject to a three month time limit from the date of dismissal.)

This piece of legislation is designed to prevent discrimination in industrial action but no more. The tribunal is not concerned with the reasons or the merits of industrial action.

It should be remembered that whilst there is a right to withdraw one's labour, this immediately makes the employee in breach of contract which is a ground for terminating his contract at common law. No statutory protection is available to such a person unless he has been victimised and the above sections apply.

Health and safety cases: S.44 Employment Rights Act 1996

Section 100 of the Employment Rights Act now gives employees the right not to be dismissed in certain 'health and safety' situations. Thus, they should not be dismissed after carrying out or proposing to carry out health and safety activities to which they had been designated by their employer. Nor should they be dismissed for bringing a health and safety matter to the attention of the employer. Furthermore, no employee should be dismissed if he was in serious and imminent danger and had left his post refusing to return, unable to avert the danger.

Pregnancy: S.99 Employment Rights Act 1996

Dismissing a woman solely because she is pregnant is unfair, however short her time of employment has been.

Dismissal of temporary 'maternity leave' replacement: S.99 Employment Rights Act 1996

It must be made absolutely clear in writing at the beginning of the contract that the job is only temporary and subsequently the dismissal must be carried out fairly; otherwise the person dismissed will have a good case for unfair dismissal.

Sex and race discrimination

See p. 232. Anyone dismissed on grounds of sex or race discrimination is deemed to have been unfairly dismissed (even within the first year of employment).

Redundancy: S.105 ERA 1996

If only a particular employee is made redundant, yet the surrounding circumstances for other employees are the same and they have not been dismissed by reason of redundancy, then if the main or only reason for selecting that worker for redundancy is inadmissible or if he had been selected in a way that was against a customary arrangement or agreed procedure, the dismissal shall be regarded as unfair. In a decision of the Employment Appeals Tribunal, an employer who required six carpenters to go to Cornwall when they normally worked in London was considered to have unfairly dismissed them. The carpenters had no contracts determining where they should work and the offer of alternative work in Cornwall was not a suitable offer.

Employers are under a statutory duty to consult with trade unions and in certain cases to inform the Department of Employment before making people redundant. Also the Industrial Relations Code of Practice, Paras. 44 and 45, gives advice on dealing with redundancy, and if this advice is not followed it may influence a tribunal's decision that the dismissal was unfair.

Remedies of industrial tribunals

Re-instatement and re-engagement orders under the Employment Rights Act SS. 114–15

Where the employee has been unfairly dismissed, the tribunal can, if asked, order his re-instatement or re-engagement.

Re-instatement means that on return to work, the employee is treated as if nothing had happened. Thus, he will get his back pay and will lose none of his pension rights. If he would have been upgraded or got better conditions had he still been working there, then these must be awarded to him.

Re-engagement means that the employee returns to work for the employer or his successor in a job similar to that from which he was dismissed.

These orders are discretionary as in many cases they would be totally inappropriate and in some cases impracticable.

If such an order is ignored then an additional award of between 13 and 26 weeks' pay can be made and an even bigger award of between 26 and 52 weeks' pay if the dismissal was for sex or race discrimination.

Compensation under Employment Rights Act 1996

There are three types of compensation awards: the basic award, the compensatory award and the special award.

1. The basic award: Ss. 119–22 ERA 1996. The basic award is calculated in much the same way as under awards for redundancy (see Figure 9.1).

The basic award for unfair dismissal under Ss. 112(4) or S.117(3) of the 1996 Act, i.e. dismissal for an inadmissible reason, is at least £2,770.

2. Compensatory award. Subject to a maximum of £11,000 (in 1996) a compensatory award is paid such as the tribunal thinks is just and equitable in all the circumstances to compensate for loss directly attributable to the employer's behaviour.

This parallels the rule in *Hadley* v. *Baxendale* in basic contract (see p. 79). Thus, compensation may be awarded for such varied losses as loss of pension rights, future loss of earnings, loss of pay in lieu of notice etc.

3. Special award: Ss. 118 and S.125 ERA 1996. Those employees who have been unfairly dismissed because of their trade union membership or activities, refusal to join a union or unfair selection for redundancy can ask for an additional

special award provided they had asked for re-instatement or re-engagement.

If no such order to re-instate or re-engage is made by the tribunal then the employee receives 104 weeks' pay or £13,775 whichever is the highest amount subject to a limit of £27,500.

If an order for reinstatement **was** made by the tribunal but the employer failed to comply then the award will be 156 weeks' pay or £20,600, whichever is the higher amount.

Industrial tribunals

Origins

These were originally set up by the Industrial Training Act 1964. They are independent bodies which make decisions on questions or claims concerning aspects of employment under a number of Acts.

Relevant legislation

The most important relevant Acts are: Industrial Tribunals Act 1996; Employment Rights Act 1996; Race Relations Act 1976; Sex Discrimination Act 1975; Health and Safety at Work etc. Act 1974; Equal Pay Act 1970; Disability Discrimination Act 1995; Trade Union Reform and Employment Rights Act 1993; Trade Union and Labour Relations (Consolidation) Act 1992; Industrial Tribunals (Constitution and Rules of Procedure) Regulations 1993; and Industrial Tribunals (Constitution and Rules of Procedure) (Amendment) Regulations 1996.

Application

Forms of originating application (IT1) are obtained from the tribunal offices which are found in most big towns or from the Department of Employment. On completion, the form must be sent to the Central Office of the Industrial Tribunals in London or to a regional office.

The application is scrutinised to see whether it asks for relief which the tribunals are capable of giving under the relevant Acts. On registration of the application, a hearing at a local tribunal is arranged. Copies of the claim are sent to the respondent who is asked to file a notice of appeal and a defence, if any.

Conciliation

In some cases, copies of all relevant documents are sent to a **conciliation officer** from ACAS, i.e. the Advisory, Conciliation and Arbitration Service. The officer attempts to come to some agreement between the parties before the hearing. This is designed to save time and money and a more harmonious relationship, especially if re-instatement to the job itself is sought. It is interesting to note that, because of conciliation, approximately 50 per cent of all cases never actually reach a hearing by the industrial tribunal.

Composition of tribunals

As the word tribunal indicates, each tribunal consists of three people, the chairman who is a lawyer and two lay members who are chosen from employers' and employees' associations. All have relevant experience but the lay members are not intended to represent a particular point of view, although the Department of Employment has been trying to recruit younger people, women and members of ethnic minorities.

Pre-hearing reviews

In order to reduce costs, a pre-hearing review is often held in which the case is scrutinised to see whether the applicant has a good case to go before the tribunal: S.9 Industrial Tribunals Act 1996.

The hearing

They are less formal than court hearings, although they are usually heard in public, and legal representation is not necessary. (It may be advisable, however.) Most trade unions have officers who will present a member's case for him. The applicant must provide sufficient evidence to support his allegations. Witnesses can be compelled to attend and the production of documents can be ordered. In such cases, this is best ordered at the pre-hearing (if any). Otherwise, an adjournment may be required.

Despite the informality of the proceedings, evidence of a cogent nature must be presented, i.e. it must be believable and support the claims made, but they are not bound by the normal rules of evidence. The burden of proof is the normal civil 'balance of probabilities' and hearsay evidence may be accepted.

Previous conduct, whilst not acceptable in criminal cases, can be used in tribunal cases. For example, if the employee was dismissed for dishonesty, evidence of earlier dishonesty would be accepted. The hearing is in public.

Reviews (not appeals)

Where either party thinks the decision made was **wrong** they can make an application to the industrial tribunal for a **review** of the proceedings within 14 days of the decision (or at the hearing).

A review may be requested where:

- the decision was made through an **error** on the part of staff at the tribunal.
- where a party received **no notice** of the proceedings. (This may occur through not sending a notice to a company's registered office where all official letters should be sent.)
- where a decision was made in the absence of one of the parties. This will usually have arisen through illness.
- where new evidence has been found since the decision whose existence would not have been known or not reasonably foreseen. Such evidence should be important, relevant to the case, not have been available before and must be believable.
- *where it is in the interests of justice.* This is a 'catch all' for the tribunal and can cover many types of situations: R.11(1) 1993 Regulations.

The procedure is simple and cheaper than applying for an appeal on a point of law (if there is one involved) to the Employment Appeals Tribunal. The application can be made by mere letter setting out a request for review, giving the reasons for the application and his reasons why the decision was wrong.

Appeals

Appeals on points of law only are taken to the Employment Appeals Tribunal (EAT). This is presided over by High Court judges assisted by members appointed by the Lord Chancellor and the Secretary of State for Employment, who have specialist knowledge of industrial relations. Like the industrial tribunals, the lay members are appointed by the Trades Union Congress and the Confederation of British Industry.

Each hearing is heard by a judge plus either two or four members. The numbers must, of course, remain equal so that there can be no bias towards workers or employers. The judge has no greater power than the lay members and thus it is possible for him to be outvoted.

Most appellants, if employees, are represented by counsel, some by solicitors and a small minority represent themselves or get a friend or trade union representative to do this. Employers are usually represented by counsel.

Especially difficult cases are heard by the President of the Employment Appeals Tribunal in a special procedure list who first listens to agreements for approximately 30 minutes before deciding whether to allow the appeal to be made to the full tribunal.

The Construction Industry Training Board (CITB)

Industrial training boards are governed by the Industrial Training Act 1982 as amended and the Secretary of State for Employment. There has been a Construction Industry Training Board since 1964 authorised under previous legislation.

Training boards are designed to:

- provide or arrange training courses for entrants into an industry;
- act as a watchdog over such courses;

- provide advice for those involved in training and encourage research;
- make recommendations and public reports *vis-à-vis* standards to be achieved as a result of training and whether they are being achieved.

In achieving the above ends, the boards may provide financial assistance for persons undergoing training or to employers whose personnel are going to undergo training courses.

In order to be able to monitor an industry an ITB can require employers to provide such information as may be necessary and to keep certain records for examination. An employer who fails to comply with these requirements is guilty of an offence.

To pay for the training system, levies are imposed on employers based on the number of their employees. At present the levy rates are 0.25 per cent of their annual payroll and 2 per cent of payments to labour-only subcontracts. A maximum ceiling of 1 per cent of the total cost of remuneration to employees is levied. 'Small' builders whose total payroll plus 'lump' payments did not exceed £61,000 pa for the financial year up to 5 April 1996 are exempt.

If an employer arranges his own approved training courses, e.g. through a further education establishment, he can apply for exemption. The exemption certificate can be made subject to many conditions and shall not exceed three years from its issue. If the application is turned down by the CITB an appeal can be made to referees appointed under the Act.

The board also operates a training centre specializing in certain aspects of construction work.

Trade unions

Reasons for setting up

These are groups of workers who gather together on a formal basis with the intention of improving their pay and other working conditions. In the nineteenth century they were illegal, classed as unlawful conspiracies, e.g. the Tolpuddle Martyrs, the centenary of whose trial was celebrated in 1984.

Nowadays, they are governed by the Trade Union Act 1984, Trade Union and Labour Relations (Consolidation) Act 1992 (TULRCA) as amended by the Trade Union Reform and Employment Rights Act 1993.

Not corporations

Generally trade unions do not have corporate status. However, a series of statutes have given them some of the outward characteristics of a corporation. For example, they can sue and be sued and make contracts in their own name. Their property, however, cannot be vested in the union itself like a normal corporation. It must be transferred to trustees to hold for the benefit of the union.

Also the members can be prosecuted in the union's name and the union's property can be taken to satisfy any judgment made against it in court.

Possible causes of action against trade unions or unionists in common law

A union or a member of a union could possibly be sued for the torts of inducing a breach of contract, intimidation and unlawful interference with trade or employment contracts. Under S.219 TULRCA 1992, a trade union may be liable for inducing a person to break his contract of employment or a commercial contract if no secret ballot was first taken which supports industrial action. A ballot need not be taken but, if so, statutory immunity under S.219 below would be lost. Furthermore, a number of unionists could be prosecuted for criminal conspiracy if they had conspired to perform a criminal act or could be sued for civil conspiracy if they conspired to commit a tort.

Statutory immunity of individuals: S.219 TULRCA 1992

Whilst no protection should be given to unions or unionists to enable them to commit crimes, a limited amount of protection has been given to individuals in situations in which they try to pursue their legitimate interests. Thus, by S.219(1), an act committed by a person in contemplation or furtherance of a trade dispute shall not be actionable in tort if it only:

- induces another to breach a contract *or*
- induces any other person to interfere with its performance, *or*
- consists of threatening that a contract (to which he may or may not be a party) will be broken or its performance interfered with, *or*
- that he will induce another person to break a contract or interfere with its performance.

As these actions are the direct result of a strike call, S.219(1) protects the right of an **individual** to be able to strike or encourage others to do so, in that they cannot be sued for such action. It should be noted that striking is still a breach of contract and would be a ground for dismissal.

The possibility of being sued for civil conspiracy has also been restricted provided that the agreement between two or more persons to do or procure the doing of an act in contemplation of a trade dispute would **not be actionable** if done by **one person**. It should be noted that such immunity only applies to torts. There is no immunity against criminal conspiracy: S.219(2).

Statutory immunities of trade unions

Probably this will be dealt with by applying S.219 to the larger membership as union immunity *per se* was repealed by the Employment Act 1982.

Thus, a trade union can be liable in tort but only if the act which resulted in the tort being committed was authorised by a 'responsible person', e.g. employed officials or executive committee.

Section 10 of the Trade Union Act 1984 deprives a union of such immunity if it failed to gain the support of a secret ballot of its members.

Any damages recovered are limited in proportion to the size of the membership of the union. The damages are restricted to each claim and not per dispute and there are no restrictions in personal injury claims.

Permitted peaceful picketing: S.220 Trade Union and Labour Relations (Consolidation) Act 1992

Peaceful picketing which is in contemplation or furtherance of a trade dispute is permitted:

- by a person at or near his **own** place of work, *or*
- by an official of a trade union, at or near the place of work of a member of that union, whom he is supporting or whom he represents.

There is a Code of Practice on Picketing issued by the Secretary of State for Employment.

Secondary action: S.224 Trade Union and Labour Relations (Consolidation) Act 1992

Unions used to seek to increase the effect of their industrial action by taking action against companies and individuals who are not in direct contact with the dispute.

Secondary action can be described as consisting of inducing or threatening to break or interfere with a contract of employment where the employer in that contract is **not** a party to the dispute. Such actions are now no longer protected under S.224 TULRCA.

10 Safety law

Introduction

Obviously unsafe work situations are affected by the common law rule relating to negligence and nuisance, and statutes such as the Occupiers Liability Act 1957 and the Defective Premises Act 1974. However, the main object of these rules is to **compensate** victims of incompetence rather than to prevent accidents happening in the first place. It has therefore been left to Parliament to introduce legislation to encourage safe working conditions.

Unfortunately, whilst the majority of accidents occur in the home, it would be impossible to monitor domestic safety conditions. For this reason, legislation has concentrated on trying to control accidents at **work**.

Originally, legislation was introduced in a piecemeal way covering at first only the most dangerous of occupations and later the most common, such as factories, offices and shops. In the building industry, the multiplicity of processes and varied places of work have meant that building work was (and still is) covered by a number of different Acts and regulations. For example, a site office came under the Offices, Shops and Railway Premises Act 1963, as did the contractor's permanent office, but the building work itself was covered by the Factories Act 1961 and the Construction Regulations. This piecemeal approach to legislation has now been 'tied' together by the Health and Safety at Work etc. Act 1974, which covers all work situations. It takes under its 'umbrella' all the old Acts and regulations which continue in force until replaced by new regulations made under the Act (see Figure 10.1). (Note that the Building Regulations are more concerned with the quality of the work, whereas the Construction Regulations are designed to protect the worker whilst doing the job.)

Construction work is now covered by many different regulations (see the end of the chapter). The most relevant are the Construction (Health, Safety and Welfare) Regulations 1996 and the Construction (Design and Management) Regulations 1994. Main offices are covered by the Workplace (Health, Safety and Welfare) Regulations 1992. All these regulations have been made under the main Health and Safety at Work etc. Act.

As stated before, the Act does **not** set out to enable compensation to be recovered from wrong-doers, although compensation may be awarded if there is a breach of a **specific** duty under the Act (see p. 209). The Act's prime responsibility is to impose **criminal** liability on those breaching the Act or regulations made under it.

Reasons for enactment

This came into effect in 1975 and was designed to provide, for the first time, safety, health and welfare protection for **all** people at work (except domestic servants) **and** also for people **affected** by the dangers of work.

The Act thus provides a framework on which complex technical and scientific regulations can be formulated to cover all work processes, but which will be administered and enforced in a standardised way.

The Health and Safety Commission

This is a body having corporate status (see p. 148) which was set up by the Act with the purpose of achieving better health and safety of working

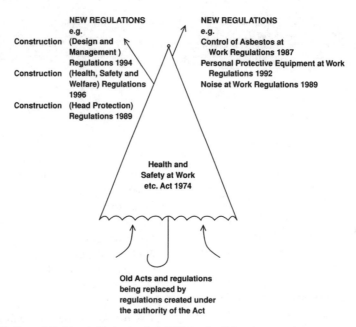

NEW REGULATIONS
e.g.

Construction (Design and
Management)
Regulations 1994

Construction (Health, Safety and
Welfare) Regulations
1996

Construction (Head Protection)
Regulations 1989

NEW REGULATIONS
e.g.

Control of Asbestos at
Work Regulations 1987

Personal Protective Equipment at Work
Regulations 1992

Noise at Work Regulations 1989

Health and
Safety at Work
etc. Act 1974

Old Acts and regulations
being replaced by
regulations created under
the authority of the Act

Figure 10.1 The Health and Safety at Work etc. Act 1974 'umbrella'

people, by undertaking research and training, formulating policy, providing advice and information and acting in close liaison with appropriate government departments. By S.15 of the Act the Commission may with the appropriate Minister of State issue regulations which form part of the body of health and safety legislation. Examples include Control of Asbestos at Work Regulations 1987, Personal Protective Equipment at Work Regulations 1992, Construction (Design and Management) Regulations 1994 and the Construction (Health, Safety and Welfare) Regulations 1996. The commission may also issue codes of practice which, whilst **not** actually forming part of the legislation or general law, provide good evidence in court if any recommended practice has not been followed, e.g. the Code of Practice on p. 256.

The commission may also approve existing codes of practice issued by other specialist bodies or government departments. It must have the consent of the Secretary of State before approving a code and must first consult appropriate bodies.

The commission may also hold informal investigations and formal inquiries into accidents or situations when it thinks necessary.

The Health and Safety Executive

Prior to the 1974 Act, there were a number of different inspectorates having similar responsibilities, but whose authority stemmed from their controlling Act, e.g. the factory inspector was empowered by the Factories Act 1961. All the inspectorates have now been brought under the control of the executive, which is the health and safety law enforcement agency and which also has corporate status. Within the executive, their titles indicate their responsibilities, e.g. inspectors of nuclear safety, pollution inspectors, inspectors of mines, inspectors of factories, agriculture and quarries and inspectors of pipelines.

Some of the inspector's powers may be transferred to other enforcement agencies. For example, activities involving catering, offices, the sale or storage of goods in shops and warehouses and launderettes are dealt with by the local authorities' environmental health departments. They have exactly the same powers (under the Act) as the other health and safety inspectorates (Health and Safety (Enforcing Authority) Regulations 1989).

The Act is an enabling Act

Therefore, no new Acts need be passed covering health and safety at work (although it is of course possible). New regulations containing new laws will be made by the appropriate Minister of State in liaison with the Health and Safety Commission.

Powers of the inspectorate

Investigative powers

An inspector may, at any reasonable times (or at any time, if it is believed that there is something dangerous going on), enter premises and may take a police constable with him to prevent being obstructed in the course of his duty. Once on the premises, he may, *inter alia*:

- examine and investigate
- take samples
- take photographs
- make measurements
- dismantle or remove something dangerous
- take possession of something to prevent it being tampered with
- inspect documents
- and require **anyone** to give such information as he requires.

The above powers are of an investigative nature. If he finds that there has been a breach of the Act or regulations he may take the following enforcement steps.

Enforcement powers (mnemonic PIPS)

Issue of improvement notice: S.21
Such a notice informs the recipient of the breach, and requires that the situation be remedied within a specified period.

The measures required to remedy may be specified. However, in the case of an improvement notice served in relation to a building, then any requirements made cannot be greater than those imposed under the Building Regulations.

Issue a prohibition notice: S.22
If there has been a breach of a statutory provision and the inspector is of the opinion that if the situation continues, there is a likely risk of **serious personal injury**, then he may serve a **prohibition notice**. This prohibits the behaviour amounting to the breach, immediately if the risk appears imminent or it may be deferred until the expiry of a specified period. Prohibition notices can thus be used to stop a particular practice, e.g. a worker who fails to use protective clothing, or it can be used to close a whole workplace, e.g. where toxic fumes are not being effectively removed from a factory. The service of a prohibition notice in building work is to be avoided at all costs, as the responsibility for any delay would invariably fall on the shoulders of the building contractor unless the liability for the notice was due to the fault of someone else, e.g. the employer.

(Note: improvement and prohibition notices cannot be enforced against the Crown. Nevertheless, the commission allows a **Crown notice** to be served on government departments etc. which specifies the breach and imposes a moral duty to comply. The National Health Service (Amendment) Act 1986 has removed Crown immunity from National Health Service Hospitals in relation to health and safety and food regulations.)

Seizing, rendering harmless or destroying articles or substances: S.25
Sometimes the inspector may find articles or substances which he believes could cause imminent personal injury, e.g. gelignite which has deteriorated and has become unstable. In such cases, the inspector may remove the goods in order to destroy them or to have them made safe.

Prosecution: S.33
A person who contravenes a relevant statutory provision, including ignoring prohibition or improvement notices, can be prosecuted in the magistrates' court and in certain specified cases in the Crown Court. Fines and even imprisonment can be imposed. Furthermore, if a notice is ignored, the court can order that work should be carried

out or some step be taken which if also ignored will amount to contempt of court. If a recipient of a notice fails to carry out the requirements contained in any notice or order, then each day he continues to be in contravention, he will be fined a further fixed sum.

Anyone, including a corporation such as a company, may be prosecuted if in breach of the Act. However, it is interesting to note that under S.37, where it appears that an offence committed by a corporation (an inanimate person) has been committed with the consent or connivance of or through the neglect of a director, manager or other officer, then he may also be prosecuted: *Armour* v. *Skeen* 1977. Remember, ignorance of the law is **no** defence. It is interesting to note that although the Crown cannot be prosecuted (see p. 156) the inspectorates have informed government departments and other Crown departments such as health authorities that they are still entitled to use Ss. 7 and 8 (see p. 249) against individuals.

These sections concern an employee's duty to take care of himself and others and the duty of anyone not to intentionally or recklessly interfere with anything provided to comply with health, safety and welfare provision.

Appeals against notices: S.24

Of course, the recipient of a **notice** may think he has grounds for an appeal and he may do so to an **industrial tribunal**. The grounds include insufficient time being given to remedy the situation. Also, the alleged breach may in fact not be a breach of a statutory duty. These rules and regulations must therefore be scrutinised to check whether there has been a breach and it is generally no defence to say that there have been no accidents in the past despite the breach. That might have been pure luck. The degree of likely risk is also a ground for appeal. If the risk is likely then all steps must be taken to remedy the situation. But if the risk is less likely, then the tribunal may be satisfied with fewer safeguards than are required by the notice. This also affects the financial aspects which may influence a tribunal to lessen the requirements of a notice. Thus, a firm may successfully argue that the requirements will prove

financially ruinous, especially as the risk is not reasonably likely. Should the risk be likely, however, no-one will relieve anyone of their duty to guard against that risk.

The tribunal may at its discretion award costs against a party who is unsuccessful in bringing an appeal.

An inspector has only those powers set out in a document given to him on his appointment, and this should be produced at any time he chooses to exercise his rights. Should he overstep those powers, he can be sued by the person suffering as a result. Under S.26 of the Act, the employers have a power (not a duty) to indemnify him for any claims made against him.

Civil actions for breaches of the Act

By S.47, a breach of the **general** duties under the Act does **not** give rise to civil action in tort for a breach of a statutory duty (see p. 209). Thus, if there is a breach of a general duty, only criminal proceedings may be taken *vis-à-vis* **the Act**. However, this does not prevent civil action being taken in negligence, but all the requirements of that tort will have to be met.

On the other hand, if there is a breach of one of the **particular** Acts, e.g. the Factories Act 1961 or one of the regulations made thereunder, a civil action for breach of a statutory duty may be taken in addition to (or instead of) criminal proceedings **unless** the Act/regulation specifically forbids this, e.g. as under the Management of Health and Safety at Work Regulations 1992.

Should someone wish to sue as a result of a dangerous occurrence which has been investigated by an inspector, then a party in the action may require the inspector to provide a written statement concerning the occurrence.

The general duties under the Act

Under S.2(1) an employer is under a duty 'to ensure, so far as is reasonably practicable, the health, safety and welfare at work of all his employees'. This duty is expanded by S.2(2) which specifies five areas of responsibility.

1. The employer must provide and maintain plant and systems of work which are, as far as is reasonably practicable, safe and without risks to health.
2. He must ensure the safety and absence of risks to health in the use, storage, handling and transport of substances and articles.
3. He must provide sufficient information, instruction, training and supervision in order to ensure the health and safety of his employees.
4. He must maintain the place of work in a condition which is safe and without risk to health and ensure safe access to and egress from it.
5. He must provide a safe working environment with adequate welfare facilities.

Furthermore by S.3 every employer and every self-employed person is under a duty to ensure so far as is reasonably practicable that persons not in his employment are not exposed to risks to their health or safety by the conduct of the business. Thus passersby and visitors to a site would be owed a duty under S.3.

Independent contractors on a building site would come within the scope of S.3 both as owing and being owed such a duty. Furthermore by S.3(3) in certain prescribed situations, employers and self-employed persons must give the required prescribed information to non-employees as to how the work affects their health and safety. For example in *R. v. Associated Octel Co. Ltd* 1994 an employer contracted with a specialist independent firm of contractors to repair a tank in a chlorine plant. The independent contractor's employee was injured in a fire caused by an explosion after his lamp broke igniting acetone vapour in the tank. The defendant company was found to be in breach of S.3 and should have provided sufficient information on how the work ought have been carried out in order to keep the independent contractor's employee safe.

By S.4, a person in control of non-domestic premises which are used by people who are not in his employment is under a duty to ensure, so far as is reasonably practicable, that the premises, the means of access and egress and any plant or substance used there are safe and without risks

to health. This section would cover places such as launderettes or self-service petrol stations. In *Westminster City Council* v. *Select Managements Ltd* 1985 the common parts of a block of flats were deemed to be 'non-domestic' premises so that a duty under the Act was owed to repairmen who went to repair and maintain lifts and electrical installations (cf. the civil liability of occupiers on p. 184).

By S.4(3) a person who has by a covenant or tenancy an obligation to repair or maintain the premises or the means of entrance or exit (access and egress), or to see that there are no risks to health from plant or equipment or substances on the premises, is treated as one in control of those premises. So commercial repair and maintenance contracts carry with them statutory duties under the Act which cannot be ignored merely because the premises do not belong to the contractors.

From S.4(4) it is quite clear that the duties imposed by S.4 are in relation to business premises, whether a profit is made or not, and not to domestic premises (subject to the interpretation placed by Westminster City Council above).

By S.7 an employee is under a general duty to take reasonable care at work for the health and safety of himself and others such as his workmates and he must co-operate with his employer and others in complying with any relevant statutory provision.

Under S.8 a person who wilfully or recklessly interferes or misuses anything provided in pursuance of a relevant statutory provision, such as safety apparatus, is guilty of an offence.

By S.9 no employer is allowed to charge his employees for things done or provided by virtue of complying with specific requirements of the Act or regulations made under the Act.

The Construction Regulations

Like all areas of society, construction work is being increasingly bound by new rules and regulations. The main regulations that the builder should be concerned about are the Construction (Head Protection) Regulations 1989, the Construction

(Lifting Operations) Regulations 1961 (due for replacement shortly), the Workplace (Health, Safety and Welfare) Regulations 1992, the Construction (Design and Management) Regulations 1994 and the Construction (Health, Safety and Welfare) Regulations 1996. These are outlined below.

Other regulations such as the Provision and Use of Work Equipment Regulations 1992, the Personal Protective Equipment at Work Regulations 1992, the Control of Substances Hazardous to Health Regulations 1994 and the Management of Health and Safety at Work Regulations, whilst very important, are beyond the scope of this book.

Construction (Health, Safety and Welfare) Regulations 1996

These have *inter alia* revoked and replaced the Construction (Working Places) Regulations 1966 and the Construction (Health and Welfare) Regulations 1966, partially revoked the Construction (General Provisions) Regulations and amended the Construction (Lifting Operations) Regulations 1961, Workplace (Health, Safety and Welfare) Regulations 1992 and the Construction (Design and Management) Regulations 1994.

They impose health and safety requirements on those people undertaking 'construction work' and other people who may be affected by such work. They came into effect on 2 September 1996.

Regulation 2 contains a number of statutory definitions.

'Construction site' means any place where the principal activity being carried out is construction work.

'Construction work' means the carrying out of any building, civil engineering or engineering construction work and includes any of the following:

(a) the construction, alteration, conversion, fitting out, commissioning, renovation, repair, upkeep, redecoration or other maintenance (including use of high pressure water or abrasive or classified corrosive or toxic substances), decommissioning, demolition or dismantling of a structure;

(b) the preparation for an intended structure, including site clearance, exploration, investigation (but not site survey) and excavation, and laying or installing the foundations of the structure;

(c) the assembly of prefabrication elements to form a structure or the disassembly of prefabricated elements which, immediately before such disassembly, formed a structure;

(d) the removal of a structure or part of a structure or of any product or waste resulting from demolition or dismantling of a structure or from disassembly of prefabricated elements which, immediately before such disassembly, formed a structure; and

(e) the installation, commissioning, maintenance, repair or removal of mechanical, electrical, gas, compressed air, hydraulic, telecommunications, computer or similar services which are normally fixed within or to a structure,

but does not include the exploration for or extraction of mineral resources or preparational activities for such work.

By R.3(1) the regulations only apply to construction work carried out by a person at work. So DIY is not covered. Also the regulations **do not** apply to parts of the construction site which is set aside for other **non-construction** purposes: S.3(2) (in such cases the Workplace (Health, Safety and Welfare) Regulations 1992 will apply; see below).

Regulations 15, 19, 20, 21, 22 and 26(1) and (2) apply only to and in relation to construction work carried out by a person at work on a construction site: R.3(3).

Persons upon whom duties are imposed: R.4

This follows along the lines of the main Act. So employers and the self-employed must carry out the provisions of the regulations in so far as they affect their employees, themselves or any person at work under their control: R.4(1).

Any other person who has control of people carrying out construction work must comply with the provisions if matters are within his control: R.4(2).

Every employee is under a duty to comply with the requirements of the regulations: R.4(3).

Every person at work must co-operate with others to enable duties or requirements of the regulations to be carried out and to report to those persons defects which he is aware may endanger the health, safety and welfare of himself or others: R.4(4).

Safe places of work: R.5

So far as is reasonably practicable there should be safe access to and egress from every place of work. The place itself should be safe and, importantly, suitable and sufficient steps should be taken that, so far as is reasonably practicable, no person gains access to any place which does not comply with the above.

Persons **making** the place safe are not included in the above, provided all practicable steps have been taken to ensure their safety.

Every place, so far as is reasonably practicable, should have sufficient working space having regard to the type of work being carried on.

Falls: R.6

Suitable and sufficient steps shall be taken to prevent, so far as is reasonably practicable, any person falling. Such measures include the provision of guard-rails, toe-boards, barriers etc. or any working platform. (These must comply with the requirements contained in Schedules 1 and 2 to the Regulations which give more detailed requirements and specifications. For example the main guard-rail shall be at least 910 mm above the edge from which any person is liable to fall.)

Apart from this general principle, where anyone is to carry out work or use as a means of access or egress, where he is liable to fall **2 m or more**, suitable and sufficient guard-rails etc. should be provided which should comply with the provisions of Schedule 1 (Requirements for guard-rails): R.6(3)(a). Should a working platform be required, this should be provided in sufficient number and must comply with Schedule 2 (Requirements for working platforms): R.6(3)(b).

If it is not practicable to comply with (a) or (b) above or if the work is going to be of short duration so that compliance with these paragraphs is not reasonably practicable, then suitable personal suspension equipment should be used which must comply with the requirements of Schedule 3 (Requirements for personal suspension equipment).

If it is not practicable to comply with (a) or (b) or (c) above or if the work is going to be of short duration so that compliance with these paragraphs is not reasonably practicable, then a suitable and sufficient means of arresting the fall must be used which must comply with the provisions of Schedule 4 (Requirements for arresting falls).

When materials need to be moved, the safety equipment mentioned above may be removed but must be replaced as soon as practicable: R.6(4).

A ladder shall not be used as a means of access or egress to or from a place of work unless it is reasonable having regard to

(a) the nature of the work and its duration and
(b) the risks to the safety of anyone using the ladder: R.6(5).

If a ladder is used it must comply with Schedule 5 (Requirements for ladders) and the provision of paragraph (3) above does not apply (see above: guard-rails etc., working platforms, personal suspension equipment and methods of arresting falls): R.6(6).

By paragraph (7) any equipment used in this connection shall be properly maintained.

Paragraph (8) is concerned with scaffolds. The installation or erection of any scaffold in relation to paragraph (1) or paragraph (3)(b) and any substantial addition or alteration shall only be carried out under the supervision of a competent person: R.6(8)(a).

Similarly if personal suspension equipment or a means of arresting falls is being installed or erected then this must also be under the supervision of a competent person. Installation does not mean the personal attachment of any equipment: R.6(8)(b).

By R.6(9) no toe-boards shall be required in respect of any stairway or rest platform forming part of a scaffold if they are used solely as a means of access to or egress from a workplace, provided that the stairway or platform is not being used to store or keep materials.

Fragile material: R.7

Suitable and sufficient steps shall be taken to prevent anyone falling through any fragile material.

In particular, no-one shall pass across or work on or from fragile material through which he is likely to fall 2 m or more, unless suitable and sufficient platforms, coverings or similar are used to support his weight and guard-rails, coverings etc. are provided to prevent his fall through that material. Furthermore, in such situations prominent warning notices must be fixed at the approach to such material.

Falling objects: R.8

Suitable and sufficient steps must be taken to **prevent** the fall of any material or object so far as is reasonably practicable. This can include the use of guard-rails, toe-boards etc. or working platforms. If this is not reasonably practicable, then steps must be taken to to prevent people being injured by falling objects.

No material or object shall be thrown or tipped from a height in circumstances where it could cause injury.

Materials and equipment must be stored in such a way as to prevent danger to any person arising from their collapse, overturning or unintentional movement.

Stabilty of structures: R.9

All practicable steps must be taken to prevent danger, by ensuring that any new or existing structure which may become unstable or temporarily weak due to the carrying on of construction (including excavation work) does not collapse accidentally.

Structures must not be so loaded as to render them unsafe.

Buttresses, temporary supports or structures used to support a permanent structure shall be erected or dismantled only under the supervision of a competent person.

Demolition or dismantling: R.10

This must be planned and carried out in such a manner as to prevent, so far as is reasonably practicable, any danger and under the supervision of a competent person.

Explosives: R.11

An explosive charge shall only be fired or used if suitable and sufficient steps have been taken to prevent exposure to risk of injury from the explosion or from flying material.

Excavations: R.12

All practicable steps shall be taken to prevent danger, by ensuring that any new or existing excavation which may be temporarily weak or unstable due to construction work (including other excavations) does not collapse accidentally.

Suitable and sufficient steps must be taken, so far as is reasonably practicable, to prevent anyone being buried or trapped by a fall or dislodgement of any material.

Additionally, in order to prevent such dangers, as early as practicable the excavation work must be sufficiently supported using appropriate equipment. The support work must be carried out only under the supervision of a competent person.

Proper steps must be taken to prevent any person, vehicle or plant and equipment or any accumulation of earth falling into any excavation.

If there is a likelihood of collapse of the excavation then no person or thing shall be placed or moved near.

No excavation work shall be carried out unless proper steps have been taken to identify and, so far as is reasonably practicable, prevent injury from underground cables or service.

Cofferdams and caissons: R.13

These must be of suitable design, construction and material and be of sufficient strength and capacity for their purpose and must be properly maintained. The construction, installation, alteration or dismantling of these must only take place under the supervision of a competent person.

Prevention of drowning: R.14

If there is a likelihood of drowning in water or other liquid, proper steps must be taken to prevent falls, so far as is reasonably practicable, minimise the risk of drowning and provide suitable rescue equipment.

If workers have to be transported by water then the transport provided should be safe, of suitable construction, properly maintained and controlled and not be overcrowded or overloaded.

Traffic routes: R.15

Every construction site shall be organised in such a way that, so far as is reasonably practicable, pedestrians and vehicles can move safely and without risks to health.

Traffic routes must be appropriate for the persons or type of vehicle using them. They will not be suitable unless steps are taken to ensure that pedestrians or vehicles may use a route without causing a danger to others nearby. Gates and doors for pedestrians leading onto vehicle routes must be sufficiently separated from that traffic route to enable the pedestrians to see approaching vehicles from a place of safety.

There should be sufficient separation of pedestrians and vehicles to ensure safety and if this is not practicable other means of protecting the pedestrians should be used and effective arrangements should be made to warn anyone likely to be crushed or trapped by a vehicle of the approach of that vehicle.

Loading bays must have at least one exit point for pedestrians only.

If it is unsafe to use a gate intended for vehicles, a clearly marked door should be provided close by and kept free of obstruction.

No vehicles shall be driven on a route unless, so far as is reasonably practicable, it is free from obstruction and allows sufficient clearance. If that is not practicable, then suitable warnings must be given of obstruction or lack of clearance.

Every traffic route shall be properly indicated for health and safety reasons.

Doors and gates: R.16

In order to prevent injury, doors, gates and hatches must be fitted with safety devices. Sliding doors etc. must have devices to prevent them coming off their tracks, upward opening doors etc. to prevent them falling back and powered doors etc. from trapping people; these must also be capable of being opened manually if the power fails.

Vehicles: R.17

The unintended movement of vehicles must be controlled or prevented.

Should anyone be in danger from the movement of a vehicle, then the person in control must give a suitable warning.

Vehicles must be driven, operated, towed and loaded safely.

No-one should ride on a vehicle except in a properly provided safe place.

No-one shall remain on a vehicle while it is being unloaded or loaded unless they are in a properly provided safe place.

Vehicles used for excavating, handling or tipping materials should be properly prevented from falling into excavations, pits, water or overrunning edges of embankments or earthworks.

Suitable equipment must be provided for replacing derailed rail vehicles onto their tracks.

Prevention of risk from fire etc.: R.18

Suitable steps shall taken to prevent, so far as is reasonably practicable, risk of injury caused by fire or explosion, flooding or substances causing asphyxiation.

Emergency routes and exits: R.19

A sufficient number of suitable direct emergency routes must be provided to enable anyone to reach an identified place of safety in case of danger. These must be kept clear of obstructions, indicated by suitable signs and provided with emergency lighting, if necessary. Provisions for such routes must take account of the type of work on site, the characteristics and size of the site, the number and locations of places of work there, plant and equipment used, the number of people present, and the physical and chemical properties of any substances or materials on or likely to be on the site.

Emergency procedures: R.20

Suitable emergency procedures (including evacuation) to deal with foreseeable emergencies must be prepared and if necessary implemented. The same factors as under R.19 should be taken into account. These should be brought to the attention of anyone affected and the procedures should be tested at suitable intervals.

Fire detection and fire-fighting: R.21

Suitable fire-fighting equipment should be kept on site and properly indicated, as also fire detectors and alarms which should be suitably located. The same factors as under R.19 should be taken into account. The equipment should be properly maintained and tested. If the equipment does not come into use automatically, like sprinklers, then it must be easily accessible. Everyone on site must be instructed in the use of the equipment. If there is a particular risk of fire in a particular job, then no-one must work unless suitably instructed to prevent that risk.

Welfare facilities: R.22

The person in control of the site is under a duty to see that the requirements of this regulation are complied with in relation to that site. Employers and the self-employed are also under a duty to see that the following provisions are complied with in relation to those under their control.

Suitable and sufficient sanitary conveniences shall be provided or made available at readily accessible places. They must comply with the provisions of Schedule 6. Thus, they must be adequately ventilated and lit; they (and the rooms) must be clean and in an orderly condition; separate rooms must be provided for men and women, unless the cubicles can be locked from the inside.

Suitable and sufficient washing facilities, including showers if necessary, shall be provided or made accessible. They must comply with the provisions of Schedule 6. Thus, washing facilities shall be provided close to each sanitary convenience and in the changing rooms. They shall include clean hot and cold or warm running (if possible) water, soap and towels or other means of drying. The rooms must be sufficiently ventilated and

lit and kept in a clean and orderly condition. Separate facilities shall be provided for men and women unless provided in a room which can be locked from the inside and intended to be used one at a time. Separate facilities are not required in relation to the washing of hands, forearms and face.

An adequate supply of wholesome drinking water must be provided. It should be readily accessible and be conspicuously marked. Sufficient cups or a drinking fountain should be provided.

Suitable accommodation shall be provided for non-work clothes and for work clothes which are not taken home. Such accommodation shall allow for facilities to dry clothes.

Also, suitable accommodation must be provided to change clothing where workers have to wear special clothing for work and they cannot for health or propriety reasons be expected to change elsewhere. Separate facilities should be provided for men and women where necessary.

Facilities for rest should be made available at readily accessible places, so far as is reasonably practicable. Schedule 6, paragraph (14), states that such facilities shall include those provided in more than one area, protection of non-smokers from smoke, provision for pregnant women or nursing mothers to rest, and suitable arrangements for the preparation and eating of meals and the boiling of water.

Fresh air: R.23

Workplaces and their approaches must be properly ventilated by fresh or purified air. Plant used should have a visible or audible warning device to indicate failure.

Temperature and weather protection: R.24

Suitable steps shall be taken to ensure, so far as is reasonably practicable, that during working hours the temperature in an indoor place of work is reasonable regarding the purpose for which it is being used.

Outdoors, so far as is reasonably practicable, protective equipment and clothing must be provided to guard against adverse weather.

Lighting: R.25

Suitable and sufficient lighting, preferably natural, must be provided for each workplace, approach and traffic route. The colour of any artificial lighting must not distort the perception of signs or signals. Emergency lighting should be provided in the event of failure of primary artificial lighting if this would cause risk to health or safety.

Good order: R.26

Every part of a construction site should be kept clean and in good order, so far as is reasonably practicable. Where necessary the perimeter of the site must be identified by suitable signs which also indicate its extent.

No timber or other material with projecting nails must be used or allowed to remain so as to be a danger.

Plant and equipment: R.27

All plant and equipment shall, so far as is reasonably practicable, be safe and without risks to health, be of sound construction, sufficiently strong for its purposes, and properly maintained and used.

Training: R.28

Where training, technical knowledge or experience is required to prevent risks of injury, then those people should have such a background or be supervised by someone with the necessary expertise.

Inspection: R.29

Certain works require inspections as set out in Schedule 7 before work can start or continue, e.g. a cofferdam or caisson must be inspected before any person carries out work at the start of every shift and after any event likely to have affected its strength or stability. The inspection will include inspecting plant, equipment and materials if it could affect the safety of the workplace.

Apart from the above, where the workplace is part of a scaffold, excavation, cofferdam or caisson, any employer or person controlling construction work must ensure that the workplaces are stable, of sound construction and that any required safeguards are in place before the workplace is first used.

If, after a required inspection, the workplace is unsatisfactory then the appropriate person, e.g. the person in control of the site, shall be informed and that workplace shall not be used until the problems have been remedied.

Reports: R.30

Where an inspection is required under the previous regulation, the person inspecting must prepare a report before the end of the working period containing certain particulars set out in Schedule 8. These particulars are:

1. Name and address of person for whom inspection was done
2. Location of workplace
3. Description of workplace or part inspected (including plant and equipment)
4. Date and time of inspection
5. Details of problems found
6. Details of action taken
7. Details of further action necessary
8. Name and position of person inspecting

Within 24 hours the report, or a copy, must be given to the appropriate person.

The report or copy must be kept at the site and must be retained at the office of the person for whom the inspection was carried out, for three months from the date of completion of the work.

Such a report must be available for inspection by any (health and safety) inspector and the keeper of the report must send such extracts or copies as may be required by an inspector.

No reports are required in respect of working platforms or similar from which someone would fall less than 2 m.

No reports are required regarding mobile tower scaffolds unless erected for seven days or more. Only one report per 24 hours is required in respect of a working platform which has been substantially added to, dismantled or altered. Only one report per seven days is required in respect of **normal pre-shift** inspection of excavations or cofferdams or caissons.

Workplace (Health, Safety and Welfare) Regulations 1992

These regulations came into effect on 1 January 1996. They gave effect to EEC Council Directive 89/654 regarding minimum safety and health requirements for the workplace. These regulations do not apply to construction works (R.3(1)(b)) and are modified in relation to temporary work sites (R.3(2)). Builders need to know about them because they will apply to non-construction work bases such as main offices and long-term site offices. In the case of temporary work sites the requirements to comply with Regulations 20–25 are so far as is reasonably practicable. An Approved Code of Practice 'Workplace, health, safety and welfare' has been issued.

Employer's obligations: R.4

The employer must ensure compliance with the regulations.

Maintenance of workplaces and of equipment, devices and systems: R.5

They must be maintained and cleaned in an efficient state, be in efficient working order and in good repair.

Ventilation: R.6

Workplaces must be properly ventilated by fresh or purified air. Plant used should have a visible or audible warning device to indicate failure.

Temperature: R.7

Workplaces should be kept at a reasonable temperature without the presence of injurious fumes caused by heating or cooling devices. Sufficient thermometers should be provided to check the temperature.

Lighting: R.8

Workplaces must be suitably and sufficiently lighted by natural or, if not reasonably practicable, artificial lighting. Emergency lighting should be provided if there is a special risk of danger.

Cleanliness: R.9

Every workplace and furniture, furnishings and fittings must be kept sufficiently clean. Waste materials should not be allowed to accumulate there.

Room dimensions and space: R.10

Every room where persons work should have sufficient floor area, height and unoccupied space for purposes of health, safety and welfare. The Approved Code of Practice suggests that each person should have 11 m^3 of breathing space.

Workplaces and seating: R.11

Every workstation shall be so arranged that it is suitable for the person working there and the work he is doing. A workstation out of doors shall be arranged so that, as far as is reasonably practicable, it is sheltered from adverse weather, allows easy egress in the event of an emergency and prevents a person slipping. A suitable seat must be provided.

Conditions of floors and traffic routes: R.12

They must be suitable for the purpose used. No floor or surface shall have a hole or slope, be uneven or slippery to expose people to risks. Floors should be effectively drained. Floors and traffic routes must be free of obstructions. The Approved Code of Practice suggests that floors and indoor traffic routes are cleaned at least once a week.

Falls or falling objects: R.13

Suitable and effective measures must be taken to prevent people from falling a distance or being struck by a falling object. Such measures should be taken, where possible, not by merely providing personal protection equipment. If there is a pit or tank etc. these should be covered.

Windows and transparent doors: R.14

These should where necessary be of safety material and be marked appropriately.

Windows, skylights and ventilators: R.15

None should be opened, closed or adjusted or be in such a position as to expose anyone to risk.

By R.16 they must be capable of being cleaned safely.

Organisation of traffic routes: R.17
Pedestrians and vehicles must be able to circulate in a safe manner.

Doors and gates: R.18
These must be suitably constructed, e.g. sliding doors or gates must have a device to prevent them from coming off their tracks. Powered doors must have a device to stop anyone being trapped in them.

Escalators and moving walkways: R.19
They must function safely and be equipped with any necessary safety device and an emergency stop control which is easily identifiable and readily accessible.

Facilities for changing clothing: R.24
If someone has to wear special clothing, suitable and sufficient facilities must be provided.

Facilities for rest and to eat meals: R.25
Suitable and sufficient places should be provided.

Construction (General Provisions) Regulations 1961

The sections of the former Construction (General Provisions) Regulations 1961 relating to dangerous and unhealthy atmospheres, ventilation of excavations and internal combustion engines have been revoked and are now covered by the Control of Substances Hazardous to Health Regulations 1994 which are beyond the scope of this book.

The sections of the former Construction (General Provisions) Regulations 1961 relating to the fencing of machinery have been revoked in relation to new use by the Provision and Use of Work Equipment Regulations 1992 which are beyond the scope of this book.

The section of the former Construction (General Provisions) Regulations 1961 relating to the protection of eyes is revoked by the Protection of Eyes Regulations 1974.

Construction (Lifting Operations) Regulations 1961

These have been amended by the Construction (Health, Safety and Welfare) Regulations 1996.

Part III: lifting appliances
A lifting appliance is defined in R.4 as a crab, winch, pulley, block or gin-wheel used for raising or lowering, and a hoist, crane, sheer legs excavator, dragline, piling frame, aerial cableway, aerial ropeway or overhead runway.

- These should be properly constructed, maintained and inspected: R.10.
- They should be supported, anchored, fixed and erected: R.11.
- Precautions should be taken where the lifting appliance has a travelling or slewing motion: R.12.
- Platforms should be provided for crane drivers and signallers: R.13.
- Cabins should be provided for drivers: R.14.
- Drums and pulleys should be of suitable diameter and construction: R.15.
- Brakes, controls and safety devices should be provided to prevent the load falling or accidentally moving: R.16.
- A safe means of access should be provided: R.17.

The regulations also concern:

- poles or beams supporting pulley blocks or gin wheels: R.18
- stability of lifting appliances: R.19
- rail mounted cranes: R.20
- mounting of cranes: R.21
- cranes with derricking jibs: R.22
- restriction on use of cranes: R.23 as amended by the Lifting Plant Equipment (Records and Test and Examination etc.) Regulations 1992
- use of cranes with timber structural member prohibited: R.24
- erection of cranes under supervision: R.25

- competent persons to operate lifting appliances and give signals: R.26
- giving of signals: R.27
- testing and examination of cranes: R.28 as amended as above
- marking of safe working loads: R.29
- indication of safe working load of jib cranes: R.30
- load not to exceed safe working load: R.31
- precautions on raising or lowering loads: R.32
- scotch and guy derrick cranes: R.33

Part IV: chains, ropes and lifting gear
These regulations concern:

- construction, testing, examination and safe working load: R.34
- testing rings etc. altered or repaired by welding: R.35 as amended as above
- hooks: R.36
- slings: R.37
- edges of load not to come into contact with sling etc.: R.38
- knotted chains etc.: R.39
- examination of chains, ropes and lifting gear: R.40
- annealing of chains and lifting gear: R.41

Part V: special provisions as to hoists
These concern:

- safety of hoistways, platforms, cages: R.42
- operation of hoists: R.43
- winches: R.44
- safe working load and marking of hoists: R.45
- test and examination of hoists: R.46

Part VI: carriage of persons and secureness of loads
These concern:

- carrying persons by means of lifting appliances: R.47
- hoists carrying persons: R.48 as amended
- secureness of loads: R.49
- suspended scaffolds (not power operated): R.48(A) and (B) introduced by the Construction (Health, Safety and Welfare) Regulations 1996

Part VII: keeping of records as amended by the Lifting Plant Equipment (Records and Test and Examination etc.) Regulations 1992
Under the regulations, certain reports must be kept and these shall be kept either at the site or at the contractor's or employer's office, depending on whether the work is carried out on site or not. If the work is to take less than six weeks, then they can be kept at the contractor's office. They shall be available for inspection by the health and safety inspectors.

The reports are the following:

Form 91	Part I	Inspection report of lifting appliances
	Part II	Testing of crane and load, radius of jib anchorage
	Part IV	Results of tests of hoists for carrying persons

Former provisions of the Offices, Shops and Railways premises Act relating to the handling of heavy loads are now governed by the Manual Handling Operation Regulations 1992 which are beyond the scope of this book.

Construction (Design and Management) Regulations 1994

These came into effect in March 1995 and were made under the Health and Safety at Work etc. Act 1974. There is also an Approved Code of Practice.

Interpretation: R.2
Construction work is defined as the carrying out of any building, civil engineering or engineering construction work and includes, *inter alia*, alteration, conversion, fitting out, commissioning, renovation, repair, upkeep, redecoration or other maintenance work, high pressure water cleaning;

site clearance and foundation; assembly and dis-assembly of prefabricated structures; removal of waste from demolition; and installation, mainten-ance or removal of electrical, gas, telecommun-ications or similar services normally fixed to a structure.

The contractor is anyone who carries on a trade, business or other undertaking (whether for pro-fit or not) in connection with which he under-takes construction work or arranges for others to do so.

A domestic client is one for whom the project is being carried out and who is not in trade, busi-ness etc.

Application of regulations: R.3
The regulations apply to construction work as defined above, except where the client has rea-sonable grounds to believe that the project is

- not notifiable and
- no more than five people will be involved in the work at any one time.

Regardless of the above exceptions, the regula-tions will apply to demolition work.

Clients and agents of clients: R.4
Clients may appoint an agent (including another client) to act on their behalf as the **only** client in respect of a project. In such cases the following applies.

- The client must be reasonably satisfied that his agent has the competence to perform the duties required by the regulations.
- Where the agent makes a declaration under paragraph 4 below then, from the date of receipt by the Health and Safety Executive, all the requirements and prohibitions of these regulations apply to the agent as if he were the only client, providing he remains so appointed.
- Paragraph 4 states that such a declaration
  (a) is in writing, signed by or on behalf of the client, that the agent will act as the client under the regulations;

(b) includes the name of the person giving the authority, the address for service of documents and address of the construc-tion site;
(c) is sent to the Executive.

The reasoning behind this is to have only one party for the Executive to deal with. In practice serious decisions will have to be taken over who adopts this responsibility and funding must be provided to cover these duties.

Requirements on developer: R.5
If the project is for a domestic client and the client enters into an arrangement with a developer who carries on a trade, business or other undertaking (for profit or not) in connection with which

(a) land or an interest in land is granted or trans-ferred to the client and
(b) the developer undertakes that construction work will be carried out on the land and
(c) after the work, the land will have premises intended to be occupied as a **residence**,

then from the time of such an arrangement the requirements of Regulations 6 and 8–12 shall apply to the **developer** as if **he** were the client.

Appointments of planning supervisor and principal contractor: R.6
Every client must appoint a **planning supervisor** and a **principal contractor** in respect of each project. Only contractors may be appointed as principal contractors.

The planning supervisor must be appointed as soon as is practicable after the client has suf-ficient information about the project and work involved as will enable him to comply with the requirements of Regulations 8(1) and 9(1) (com-petence of such people and intention to allocate adequate resources to enable him to carry out his functions).

Similarly, the principal contractor must be appointed as soon as is practicable after the client has sufficient information about the project and work involved as will enable him to comply with the requirements of Regulations 8(3) and 9(3).

The appointments may be ended, changed or renewed to ensure the posts are consistently filled until the end of the construction phase.

The same person may hold the post of planning supervisor and principal contractor providing he is suitably competent.

Notification of project: R.7

The planning supervisor must give notice of the project to the Executive unless he reasonably believes the project is not notifiable.

The notice must be in writing or in any other manner required by the Executive. Certain particulars must be given as specified in Schedule 1 as soon as practicable after appointment of the planning supervisor or, if not possible, before the start of the construction work.

Competence of planning supervisor, designers and contractors

No client shall appoint anyone to act as planning supervisor for a project unless he is satisfied that he is competent to act as such. Similarly, no-one should appoint a designer to prepare a design or a contractor to carry out or manage construction work unless he is reasonably satisfied that they have the competence to do so. Competence in this respect relates only to the work involved and requirements of the regulations.

Provision for health and safety: R.9

No-one must appoint a planning supervisor, designer or contractor unless they are satisfied that they have allocated or will allocate adequate resources to enable them to perform their appropriate functions under these regulations.

Start of construction phase: R.10

Every client must ensure that, as far as is reasonably practicable, the construction phase does not start unless a health and safety plan has been prepared. (Failure to do this can lead to civil liability as well as being a criminal offence.)

Client to ensure information is available: R.11

Every client must ensure that the planning supervisor is provided with all the information he needs for his job regarding the state or condition of the premises where the work is to be carried out. Such information is that which is reasonable for a client to actually have or to make enquiries about.

Client to ensure health and safety file is available for inspection: R.12

Any health and safety file in the client's possession should be available for inspection by anyone needing information in order to comply with the regulations.

Requirements on designer: R.13

Except where a design is prepared in-house, no employer shall allow an employee to prepare a design unless he has taken reasonable steps to ensure the client is aware of the duties under these regulations and of any guidance issued by the Commission in their connection.

Every designer shall ensure that his design shall have regard to the need to avoid foreseeable risks, to combat risks at source and to give priority to protective measures in respect of construction site workers, cleaners or anyone who could be affected by the work.

The designer must ensure that the design includes adequate information about aspects of the work which might affect the health or safety of such people. Furthermore, he must co-operate with the planning supervisor and with any other designer in order to comply with the regulations.

Requirements on planning supervisor: R.14

The planning supervisor shall ensure, so far as is reasonably practicable, that the design of any structure includes among the design considerations those needs specified in R.13 above. He must also take reasonable steps to ensure co-operation between designers so that they can comply with the requirements of R.13. He must be in a position to give adequate advice to any client and contractor to enable them to comply with R.8(2) and R.9(2) and to any client to comply with R.8(3), R.9(3) and R.10.

He must ensure that a health and safety file is prepared in respect of each structure in the project containing required information and any other information which it is reasonably foreseeable would ensure the health and safety of workers, cleaners and those affected by such work.

He must review, amend or add to the file as necessary.

He must deliver the health and safety file to the client at the end of the project.

Requirements relating to the health and safety plan: R.15

The planning supervisor shall ensure that such a plan has been prepared and provided to any contractor prior to the arrangements being made for the carrying out of or management of the work. (The Approved Code of Practice suggests that it is available at pre-tendering stage so that contractors can incorporate its requirements and allow the finance for it in their tenders.)

Such a plan must contain the following information:

(a) a general description of the project construction work;
(b) details of date for completion and intermediate stages;
(c) details of reasonably foreseeable risks as are known to the planning supervisor;
(d) any other information which the planning supervisor knows or could reasonably find out and which would be necessary to show that he is competent under R.8 (see above) or that he has allocated sufficient resources to satisfy R.9 (see above);
(e) such information as the planning supervisor knows or could reasonably find out and which he would expect the principal contractor to need in order to comply with paragraph (4) below;
(f) similarly, information that any contractor would need to know so he could comply with welfare requirements of statutory regulations.

By paragraph (4) the principal contractor must take reasonable measures to ensure that the health and safety plan has the following features until the end of the construction phase:

(a) project arrangements ensuring the health and safety of workers and those affected by the work, to include, if necessary, management and monitoring and to take account of construction risks and paragraph (5) activities;
(b) sufficient information about welfare arrangements to enable any contractor to understand how to comply with any statutory welfare requirements.

Paragraph (5) activities are:

(a) those carried out by persons at work and
(b) those carried on at the construction site premises, and either
   (i) are such as will affect the health or safety of workers carrying out the construction work or those affected by such work, or
   (ii) are such that the health and safety of workers carrying on a non-construction activity may be affected by construction workers' work.

Requirements on and powers of principal contractor: R.16

(a) He must take reasonable steps to ensure co-operation between all contractors sufficiently to enable them to comply with their own statutory requirements. (They may be sharing the construction site under R.9 of the Management of Health and Safety at Work Regulations 1992. This is beyond the scope of this book.)
(b) He must ensure that every contractor and every employee complies with the rules in the plan.
(c) He must take reasonable steps to ensure that only authorised persons are on site.
(d) He must ensure that R.7 particulars are prominently and legibly displayed to workers.
(e) He must promptly give information in his possession to the planning supervisor regarding contracting work and any information the planning supervisor would require to comply with R.14 for the file and which the planning supervisor does not have.

The principal contractor may give reasonable directions to any contractor so that he (the principal contractor) can comply with his duties. Furthermore he may make such rules as are reasonably required in the plan to achieve health and safety. Such rules must be in writing and must be brought to the attention of anyone who could be affected by them.

It should be noted that breach of R.16 could invoke civil liability (see R.21 below).

Information and training: R.17

Sufficient information on the risks of that work to employees, persons under their control or those affected by the work must be provided by the principal contractor for every contractor.

He must ensure that each contractor employer provides his employees with any information and training required to be provided by R.8 and R.11(2)(b) respectively of the Management of Health and Safety at Work Regulations 1992.

Advice from, and views of, persons at work: R.18

The principal contractor must be open to discussion and advice from employees and self-employed persons on matters that could affect their health and safety. He must also ensure that such views can be co-ordinated.

Requirements and prohibitions on contractors: R.19

Every contractor must co-operate with the principal contractor to enable him to carry out his statutory duties and promptly provide the principal contractor with any information (including risk assessments made under the Management of Health and Safety at Work Regulations 1992) which might affect construction workers etc. or which might justify a review of the plan. Furthermore he must comply with directions given by the principal contractor, comply with health and safety plan rules, promptly provide information on accidents as required by the Reporting of Injuries, Diseases and Dangerous Occurrences Regulations 1996 and promptly provide the principal contractor with the sort of information he would require under R.16 and which only the contractor would know or could find out about.

No employer must permit anyone to do construction work nor must a self-employed person commence such work unless he has been provided with the following information: the names of the project planning supervisor and project principal contractor and such contents of the health and safety plan as are relevant to them (paragraphs (2), (3) and (4)). It will be a defence in proceedings for contravention of these paragraphs for the employer or self-employed person to show that he had made enquiries and reasonably thought he had got the information or that this regulation did not apply to construction work.

It should be noted that under R.21 only breaches of R.10 and R.16 impose civil liability.

The enforcing authority is the Health and Safety Executive: R.22.

Construction (Head Protection) Regulations 1989

'Suitable head protection' is defined by R.1.2 as head protection which

(a) is designed to provide protection, so far as is reasonably practicable, against **foreseeable** risks of head injuries;

(b) fits the wearer after adjustment; and

(c) is suitable for the particular job.

Application: R.2

The regulations as amended apply to 'construction work' as defined by R.2(1) of the Construction (Design and Management) Regulations 1994 (see p. 258).

Provision, maintenance and replacement of suitable head protection: R.3

This must be provided by every employer for each of his employees who are working on relevant operations or works. The head protection must be maintained and replaced whenever necessary.

Self-employed persons should provide their own head protection, maintaining and replacing it similarly.

Such head protection must comply with the Personal Protective Equipment at Work Regulations 1992.

What is suitable must be assessed before deeming the head protection to be suitable. The assessment must define the characteristics of suitability required and must compare different types available with what is required. The assessment must be reviewed if one suspects it is no longer valid or there has been a significant change in the relevant work. Appropriate accommodation must be made available for the head protection when not being used.

Ensuring suitable head protection is worn: R.4

Employers must ensure, so far as is reasonably practicable, that his employees at work wear their head protection, unless there is no foreseeable risk of injury to their heads other than by falling.

Similarly, employers, self-employed persons and employees who have control over others at work to whom the regulations apply must ensure that, so far as is reasonably practicable, those people wear their head protection unless there is no risk, as above, except from falling.

Rules and directions: R.5

The person in control of the site where construction work is being carried out may make rules regulating the wearing of head protection.

Such rules must be in writing and must be brought to the attention of those affected.

In order to comply with R.4, an employer may give directions requiring his employees to wear suitable head protection.

Similarly, employers, self-employed persons or persons in control of other self-employed persons may give directions requiring the others to wear their head protection. This would be important in relation to the supervision of labour-only subcontractors.

Wearing of suitable head protection: R.6

Employees must wear their head protection provided when required to so by the rules of R.5, as must self-employed people. Full and proper use must be made of the head protection and all reasonable steps must be taken to return it to the accommodation provided for its storage after use.

Reporting the loss of, or defect in, suitable head protection: R.7

Employees must take reasonable care of their head protection and report any loss or obvious defect.

Certain exemptions may be granted. For example Sikhs who wear turbans do not have to wear head protection.

11 Insurance

Introduction

The law of insurance is found principally at common law and equity and in some statutes mentioned in the text.

Insurance is merely another branch of contract. In this case, the contract is between the **insured** who pays a **premium** to an **insurer** in return for a sum of money or some other compensating action, such as replacing an article of property if a certain specified event occurs, e.g. death, illness, having had judgment awarded against the insured in a civil action, or the loss of property through theft or destruction (see Figure 11.1).

The cost of paying the insured is met by the insurer out of money made by investing all the other premiums collected from persons insuring through the company. Such payments may be by way of an **indemnity**, i.e. they make good any financial loss the insured had suffered as a result of, for example, a fire or a court judgment.

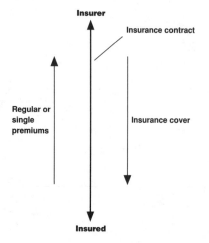

Figure 11.1 An insurance contract

Alternatively, the sum paid may be a **benefit**, payable in the case of death, injury or illness, whereby the insured pays a sum specified in the insurance policy on the happening of such events. Benefit insurances cannot be indemnity insurances, as it is impossible to revert the insured person into the same financial position as he was prior to the accident, illness etc. In both cases, there must be an element of **uncertainty** which is beyond the control of the insured. (In the case of death, the uncertainty concerns the date of death.) Often, the first type of benefit insurance is called **assurance** as the fact of death is **assured**, and the second and third types are called **insurance** which is needed **in** case the event happens. Such terminology is not standard, however, and the word insurance is perfectly acceptable at all times.

Form of contract

Insurance contracts are not like ordinary contracts in that the greater bargaining power is in the hands of the insurance companies, and a potential insurer must either accept their standard terms and conditions or go without insurance.

There is no statutory requirement for the contract to be in any particular form. Thus, it can be in writing, made orally, implied from behaviour or a combination of any form. Often, when asking for temporary cover for a new car, this is granted orally over the telephone and subsequently confirmation will be made in writing. Normally, however, contracts will be entered into by a combination of completing a **proposal form** which may or may not be accepted by the insurance company, who then send a written **policy**, which in reality is confirmation of what has already been agreed between the parties, i.e. an acceptance of the proposal form offer. A copy of the standard policy, therefore, should have already been made available on request for reference by the prospective insured. Thus, all the peripheral details may not necessarily have been expressly agreed upon by the insured and his agreement to such details will be presumed. However, there must have been agreement on the main terms, such as the **subject-matter** of the insurance, e.g. a dwelling house, the **risks** covered, e.g. fire, the **premium** paid, and the amount of compensation, e.g. full replacement value or the insured to pay the first £50 of loss.

On receipt of the policy, it should be minutely checked to see that it complies with what was agreed, and sent back for alteration if it is incorrect. Otherwise a claim may be prevented at a later date.

The proposal form and the *uberrimae fidei* rule

In most cases nowadays, a proposal form must be completed by the proposer. The form contains many questions, all of which must be answered truthfully. This is of vital importance as insurance contracts are *uberrimae fidei*, i.e. of the **utmost good faith**. This is so because the insurer cannot properly assess the degree of risk involved unless he knows **all** the material facts. For example in the case of life insurance, the insurer will want to know about the health of the proposer including details of past illnesses.

Unfortunately, the concept of *uberrimae fidei* takes the giving of information much farther than one would expect.

It is not enough to answer all the questions on the proposal form truthfully. If there are **any other facts** which could be of **material** importance, then they must be disclosed. The consequences of failing to do this is that the contract will be **voidable** at the option of the insurer. Voidable means that, whilst the contract was previously valid, it can be set aside and treated as null and void, giving the insured no rights to payment.

This duty to provide information falls on both parties, but of course, in the majority of cases, it is the proposer who knows more of the surrounding circumstances than the insurer. Thus, there have been many unhappy cases where someone has claimed on an insurance policy only to find that, on investigation of the events, undisclosed material facts have come to light which cause the insurer to avoid the policy.

Certain facts do not require disclosure as they are **not** material, such as where a structural survey has been made by the insurer and he fails to discover obvious facts.

The insurance contract is only *uberrimae fidei* at the pre-contract stage. Once the insurance contract is in being, it becomes a normal sort of contract subject only to the usual requirement of good faith. Nevertheless, most policies contain terms which require the insured to tell the insurer of any change in circumstances. Failure to do so may again lead to the contract being set aside. The agreement may also require the insured to promise (warrant) that certain facts are true at the time of making the contract and that they will continue to remain so. Such a clause in the contract must be made quite clear to the proposer if the insurer is going to be able to rely on it at a later date (see statement of practice post).

An ordinary householder's policy may possibly be avoided if work is being carried on at the premises which would increase the risks undertaken by the insurer (and thus be a material fact). Whilst this seems a pedantic point of view, claims are only made *after* damage or injury has been suffered, and if this was caused by building contractors working on the house, the insurer may very well try to avoid the policy because they were *not* told of the builder's presence.

In an Association of British Insurers and Lloyds Statement of General Insurance Practice, proposal forms, it is stated, should only be completed by the proposer according to his knowledge and belief. Thus, if he knows something and thinks it to be true, and signs the proposal form to that effect, then that should relieve him of the harshness of the *uberrimae fidei* rule.

Furthermore, facts which the insurer regards as material should be the subject of clear questions set out in the form. Neither the proposal form nor any policy must contain any provision converting the statements as to past or present facts in the proposal form into warranties. But insurers may require specific warranties about matters which are considered material to the risk. So, for example, cover may be provided where the insured has antiques on the premises, provided there is a working burglar alarm and the insured warrants that this is so.

It should be noted that this statement only applies to contracts of insurance entered into by the insured in his **private capacity** and for **non-life** insurance. As much insurance is taken out to cover both a person's working life and his private life, this may mean that the strict *uberrimae fidei* rule applies to 'mixed contracts' or it could be construed to cover only the private part of the contract. Thus, a builder's professional insurance cover, insurance of the works etc. will all be subject to the *uberrimae fidei* rule.

Relaxation of *uberrimae fidei* rule

As this is very strict rule, in certain circumstances legislation has reduced its harshness, e.g. certain inaccuracies in relation to age or physical condition of a driver are not allowed to deny compensation under a policy under the Road Traffic Acts.

Also in the Association of British Insurers and Lloyds Statement of General Insurance Practice mentioned above, it was clearly set out that the insurer must not unreasonably refuse to indemnify a policy-holder in three situations:

(i) where the insured failed to disclose a material fact which a policy-holder could not reasonably be expected to have disclosed;
(ii) where the insured made a misrepresentation unless it is a deliberate or negligent misrepresentation of a material fact;
(iii) where there has been a breach of a warranty or condition in the agreement, if the loss is unconnected with such a breach, the insurer must still pay out, unless fraud is involved.

(Note: the statement only covers private non-life insurance. It does not cover commercial or business contracts.) (See above.)

An insurable interest

Generally, no insurance cover may be given to anyone unless they have an **insurable interest** in the subject-matter of the contract.

An insurable interest may be in a human being, property or legal liability. The proposer must have such an interest, so that if they are destroyed or damaged, or if he has to pay because of legal liability, then he will suffer financially as a result. The requirement of having a legally recognised insurable interest was introduced by the Life Assurance Act 1774, which got rid of the old practice of insuring the lives of famous people, which really amounted to gambling (gambling, despite its contractual appearance, is unenforceable in the courts).

Thus, as far as people are concerned, husbands and wives can insure each other's lives, creditors can insure their debtor's lives and so can partners for their co-partners. Main contractors often insure all the contract works on behalf of all the subcontractors.

Third-party insurance, however, is an example of one of the exceptions in which someone with no insurable interest can claim under an insurance contract.

Further points of interest in relation to insurance in general

Subrogation

We have already seen that many contracts of insurance are contracts of indemnity, i.e. the insurer pays compensation to the insured party, placing that party in the financial position he was in prior to an event, the risk of which had been covered by insurance. It follows from this right of indemnity that the insurer may take over any rights to take legal action belonging to the insured in relation to that event. This right is called **subrogation** and it is often described as a process whereby one party (the insurer) stands in the shoes of the other party (the insured) and sues the third party who caused the loss through tort or breach of contract. It was originally an equitable creation.

The effect of, and one of the reasons for, the doctrine of subrogation is to prevent the insured from profiting from his misfortune by claiming his insurance money **and** keeping any money recovered in a court action. The insurer takes the action in the name of the insured.

In certain situations, insurers may agree to waive their right of subrogation. In particular this occurs in so-called 'knock for knock' agreements whereby motor insurers will agree to indemnify their own insured drivers and not claim damages from the other insurer's driver (see Figure 11.2). It should be noted that the 'knock for knock' agreements between the insurance companies do not take away from the insured the right to sue the party in the wrong. Thus, action may be taken to prevent the indemnified innocent party from losing his 'no claims bonus'. However, on receipt of any damages, the insured must then reimburse the insurer for the amount previously paid to compensate him for his loss.

Insurers have also agreed not to use their rights of subrogation in employers' liability cases, where the employer is vicariously liable for a tort committed by an employee which injures another. Strictly speaking, the insurer would then be able to sue the negligent employee in order to recover

the money paid in compensation. This right has thus been waived by agreement.

In practice, most insurance policies are extended to go beyond the common law rights of subrogation. Thus, a clause is inserted in most policies whether they are insurances of indemnity or not.

Average

A clause called an average clause is often inserted into certain insurance policies, in particular to cover fire and accident claims. Such a clause is designed to encourage the insured to insure for the full value of his property. Otherwise, if he under-insures and there is only partial damage to his property, it would be unfair (on the insurer) to expect to receive full compensation of the partial damage. Thus, if property worth £50,000 is insured for only £25,000, and a fire causes £10,000's worth of damage, then the insurer will pay only £5,000, i.e. reflecting that the property was under-insured by half.

Warranties in insurance contracts

For historical reasons, a warranty in an insurance contract is what one would normally call a condition in ordinary contract law. Thus, if there is a breach of a warranty in an insurance agreement, this will give the other party the right to repudiate the contract. Thus, all warranties must be strictly adhered to, whether or not the warranty appears to relate only to certain types of claim. Warranties as to the existence of burglar alarms at the time the proposal form was completed are not to be judged as 'continuing' unless specific stipulations are made at the outset by the insurer: *Hussain* v. *Brown* 1995.

Insurance required by statute

Certain types of insurances **must** be obtained by the builder; otherwise he will be guilty of an offence.

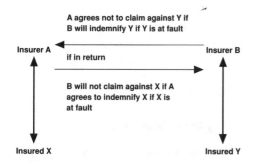

Figure 11.2 'Knock-for-knock' agreements

Employers' Liability (Compulsory Insurance) Act 1969

See p. 271. Also Cl.21.1.1.2 JCT contracts require compliance.

Road Traffic Act 1988 requirements

It is a criminal offence to use or allow the use of a motor vehicle unless there is an existing policy of insurance, covering legal liability for death or injury to third parties, hospital charges (even though we have the National Health Service, certain charges are made up to a specified limit), emergency treatment, fee for doctor or ambulance treatment.

Cover for such liability is the minimum required by the Acts, and in most cases, including the 'third-party cover' policies, the insurance provided by the companies is much greater. In *R. v Secretary of State for Transport ex.p. National Insurance Guarantee Corporation* 1996 it was held that this requirement satisfies Council Directive (EC) 90/232 which requires motor insurance to cover **all** passengers for personal injury and not just drivers.

What sort of insurance cover is required by the builder?

See Figure 11.3.

1. *Material damage insurance.* The builder should insure the building works currently under construction, his own and other people's plant and equipment, his own buildings and contents, property in transit, money and his motor vehicles against damage, destruction or theft.

2. *Liability insurance.* He should insure against his liability for injury or disease suffered by his employees in the course of their employment, personal injury or damage to property suffered by members of the public and claims caused by his professional negligence.

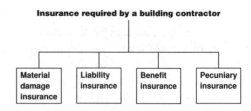

Figure 11.3 Insurance required by a building contractor

He will also need to insure against liability for injury, death or damage to property caused by the use of motor vehicles.

3. *Benefit insurance.* If he is sensible, a builder should take out insurance to cover his own life or personal injury or illness.

4. *Pecuniary loss insurance.* Such insurance is used to give cover against pure financial loss, such as building society mortgage guarantees or loss caused by dishonest employees, non-payment of debts and loss of profits following physical damage to business premises or other property.

Composite policies

Because the work of a builder is complex and interrelated, it is usual to provide insurance cover for many different types of risks in one document. The administration involved for both the insured and the insurer is easier and less likely to be overlooked.

For convenience, however, the different aspects and types of insurance will be dealt with separately.

Material damage insurance

Insuring the building works

It will be remembered from the law of contract that, where someone enters into a fixed price (lump sum or entire) contract, performance of the total contract is essential. In building work, this means that the builder will not get paid at all until he substantially completes the building. If the house is accidentally or otherwise damaged, he must bear the loss, and thus insurance is essential.

Under Cl. 22 Revised 1980 JCT standard form of building contract with quantities

By virtue of the contract the contractor agrees to carry out and complete the works in accordance with the contract drawings and bills of quantities in consideration of a certain sum.

To reduce the possibility of financial ruin for either the contractor or the employer who both stand to lose by the destruction of the works, insurance is required under Cl. 22A to cover against all risks.

It should be noted that even in an all risks policy some events are specifically excluded by the insurer. Have a look at your own household policy.

Clause 22A. 1: 'The **Contractor** shall take out and maintain a Joint Names Policy for All Risks Insurance for cover no less than that defined in Clause 22.2 for the full reinstatement value of the Works . . . and shall . . . maintain such Joint Names Policy up to and including the date of issue of the certificates of Practical Completion or up to and including the date of determination of the employment of the contractor . . . whichever is earlier.'

By Cl. 22A.2, the insurance must be placed with insurers approved by the employer and the contractor must deposit the policy and receipts for premiums with the architect. If the contractor fails to insure or to continue to insure, the employer may then himself insure in joint names and recover the cost by deducting the appropriate sums due to the contractor under the contract.

If the contractor already maintains general insurance cover under a blanket or floater policy, which would cover the contract work in this instance, then by Cl. 22A.3.1 the obligation to insure under Cl. 22A will be discharged on making

the employer the co-assured on that policy. Such policies do not, of course, have to be deposited with the employer but evidence of their existence and continued maintenance must be produced when required by the employer.

As an alternative to Cl. 22A, Cl. 22B allows the **employer** to take out insurance to cover all risks of the work and materials: 'The **Employer shall** take out and maintain a Joint Names Policy for All Risks Insurance for cover no less than that defined in Clause 22.2 for the full reinstatement value of the Works . . . and shall . . . maintain such Joint Names Policy up to and including the date of issue of the certificates of Practical Completion or up to and including the date of determination of the employment of the contractor . . . whichever is earlier.'

On discovery of such loss or damage the builder must then give notice to the employer of the extent, nature and location: Cl. 22B.3.1.

The contractor must then replace or restore the works, and it shall be treated as a variation of the contract. This means that he will be able to claim the extra money expended.

Clause 22C is concerned with existing structures owned by the employer which are going to be altered in some way by the contractor. By Cl. 22C.1 the employer must insure, in joint names with the employer, the existing structures and contents for the full cost of re-instatement, repair or replacement of loss or damage due to certain specified perils. The insurance monies are then paid to the employer. Furthermore by Cl. 22C.2 insuring the works (not being the existing buildings and contents which are dealt with above) is undertaken by the employer who must take out a joint names policy for all risks insurance for the full re-instatement value of the works.

Notice of such loss or damage must be given immediately on discovery by the contractor to the architect, giving details of the extent, nature and location of the loss or damage: Cl. 22C.4.

By Cl. 22C.4.3 the making good or re-instatement of such loss or damage will be treated as a variation required by the architect/supervising officer. However, by Cl. 22C.4.3.1, if it is just and equitable, either party may within 28 days of the occurrence of the loss or damage terminate the employment of the contractor. An opportunity to go to arbitration is made in such circumstances to decide whether it would have been just and equitable.

In the situations in which the employer should insure, the contractor is protected under the contract by being allowed to require the employer to produce evidence of the policy and, if inadequate, the contractor may take out joint names insurance and add the cost to the contract price (Cl. 22C.3).

Obviously the cost of any insurance should be reflected in the contract price and allowance must therefore be made when preparing a tender.

Insuring the builder's own plant and equipment

Generally whether a builder chooses to insure his own property is up to him as he alone suffers the loss if it is damaged or destroyed. Good sense, however, suggests that it is wise to insure; otherwise he might be placed in a difficult position and may be unable to complete a contract.

The cost varies depending on the type of equipment and whether cover is required merely whilst it is on site or if it is being transported. Most insurance of this type will be provided under a composite policy.

Insuring the builder's hired equipment

Most builders hire equipment using the Contractors' Plant Association Model Conditions for the hiring of plant. Whilst the conditions do not impose any duty to insure on the builder hiring the equipment, the conditions place liability on the builder in nearly all hiring situations, e.g. erecting and operating the plant. Once again it is implicit that insurance is necessary to give cover against any claims due to damage or loss of the equipment (fair wear and tear excepted) and claims for injury or death caused by operating

the equipment. The conditions contain a clause indemnifying the owner against any such claims by third parties.

Insurance of building contractors' own office premises or other buildings

Once again there is no compulsion for anyone to insure their own buildings, e.g. business premises and contents. However, if the builder has borrowed money to purchase such premises and has given the lender a mortgage, it is probable that a condition of the loan will be that the property shall be fully insured and the mortgagee's interest noted on the policy (this is also true of domestic mortgages). Furthermore, if the premises are leased, it will probably be a term of the lease or tenancy agreement that the premises be insured. (If the lessor is to insure, then the lessee will have to pay towards the cost of the insurance and it will be reflected in an annual maintenance payment or ground rent.)

The standard building policies cover fire, lightning and limited explosion but the policy can be extended to cover other risks such as storms, burst pipes and malicious damage.

The cover should allow for the cost of re-instatement, i.e. rebuilding.

Money

Money in this context does not just cover cash. It also includes *inter alia* bank notes, cheques, postal orders, stamps, national insurance stamps and luncheon vouchers. The loss of these must be covered on an 'all risks' basis. Thus, theft, a usual cause of loss, would be covered. However, there may be many special conditions attached to such cover, e.g. that all reasonable precautions should be taken to prevent loss, and there may be many exclusions. Thus, it is common to exclude the insurer's duty to pay if the money was stolen by a dishonest workman in the employ of the insured. (For loss caused by dishonest employees see p. 274.)

Motor insurance for damage, destruction or theft

Insurance to cover such loss is not compulsory but in most cases it is sensible to obtain cover against theft or damage caused by fire and other accidents. This is known as first-party cover as the cover is designed to protect only the insured and not the lives of third parties which is compulsory under the Road Traffic Act (see p. 268).

Comprehensive motor insurance policies would cover not only both types of risks described above but also the injury or death of the insured (not compulsory) as well as damage to the property of third parties (which is also not compulsory). Thus, a sole trader builder would find a comprehensive policy the most desirable as it would cover all the above risks. Furthermore, most comprehensive motor insurance policies provide protection for either certain named drivers of a particular motor vehicle or anyone driving with the permission of the insured.

Building companies, if they have a large workforce using company motor vehicles or even their own private vehicles, may have group or block insurance covering all or any of the risks described above.

Motor insurance is subject to many exclusions and limitations. For example, a company's block insurance policy may provide cover for a driver driving his own car whilst on company business only. Reliance must be made on the driver's own insurance when driving the rest of the time. Also, it is common for private motor insurance to be restricted to social, domestic and pleasure purposes and thus a driver of a private car would not have cover whilst on company business.

The 'small' builder should also have cover for any business equipment carried in his motor vehicle.

Liability insurance

The concepts of liability insurance, i.e. insurance against claims made as a result of legal liability, are, of course, the principles of tort (see pp. 160–211) and also contract (see p. 47). It may therefore be beneficial to refresh one's memory at this point.

Employers' liability

Insurance to cover this is compulsory. Failure to insure amounts to an offence and copies of the insurance certificate must be displayed.

It should also be noted that this insurance is required in order to cover claims for liability against employers for personal injury and disease sustained by employees in the course of their employment. It is not public liability insurance.

If the work is covered by the JCT contract, then under the indemnity to the employer required by this contract, the contractor's employer's liability to his own workers would be within the scope of this liability. Clause 21 states that the contractor must insure for his own liability under Cl. 20 (see p. 272). By Cl. 21.1.2 the insurance required *vis-à-vis* the contractor's own workers or subcontractors must comply with the Employer's Liability (Compulsory Insurance) Act 1969.

However, there are some other points of interest to note. Certain categories of employers are exempt under the Employers' Liability (Compulsory Insurance) Act 1969 from having to maintain compulsory insurance, e.g. local authorities and regional health authorities. Also, there are no limitations on the insurers' right to avoid the policy for such things as non-disclosure of material facts.

Furthermore, if the employer is in breach of the Act and has **no** insurance or only insurance which does not cover his liability, then there is no equivalent to the Motor Insurers' Bureau to fall back on (see p. 273).

Liability for injury to public or their property

By Ss. 68–71 of the Finance (No. 2) Act 1975, building contractors must deduct income tax from the pay of subcontractors. This is so unless the subcontractor can produce an exemption certificate from the Inland Revenue. One of the pre-conditions for obtaining this certificate is that the subcontractor must have personally taken out insurance to cover his liability against claims by the public for bodily injury or disease caused by the subcontractor or his employees for at least £250,000. Thus, in an indirect way, insurance against public liability for injury to the person has been imposed.

Requirements under the revised 1980 JCT standard form of contract

Under the JCT form of contract, Cl. 21, the building contractor must insure and must also cause any subcontractor to take out and maintain insurance against his own liability under Cl. 20.

Clause 20: injury to persons and property and employers' indemnity

Death and personal injury

'The Contractor shall be liable for and shall **indemnify** the Employer against any expense, liability, loss, claim or proceedings whatsoever arising under any statute or at common law in respect of *personal injury* to or the *death* of any person whomsoever *arising out of or in the course of or caused by the carrying out of the works*, except to the extent that the same is due to any act or neglect of the Employer or of any person for whom the Employer is responsible' (Cl. 20.1).

Whilst this seems rather onerous on the one hand, if one applies the rules of liability in tort, this clause merely restates the contractor's tortious liability in the contract presuming there is fault on his part. However, what it does do is to remove the possibility of being able to sue the employer as a joint tortfeasor with the contractor for vicarious liability, unless the injury or death is due or partly due to the employer himself or his agents, in which case both parties could be sued and the Civil Liability (Contribution) Act 1978 would apply (see p. 163). The burden of proving any liability on the part of the employer or his agents probably falls on the contractor. Because

of this possibility, the employer must also take out insurance to cover his own liability, should the indemnity not apply.

Property: Cl. 20.2

'The contractor shall, subject to Cl. 20.3 and where applicable Cl. 22C.1, be liable for and shall indemnify the Employer against any expense, liability, loss, claim or proceedings in respect of **any injury or damage ... to any property** real or personal in so far as such injury or damage arises out of or in the course of or by reason of the carrying out of the **Works** and to the extent that the same is due to any negligence, breach of statutory duty, omission or default of the Contractor, his servants or agents or of any person employed or engaged upon or in connection with the works or any part thereof, his servants or agents or of any other person who may properly be on the site upon or in connection with the works or any part thereof, his servants or agents, other than the employer or any person employed, engaged or authorised by him or by any local authority or statutory undertaker exercising work solely in pursuance of his statutory rights or obligations' (Cl. 20.2).

The wording of Cl. 20.2 makes the contractor liable where the loss has been caused by 'negligence, breach of statutory duty, omission or default'. This implies that he or his agents must have been at fault in some way. Some torts require no fault element such as *Rylands* v. *Fletcher* and nuisance (although they do require foresight of the damage – see p. 191) and in such cases the employer may be liable irrespective of Cl. 20.2. To avoid shouldering total responsibility in such situations, Cl. 21.1.1 states that 'the Contractor shall take out and maintain and shall cause any sub-contractor insurance in respect of his liability referred to in Cl. 20.1 and 20.2.'

It should be noted that the indemnity regarding property does not cover the contract works nor works executed and/or site materials up to the date of the Certificate of Practical Completion or up to the date of the contractor's determination of his employment: Cl. 20.3.1. This is so unless the employer has gone into possession of the contract work earlier or the contract works were damaged as a result of the contractor's fault after the date of practical completion or after the contractor's employment ceased. Thus, the employer should also insure against such risks.

Professional indemnity insurance

Professional negligence has already been discussed under negligence in Ch. 7. It is concerned with liability for the consequences of giving incorrect professional advice, which if acted upon results in either physical damage or occasionally in pure financial loss. It covers not just advice but also the making of plans, designs and suggestions of materials.

As we saw in Ch. 7 liability for physical damage or injury can be attributed to anyone who owes a duty of care, applying the neighbour principle or the proximity test: *Donoghue* v. *Stevenson* 1932 or *Caparo* v. *Dickman* 1990. Where negligent advice results in only financial loss, then a special relationship must be shown to exist between the giver of the advice and the injured party as in *Hedley Byrne* v. *Heller* (see p. 173).

The standard of care of the professional person will depend on his qualifications and experience and in most cases proof of this will be required before an insurer will provide cover.

Separate cover is required to indemnify professionals as the public liability policy taken out by the contractor covers only the **contractor**'s liability and **not** the employer's or his agents. Thus, it is essential that architects, quantity surveyors and consulting engineers take out sufficient insurance to cover their liability.

Also, it should be noted that public liability policies generally do not cover pure financial loss suffered by third parties nor damage to the contract works caused by professional negligence. Therefore, a contractor, taking on a contract in which he designs, provides plans or suggests materials, should make sure that he takes out professional indemnity insurance to cover loss occasioned by his own 'professional' negligence.

Motor insurance – third-party liability

Under Ss. 143 and 145 of the Road Traffic Act 1988 it is not lawful for a person to use or to cause or permit any other person to use a motor vehicle on a road unless there is a policy in force covering the required third-party risks. For the third-party risks see p. 268.

Failure to have such a policy or a security against such risks is a criminal offence for which there can be no excuse. As this is a statutory duty, anyone who suffers as a result of a breach of this can sue for breach of statutory duty.

As such insurance is a contract of indemnity it provides an exception to the usual rule of privity that only parties to the contract can derive rights and duties under it. Thus, if judgment is obtained against an insured, his insurance company will pay the third party by virtue of the indemnity in the policy and this could be enforced against the insurance company if they failed to pay.

As indicated on p. 268, most sensible people have cover against more risks than the mere compulsory third-party risks. Indeed, even in so called 'third-party policies' most are extended to cover fire and theft, which, of course, indemnify the insured directly.

The Road Traffic Act 1988 also provides additional protection for third parties by invalidating certain terms in the insurance contract which would normally give the insurer the right to avoid the policy (see p. 265). Unfortunately, the Act only renders certain specified matters as subjects of such terms, e.g. the age, physical or mental condition of the driver or the condition of the vehicle. Thus, an insurer may still include terms limiting his right to pay out or giving him the right to avoid the policy in many other ways, e.g. limitations on the type of use to which the vehicle is put.

Another area of protection for the third party is found in the Motor Insurers' Bureau.

Motor Insurer' Bureau

If someone is injured in a road accident caused by a driver who is not insured against such risks, or insured but whose cover is avoided by the insurer for some reason, or a hit and run driver, then they may claim directly against the bureau, membership of which is compulsory under the Road Traffic Act 1988 for all motor insurers.

Benefit insurance

Life insurance

There are various types of life insurance policies, the most common being **whole life**, i.e. the sum payable is only paid on death, or an **endowment** policy whereby the sum is paid either on death or after a certain number of years have elapsed, whichever is earliest.

Life insurance policies are choses in action and can be transferred by assignment, usually to raise cash or to be used as security. The latter is often used to provide additional security for a loan when mortgaging land. For this and other more obvious reasons, life insurance is essential for a builder.

Accident insurance

Life insurance covers death in any event. Accident insurance, on the other hand, covers all or any of the following: death, personal injury, illness, disease caused by an accident. The word 'accident' is used in a wide sense in insurance terms and its meaning is often the subject of litigation. For the purpose of first-party accident insurance, i.e. the insurance of the proposer or designated persons such as employees, an accident should be an event which is unexpected or unintended. Thus deliberately inflicted injuries would take someone outside the cover of the policy.

The benefits paid may be lump sums or pensions, depending on the terms of the policy.

Thus a contractor may decide to obtain insurance against his own accidents or illness and may also insure his employees. This may be considered to be a financial inducement to employees to enter that contractor's employment, as otherwise

the injured employee merely has to fall back on his own insurance (if any) or the State scheme if sick or injured at work.

Building and Civil Engineering Death Benefit Scheme

This scheme, operated by a trust company, provides death benefit for the surviving spouse or dependants of someone who dies whilst covered by the construction industry's national working rule agreement. Contributions by the workers are collected under the annual holiday with pay stamps scheme, the sum payable containing an element applicable to the death benefit scheme.

Private health insurance

The National Health Service (NHS) may not always be able to cover the health requirements of an individual quickly and many companies now maintain a private health insurance scheme as part of their employees' financial package to which the employee may or may not have to contribute. The specialist companies which offer such schemes give cover for operations, examinations etc., either privately in NHS hospitals or in their own private hospitals and clinics.

As physical fitness is generally essential in the construction industry, individuals may find all or any of the above benefit insurances desirable, if they are self-employed or if their employers do not provide such a benefit.

Pecuniary insurance

The builder may suffer pure financial loss (pecuniary or economic loss) in a number of ways. This may be due to

- non-payment of debts,
- dishonest employees,
- loss of profits or other income following physical damage to works or office or other property.

The employer may suffer in a similar way but in addition may suffer from non-performance or non-completion of a building programme undertaken by a builder.

The following insurance cover may thus be desirable.

Credit insurance

As we have seen in Ch. 5, a builder's debts (owed to him) are one of his most valuable assets. If is therefore good business sense to insure against non-payment by a debtor by taking out credit insurance.

Before cover is given, checks are made by the insurer as to the creditworthiness of the potential debtors and refusal to provide cover may be an indication to the contractor that a particular customer is not a good risk.

The insurer will also stipulate that there should be an efficient debt collecting scheme operating which, of course, will be an indirect encouragement to the contractor to improve this important part of his business.

Fidelity guarantee insurance

This can be obtained to cover the insured against his own pecuniary loss caused by his employees' dishonesty during their employment. Company fraud can be financially crippling, especially with the widespread use of computers and fidelity guarantees reduce the risk of insolvency.

Consequential loss

We have already noted that a builder should have insurance to cover loss caused by damage to his business premises and contents. The money he receives is, of course, designed to enable him to rebuild or replace the property. It can do no more. In the meantime, whilst he is rebuilding and replacing the contents, he may have to rent other premises, hire equipment and may not be able to operate as usual. He also has to pay certain continuing expenses such as ground rent, rates etc. As a result he will suffer a severe loss of income. These losses are suffered as a consequence of the main loss. A business interruption policy may therefore be desirable. This gives additional cover during a limited period following the disaster which closed down or otherwise stopped the business. Such cover will be conditional on there being a material damage policy already in existence.

Compensation for loss of liquidated damages

As we saw on p. 79, the employer often imposes a liquidated damages clause in the contract, so that if the contractor does not finish the work on the due date a set sum of money is paid by the contractor for each day of default. Under the JCT form of contract, this is found in Cl. 25. Under Cl. 25.4, however, should certain specified events occur as defined in Cl. 1.1.3, such as fire, lightning, flood etc., the architect is permitted to extend the due date accordingly and the liquidated damages clause does not become operational. Obviously, the employer then suffers consequential loss as a result. So by Cl. 22D.1 insurance may be required to cover such loss caused by the specified perils. The insurance is taken out, if required, by the contractor for the benefit of the employer.

Performance bonds

To some extent, the employer is always at the mercy of the contractor in relation to any building work undertaken. If the contractor, for one reason or another, is unable to finish the work, the employer is left with the problem of obtaining and paying new contractors to complete the job.

For this reason, the contractor may be required by the employer to obtain a performance bond from a bank, or more usually an insurance company, to act as a surety. Thus, if the contractor fails to complete the work, the surety pays a sum to cover the cost of completion by other builders. The amount paid under the bond is usually sufficient only to cover the cost of actual loss and no more. Also the bond will contain a 'ceiling' on the amount that may be paid.

A bond is a contract of guarantee, but contained in a deed (under seal, see p. 71), and strictly speaking therefore the surety need not require any consideration from the contractor. Obviously, as this is a commercial contract, the contractor must pay consideration, usually of a single premium, for the guarantee of his work. He should therefore take this additional cost into account when tendering for work requiring a performance bond.

It is usual for the surety to require an express indemnity from the contractor, so that if the surety has to pay out under the bond, then there is an express right to recover the sums paid back from the bonded contractor. This express indemnity merely affirms the common law indemnity that already exists in contracts of this nature.

Although the construction industry had, as I have stated above, traditionally treated the performance bond as a guarantee and not as a true bond, two decisions in recent years focused attention on performance bonds and for a time caused no little consternation. Although the decisions, on appeal, ultimately confirmed to some extent the traditional approach, the problems, in part caused by the archaic words used, led to two new sets of model forms being introduced.

The cases were *Trafalgar House Construction (Regions) Ltd* v. *General Surety and Guarantee Company Ltd* 1995 and *Perar BV* v. *General Surety and Guarantee Company Ltd* 1994. The first case which went to the House of Lords involved the use of a form of bond based on the one appended to the ICE Conditions of Contract, 5th Edition. In reversing the Court of Appeal's decision, the Law Lords confirmed that the bond was not a 'conditional on demand bond' but a 'conditional on default bond', a guarantee as the industry had always thought it to be. In other words, the employer had to show that the contractor was in breach of the contract and that the employer had suffered loss as a result. The time for payment occurs following the calculation of damages which would usually happen after the final account has been agreed. (A 'conditional on demand bond' on the other hand meant that the insurance company had to

pay up after it had been called on to do so, in good faith by the employer, as a result of the happening of certain specified events. Only if the employer had been fraudulent could the surety refuse to pay and the employer did not have to show that he had suffered loss.)

The Perar case concerned the insolvency of the contractor before the works were finished. The employer claimed on the bond arguing that going into administrative receivership had placed the contractor in breach of the contract, thereby activating the claim on the bond. Under the terms of the bond the employer's right to payment depended on proving that the contractor was in breach of the building contract. The Court of Appeal decided that Cl. 27.2 of the JCT 81 contract used provided that the building contract was automatically ended if the contractor became insolvent, and thus there was no duty to continue the work. They held that insolvency on its own was not sufficient to activate payment under the bond.

The two new model forms are a form of guarantee bond for use with construction contracts formulated by the Association of British Insurers (ABI) and the Institution of Civil Engineers bond. Both are conditional on default bonds, i.e. guarantees. Should the construction work actually be performed outside the United Kingdom, in order to avoid law suits under foreign jurisdictions, the bond should be arranged in this country.

Insurance to cover employer's losses caused by a contractor's bad workmanship

Poor workmanship, bad design, the use of unsuitable materials and the desire to build quickly has, in some cases, led to the existence of buildings which are now unfit for human habitation. As a result employers, usually local authorities, are left with two problems – the debts left to pay for the original building work and the costs involved in rehousing tenants. It now appears that a new insurance market is beginning to open up to cover employers who may suffer similar losses in the future.

State compensation for accidents at work or diseases or conditions contracted during the course of employment

Relevant law

The law governing this area is found principally under the Social Security Contributions and Benefits Act 1992, Social Security Administration Act 1992 and Social Security (Consequential Provisions) Act 1992 and the supplementary regulations made by the Department of Social Security.

To whom does the law apply?

It covers anyone at work. There is usually no minimum period of employment, but the applicant must be an employee or someone holding an office with a company who pays Schedule E income tax (pay as you earn). Thus, self-employed people are effectively ruled out unless they can be shown in law to have contracts of service (see p. 164). Contributions to the State scheme are made not only by the employee but also by the employer and State and the scheme is a sort of social insurance. However, the monies paid in are paid out the following week to claimants as there is no investment fund held as in ordinary insurance.

What benefits are available?

There are three main benefits.

Statutory sick pay – paid by the employer

This is paid by the employer for the first 28 weeks of absence after four days' absence. Some employees are not entitled to statutory sick pay because of lack of service, but they will still be entitled to all or any of the following benefits.

Incapacity benefit – paid by the State

This is paid for 156 working days if the applicant is unable to work as a result of an accident at work or a prescribed industrial disease contracted at work. In 1997 a single person's long-term maximum incapacity benefit amounted to £61.15.

Disablement benefit – paid by the State

This is payable to an employee who, as a result of his injury or a prescribed industrial disease, has a loss of mental or physical faculty, whether or not he is able to work.

It is paid after 90 days from the date of the accident or the beginning of the disease. It is based on a medical assessment of the person's loss of faculty as a result of his injury or disease in comparison with another person of the same age and sex. In 1997 100 per cent disablement amounted to £99.

In addition to this pension, if the person is severely disabled or has particular difficulties, he may claim other benefits, e.g. constant attendance allowance and exceptionally severe disablement allowance. There are restrictions on the total amount that an applicant may receive.

Industrial accidents

The accident must have occurred:

- at work, or
- in the course of the employee's employment, or
- as a result of no-one's fault, or
- as a result of another's misconduct or act of God

providing the employee did not cause the accident himself either directly or indirectly.

Prescribed diseases

The insured must be shown to be suffering from one of the prescribed diseases or conditions set out in the Social Security (Industrial Injuries) (Prescribed Diseases) Regulations 1985. There are at present 51 specified. The claimant must in some cases also show that he has been employed in a specified occupation for a required period. For example, claimants suffering from occupational deafness must have worked in a particular job for 10 years. The regulations are divided into four categories.

Examples

Condition	Occupation

Category A deals with conditions due to physical agents

4. Cramp of the hand or forearm due to repetitive movements | Prolonged periods of handwriting, typing or other repetitive movements of the fingers, hand or arm

5. Subcutaneous cellulitis of the hand (beat hand) | Manual labour causing severe or prolonged friction or pressure on the hand

Category B deals with conditions due to biological agents

3. Infection by leptospira | Work in places which are or are liable to be infested by rats, field mice, voles or other small mammals

Category C deals with conditions due to chemical agents

21. (a) Localised new growth of the skin, papillomatous or keratotic
    (b) Squamous-celled carcinoma of the skin | The use or handling of, or exposure to, arsenic, tar, pitch, bitumen, mineral oil (including paraffin), soot or any compound, product or residue of any of these substances, except quinone or hydroquinone

Category D – miscellaneous conditions

3. Diffuse mesothelioma | Any occupation involving the working or handling of asbestos or any admixture of asbestos; . . .

Claims must be made on the appropriate form obtainable from the local DSS office. A decision is first made by the insurance officer. If the claimant is dissatisfied, he has the right of appeal to the local medical tribunal and from there to the National Insurance Commissioner.

Index